JAAVSO

The Journal of
The American Association
of Variable Star Observers

Volume 47
Number 2
2019

AAVSO
49 Bay State Road
Cambridge, MA 02138
USA

ISSN 0271-9053 (print)
ISSN 2380-3606 (online)

Publication Schedule

The Journal of the American Association of Variable Star Observers is published twice a year, June 15 (Number 1 of the volume) and December 15 (Number 2 of the volume). The submission window for inclusion in the next issue of JAAVSO closes six weeks before the publication date. A manuscript will be added to the table of contents for an issue when it has been fully accepted for publication upon successful completion of the referee process; these articles will be available online prior to the publication date. An author may not specify in which issue of JAAVSO a manuscript is to be published; accepted manuscripts will be published in the next available issue, except under extraordinary circumstances.

Page Charges

Page charges are waived for Members of the AAVSO. Publication of unsolicited manuscripts in JAAVSO requires a page charge of US $100/page for the final printed manuscript. Page charge waivers may be provided under certain circumstances.

Publication in *JAAVSO*

With the exception of abstracts of papers presented at AAVSO meetings, papers submitted to JAAVSO are peer-reviewed by individuals knowledgeable about the topic being discussed. We cannot guarantee that all submissions to JAAVSO will be published, but we encourage authors of all experience levels and in all fields related to variable star astronomy and the AAVSO to submit manuscripts. We especially encourage students and other mentees of researchers affiliated with the AAVSO to submit results of their completed research.

Subscriptions

Institutions and Libraries may subscribe to JAAVSO as part of the Complete Publications Package or as an individual subscription. Individuals may purchase printed copies of recent JAAVSO issues via Createspace. Paper copies of JAAVSO issues prior to volume 36 are available in limited quantities directly from AAVSO Headquarters; please contact the AAVSO for available issues.

Instructions for Submissions

The *Journal of the AAVSO* welcomes papers from all persons concerned with the study of variable stars and topics specifically related to variability. All manuscripts should be written in a style designed to provide clear expositions of the topic. Contributors are encouraged to submit digitized text in MS WORD, LATEX+POSTSCRIPT, or plain-text format. Manuscripts may be mailed electronically to journal@aavso.org or submitted by postal mail to JAAVSO, 49 Bay State Road, Cambridge, MA 02138, USA.

Manuscripts must be submitted according to the following guidelines, or they will be returned to the author for correction:

Manuscripts must be:
1) original, unpublished material;
2) written in English;
3) accompanied by an abstract of no more than 100 words.
4) not more than 2,500–3,000 words in length (10–12 pages double-spaced).

Figures for publication must:
1) be camera-ready or in a high-contrast, high-resolution, standard digitized image format;
2) have all coordinates labeled with division marks on all four sides;
3) be accompanied by a caption that clearly explains all symbols and significance, so that the reader can understand the figure without reference to the text.

Maximum published figure space is 4.5" by 7". When submitting original figures, be sure to allow for reduction in size by making all symbols, letters, and division marks sufficiently large.

Photographs and halftone images will be considered for publication if they directly illustrate the text.

Tables should be:
1) provided separate from the main body of the text;
2) numbered sequentially and referred to by Arabic number in the text, e.g., Table 1.

References:
1) References should relate directly to the text.
2) References should be keyed into the text with the author's last name and the year of publication, e.g., (Smith 1974; Jones 1974) or Smith (1974) and Jones (1974).
3) In the case of three or more joint authors, the text reference should be written as follows: (Smith et al. 1976).
4) All references must be listed at the end of the text in alphabetical order by the author's last name and the year of publication, according to the following format: Brown, J., and Green, E. B. 1974, *Astrophys. J.*, **200**, 765. Thomas, K. 1982, *Phys. Rep.*, **33**, 96.
5) Abbreviations used in references should be based on recent issues of JAAVSO or the listing provided at the beginning of *Astronomy and Astrophysics Abstracts* (Springer-Verlag).

Miscellaneous:
1) Equations should be written on a separate line and given a sequential Arabic number in parentheses near the right-hand margin. Equations should be referred to in the text as, e.g., equation (1).
2) Magnitude will be assumed to be visual unless otherwise specified.
3) Manuscripts may be submitted to referees for review without obligation of publication.

Online Access

Articles published in JAAVSO, and information for authors and referees may be found online at: https://www.aavso.org/apps/jaavso/

© 2019 The American Association of Variable Star Observers. All rights reserved.

The Journal of the American Association of Variable Star Observers
Volume 47, Number 2, 2019

Editorial

So Long, and Thanks for All the Manuscripts
John R. Percy — 135

Life, Variable Stars, and Everything
Nancy D. Morrison — 137

Variable Star Research

Photometric Analysis of V633 Virginis
Surjit S. Wadhwa — 138

High Occurrence Optical Spikes and Quasi-Periodic Pulses (QPPs) on X-ray Star 47 Cassiopeiae
Gary Vander Haagen — 141

BVR_cI_c Photometric Observations and Analyses of the Totally Eclipsing, Solar Type Binary, OR Leonis
Ronald G. Samec, Daniel B. Caton, Danny R. Faulkner — 147

VR_cI_c Photometric Study of the Totally Eclipsing Pre-W UMa Binary, V616 Camelopardalis: Is it Detached?
Ronald G. Samec, Daniel B. Caton, Davis R. Gentry, Danny R. Faulkner — 157

The Long-term Period Changes and Evolution of V1, a W Virginis Star in the Globular Cluster M12
Pradip Karmakar, Horace A. Smith, Nathan De Lee — 167

A New Candidate δ Scuti Star in Centaurus: HD 121191
Roy A. Axelsen — 173

Observations and Analysis of a Detached Eclipsing Binary, V385 Camelopardalis
Ronald G. Samec, Daniel B. Caton, Danny R. Faulkner — 176

Multi-filter Photometric Analysis of W Ursae Majoris (W UMa) Type Eclipsing Binary Stars KID 11405559 and V342 Boo
Tatsuya Akiba, Andrew Neugarten, Charlyn Ortmann, Vayujeet Gokhale — 186

Observations and Preliminary Modeling of the Light Curves of Eclipsing Binary Systems NSVS 7322420 and NSVS 5726288
Mathew F. Knote, Ronald H. Kaitchuck, Robert C. Berrington — 194

Period Analysis of All-Sky Automated Survey for Supernovae (ASAS-SN) Data on Pulsating Red Giants
John R. Percy, Lucas Fenaux — 202

A Photometric Study of Five Low Mass Contact Binaries
Edward J. Michaels — 209

Medium-term Variation in Times of Minimum of Algol-type Binaries: XZ And, RZ Cas, U Cep, TW Dra, U Sge
Geoff B. Chaplin — 222

CCD Photometry, Period Analysis, and Evolutionary Status of the Pulsating Variable V544 Andromedae
Kevin B. Alton — 231

Table of Contents continued on following pages

Education and Outreach

Sky Brightness Measurements and Ways to Mitigate Light Pollution in Kirksville, Missouri
Vayujeet Gokhale, Jordan Goins, Ashley Herdmann, Eric Hilker, Emily Wren, David Caples, James Tompkins 241

Instruments, Methods, and Techniques

Automated Data Reduction at a Small College Observatory
Donald A. Smith, Hollis B. Akins 248

Variable Star Data

The Contribution of A. W. Roberts' Observations to the AAVSO International Database
Tim Cooper 254

Recent Minima of 200 Eclipsing Binary Stars
Gerard Samolyk 265

Recent Minima of 267 Recently Discovered Eclipsing Variable Stars
Gerard Samolyk, Vance Petriew 270

Abstracts of Papers and Posters Presented at the Joint Meeting of the Royal Astronomical Society of Canada (RASC) and the American Association of Variable Star Observers (AAVSO 108th Spring Meeting), Held in Toronto, Ontario, Canada, June 13–16, 2019, and Other Abstract Received

AAVSO Paper Sessions

A Robotic Observatory System that Anyone Can Use for Education, Astro-Imaging, Research, or Fun!
David Lane 275

DASCH Project Update (poster)
Edward Los 275

The Eye is Mightier than the Algorithm: Lessons Learned from the Erroneous Classification of ASAS Variables as "Miscellaneous" (poster)
Kristine Larsen 275

Contributions of the AAVSO to Year 1 of TESS (poster)
Dennis Conti, Stella Kafka 275

The Project PHaEDRA Collection: An Anchor to Connect Modern Science with Historical Data (poster)
Peggy Wargelin 276

The Eclipsing Triple Star System b Persei (HD 26961)
Donald F. Collins 276

Photometric Accuracy of CMOS and CCD Cameras (Preliminary Results)
Kenneth T. Menzies 276

Measuring and Correcting Non-linearities in ABG Cameras
Richard Wagner 277

Table of Contents continued on following pages

A Simple Objective Prism Spectrograph for Educational Purposes
John Thompson 277

Studies of Pulsating Red Giants using AAVSO and ASAS-SN Data
John R. Percy, Lucas Fenaux 277

Variable Stars in the College Classroom
Kristine Larsen 277

A Study of RR Lyrae Variable Stars in a Professor-Student Collaboration using the American Public University System (APUS) Observatory
David Syndergaard, Melanie Crowson 277

The Rotating Frame
Christa Van Laerhoven 278

Five Years of CCD Photometry
Damien Lemay 278

How-to Hour—Planning an Observing Program
Kenneth T. Menzies (Moderator/Coordinator), Michael J. Cook, Franz-Josef (Josch) Hambsch 278

RASC Paper Sessions (invited talks)

The Zooniverse
L. Clifton Johnson 278

The BRITE Constellation Mission
Gregg Wade 279

The Importance of Religion and Art in Sowing the Seeds of Success of the U.S. Space Program in the 1940s and 1950s
Catherine L. Newell 279

Public Outreach Through the Cronyn Observatory
Jan Cami 279

Canada's Role in the James Webb Space Telescope
Nathalie Ouellette 279

Helen Sawyer Hogg: One Woman's Journey with the Stars
Maria J. Cahill 279

Some Memories of the David Dunlap Observatory
Donald C. Morton 279

Canada's Role in the Maunakea Spectroscopic Explorer
Alan McConnachie 279

RASC Paper Sessions (other presentations)

Shadow Chasers 2017
David Shuman, Paul Simard 280

Table of Contents continued on next page

Toronto's Astronomical Heritage
John R. Percy — 280

On the Construction of the Heav'ns: A Multi-media, Interdisciplinary Concert and Outreach Event
James E. Hesser, David Lee, W. John McDonald — 280

Who doth not see the measures of the Moon/Which thirteen times she danceth every year?/
And ends her pavan thirteen times as soon
David H. Levy, R. A. Rosenfeld — 280

Dorner Telescope Museum
Rudolph Dorner, Randall A. Rosenfeld — 280

Science of the Stars: The National Heritage of the Dominion Observatory
Sharon Odell — 281

RASC Robotic Telescope Project
Paul Mortfield — 281

The Past, Present and Future of Youth Outreach at the RASC
Jenna Hinds — 281

Bringing Astronomy Alive—Student Engagement at John Abbott College
Karim Jaffer — 281

A Story of Spectroscopy: How Stellar Composition is Made Clear through Deceptively Simple Tools
Julien da Silva — 281

Things You May Not Know about the Apollo 11 Mission
Randy Attwood — 281

Teaching Neil to Land—Simulators Used in the Apollo Program
Ron MacNaughton — 282

RASC Public lecture

First Man—The Life of Neil Armstrong
James R. Hansen — 282

Other Abstract Received

Flare Stars: A Short Review
Kristinja Dzornbeta, John R. Percy — 282

Errata

Erratum: KAO-EGYPT J064512.06+341749.2 is a Low Amplitude and Multi-Periodic δ Scuti Variable Star
Ahmed Essam, Mohamed Abdel-Sabour, Gamal Bakr Ali — 283

Index to Volume 47 — 284

Editorial

So Long, and Thanks for All the Manuscripts[1]

John R. Percy
Former Editor-in-Chief, *Journal of the AAVSO*

Department of Astronomy and Astrophysics, and Dunlap Institute for Astronomy and Astrophysics, University of Toronto, 50 St. George Street, Toronto, ON M5S 3H4, Canada; john.percy@utoronto.ca

Received October 1, 2019

My ten-year term as Editor of *JAAVSO* ended as of October 1. It's a pleasure to introduce my successor Professor Nancy Morrison, hand the editorship over to her, reflect on my experience, thank the many people who have supported my work, and ride off into the sunset.

1. Introduction

Nancy Morrison is a professional astronomer with a very broad and distinguished career as a scientist and educator. She obtained her Ph.D. from the University of Hawaii, and then enjoyed a 32-year professorial career at the University of Toledo, where she is now Professor Emerita. Her research has been in the areas of stellar spectroscopy and photometry, hot supergiant stars—and in time-series analysis and variable stars. I've had the pleasure of co-authoring two papers with her, including one on P Cygni which used AAVSO data. At Toledo, she also served as Director of the Ritter Observatory (a research facility), the Ritter Planetarium and the Brooks Observatory, which serve students and the public. She has also been an exemplary citizen in her profession, serving in several roles in the American Astronomical Society including Treasurer (2014–2020). She has also served on the Board and Executive Committee of the Astronomical Society of the Pacific, a resource for educators and amateur astronomers and the public, as well as professionals. She brings remarkable experience and judgement, not just to *JAAVSO* but to the AAVSO's governance and operations.

2. Reflection

When I began as Editor, I had no formal training for the job, though I had edited some newsletters, and several conference proceedings, including that of the AAVSO's first European meeting. (I also had no formal training in university teaching and administration!) Fortunately, my predecessor Professor Charles Whitney and the AAVSO HQ staff had provided both guidance and guidelines; the latter can be found on the *JAAVSO* pages on the AAVSO website. They remind us that the mission of the AAVSO is "to enable anyone, anywhere, to participate in scientific discovery through variable star astronomy," and that the purpose of *JAAVSO* is "to promote the AAVSO's overall mission by serving as the pre-eminent publication venue not only for AAVSO Members, but for the entire community of variable star scientists and researchers." *JAAVSO* content is broad, ranging from variable star observation, data, and research, to astronomical history, biography, and education.

The *JAAVSO* authorship and readership are an equally diverse and unique combination of professional and amateur astronomers, astronomy instructors, and students. This affects the type and initial quality of the manuscripts that we receive, how they are refereed and edited, and how they appear in their final version. To put it another way: *JAAVSO* is not the *Astrophysical Journal*. Unlike some journals, we are happy to work with our authors, who may be new to manuscript preparation, to improve their manuscript, rather than to reject it outright. But, in the end, the content and presentation must still meet our publication criteria.

Our new Editor will be glad to know that my tenure has gone smoothly. I have had Mike Saladyga to keep me focussed, and the rest of the Council and staff to provide support. I have not, to my knowledge, had any issues of academic misconduct, or any overtly political conflicts, or any other complications to deal with. That's one of the joys of serving with an organization like AAVSO.

What of the future of *JAAVSO*? This depends on the vision of the new Editor, the Editorial Board, and the AAVSO Council and staff. I have already reflected on my own vision of its future in a previous Editorial (Percy 2017), so I will not do so again here. But I will follow *JAAVSO* with great interest, and continue to support it in any way that I can.

Citizen science is expanding and diversifying (Percy 2019). There is every reason to think that variable star observing and analysis by amateur astronomers and students (as well as by professional astronomers) will continue. *JAAVSO* provides a means to publish analyses and interpretations of those observations, and to summarize, through short or longer reviews, the role that AAVSO observations play in astronomy today. This provides motivation and feedback to observers, and the cycle of progress continues.

3. Thanks

To the AAVSO Council and staff, for their faith in and support for *JAAVSO*, and for constantly highlighting it on their website, reports and publications. It requires some financial and

---

[1] With apologies to Douglas Adams.

staff resources, but is absolutely central to AAVSO's mission as a premier "citizen science" organization for over a century.

To Associate Editor Elizabeth Waagen for advice, assistance, and good judgement in providing a second set of editorial eyes, and for ensuring that *JAAVSO* is well-connected to the Association's mission and history. Elizabeth has been my colleague and good friend for 40 years, and one of the reasons that the AAVSO continues to be my favourite organization.

To Production Editor Michael Saladyga especially, who turns a variety of input manuscripts into high-quality professional versions for electronic and print publication, aided by his long experience, expertise, and wisdom in both astronomy and academic writing. I thank him also for the extra task of dealing with manuscripts which my students and I submit to *JAAVSO*, in a timely and fair way. Most of all, I thank him for his excellent oversight and judgement in all things related to *JAAVSO*.

To the Editorial Board, past and present, who provide valuable advice and assistance in many different ways. Presently they are: Geoff Clayton, U.S.; Kosmas Gazeas, Greece; Laszlo Kiss, Hungary; Katrien Kolenberg, Belgium; Kristine Larsen, U.S.; Vanessa McBride, South Africa; Ulisse Munari, Italy; Karen Pollard, New Zealand; and Nikolaus Vogt, Chile. I'm delighted to have this network of "kindred spirits" around the world.

To the University of Toronto Department of Astronomy and Astrophysics for continuing to provide me with facilities and support in my "retirement"—which is not unreasonable, considering that I do a lot more in my "retirement" than edit *JAAVSO*, and hope to continue to do so.

To the anonymous and therefore unrecognized referees, who voluntarily read and assess the manuscripts, and provide evaluation and important feedback to the authors and editors, thereby helping *JAAVSO* to maintain its scientific standards and reputation.

To the *JAAVSO* authors, without whom there would be no *JAAVSO*. Thank you for providing the meat (and potatoes) for our publication, and for providing me with such interesting reading material over the past decade. It strengthens my belief that there is still an important role for amateurs and students, as well as professionals, in astronomical research and publication today. And thank you to the *JAAVSO* readers—without whom there would also be no *JAAVSO*—whether you are reading it today, or many years in the future.

References

Percy, J. R. 2017, *J. Amer. Assoc. Var. Star Obs.*, **45**, 131.
Percy, J. R. 2019, *J. Amer. Assoc. Var. Star Obs.*, **47**, 1.

Editorial

Life, Variable Stars, and Everything[1]

Nancy D. Morrison
Editor-in-Chief, *Journal of the AAVSO*

Department of Physics and Astronomy, and Ritter Observatory, MS 113, The University of Toledo, 2801 W. Bancroft Street, Toledo OH 43606; ndm@astro.utoledo.edu

Received November 20, 2019

1. Introduction

Thanks to John Percy for the introduction. I am excited and grateful for the invitation to assume the editorship of *JAAVSO*. I look forward to working with the Editorial Board and with the Board of Directors. I thank Stella Kafka, Bert Pablo, John Percy, Mike Saladyga, Owen Tooke, and Elizabeth Waagen for help in getting started and for handing me an active, well-run operation. Most of all, I am eager to learn a lot more than I already know about variable stars!

2. Observations

I joined the AAVSO about twenty years ago. As Director of Ritter Observatory, I subscribed to *AAVSO Alert Notices* in order to identify targets of opportunity for spectroscopic observation, in addition to our regular programs. A bright nova helps keep up observers' interest amid routine observations—although, it must be said, we could always look forward to the possibility of new features in the spectra of our regular variable-star targets each night. As a user of AAVSO data, it made sense for me to join.

3. Discussion

Since then, and especially this year, I have learned a lot about the AAVSO. "The mission of the AAVSO is to enable anyone, anywhere, to participate in scientific discovery through variable star astronomy." Publication is essential to research, as it is a major way of informing the community about results and receiving feedback. The role of the *Journal* is to enable variable-star researchers to let others know about their results.

In the aftermath of the very successful Las Cruces meeting last October, I received a letter from a senior professional astronomer (Smith 2019) about the importance of AAVSO publications to him as he began the study of variable stars. At first, there were only the meeting abstracts, which gave tantalizing hints about the contents of talks at the meetings. When the first issue of the *Journal* arrived, he began to receive an education in variable star science at an understandable level, and he was inspired to undertake his own research. He urges that the *Journal* be, or continue to be, at the top of the reading list for all AAVSO members. To make up for the lost impact of each hard-copy issue arriving in the mail, he said, the on-line issue should endeavor to convey the same impact. I embrace this challenge.

4. Conclusions

While the *JAAVSO* has evolved since those days and must continue to evolve along with the field of scientific publication, our basic goal of advancing variable-star research remains. As a first step, I am working on enlarging the pool of referees and, in the future, authors. I hope readers will encourage others who are doing research to publish in the *Journal*, where they will experience a collegial editorial process. Meanwhile, we will continue to publish quality articles on interesting research. As always, highlights will be noted each month in the Editor's Research Highlight in *AAVSO Communications*. I'll provide updates on new developments in later editorials. I'm looking forward to journeying with our readers through the universe of variable stars.

Reference

Smith, H. A. 2019, personal communication.

---

[1] Apologies to John Percy and to Douglas Adams.

Photometric Analysis of V633 Virginis

Surjit S. Wadhwa
Astronomical Society of New South Wales, Sydney, NSW, Australia; surjitwadhwa@gmail.com

Received April 14, 2019; revised May 26, 2019; accepted May 29, 2019

Abstract Photometric analysis of the ASAS-SN V-band photometry of V633 Virginis is presented. The results show an A-type system with an extreme low mass ratio of 0.14 and mid-range contact of 33%. There is good thermal contact between the components. Some astrophysical qualities are examined, and some absolute parameters of the system determined.

1. Introduction

The W Ursae Majoris (EW) group of short period contact eclipsing binary stars is an important test bed for the development of theories of stellar evolution. Numerous new contact systems have been discovered through various automated sky surveys. Most of the new discoveries remain unanalyzed even though the quality of the available photometry is such that reasonable analysis is feasible. The most difficult parameter to determine from light curve analysis is the mass ratio. It has been shown previously that in the presence of total eclipses even noisy data can yield an accurate mass ratio (Terrell and Wilson 2005), and the author has previously successfully analyzed survey acquired photometric data (Wadhwa 2005, 2006). This paper reports the Wilson-Devinney analysis of the V-band photometry from the All Sky Automated Survey for Super Novae of V633 Vir.

2. V633 Virginis

V633 Vir (= TYC 304-73-1, ASASSN-V J130226.05 +071834.2; R.A. $13^h 02^m 25.96^s$ Dec. $+07° 18' 34.38"$ (J2000)) was first reported as a variable in the Catalina Survey (Drake *et al.* 2014). The ASAS-SN variable star catalog (Jayasinghe *et al.* 2018) lists it as contact binary star with an amplitude of 0.33 magnitude and period of 0.4080688 day. Visual inspection of the ASAS-SN light curve suggests a definite total eclipse and hence the suitability of the system for photometric analysis. Very little other data apart from B–V (0.47) and J–H (0.213) magnitudes are available from the SIMBAD database. Using the color calibration presented by Noll (2014), the star is likely to be of spectral class F with an effective temperature of the primary (T_1) of 6700 K. The V-band photometry for the system is available from the ASAS-SN website (Jayasinghe *et al.* 2018) and was used in the analysis. All data points with reported errors greater than 0.05 magnitude were excluded. This yielded a light curve with over 180 data points with reasonable phase coverage.

3. Photometric analysis

The mass ratio of a contact binary system is usually determined by radial velocity studies. The mass ratio forms a critical part of the formal light curve solution which yields appropriate values for parameters such as the inclination, degree of contact, and temperature variations. However, where radial velocity data are not available, under certain circumstances, such as when the system exhibits at least one total eclipse, a systematic search of the parameter space for various values of the mass ratio can be employed to determine the correct mass ratio for the system. This is sometimes referred to as the grid search method and has previously been employed on many data sources, including data obtained through automated sky patrols (Wadhwa 2005, 2006). One of the most difficult aspects of such an approach is selecting the starting value mass ratio to begin the grid search. Ruciński (1993) through simulation experiments derived a relationship between the amplitude of the light curve and mass ratio for various degrees of fillout. Using this relationship for a mid-range fillout we estimated the likely mass ratio to be near 0.15 and we used this as the starting value for the mass ratio.

The WDWIN 5.6 package (Nelson 2009), which is a graphical front for the LC (light curve) and DC (differential corrections) components of the Wilson-Devinney model, was employed for the complete analysis.

As radial velocity data were not available for any of the systems analyzed in this paper, the grid search method as described in the above-referenced articles was employed. Ruciński (1993) demonstrated that in the case of contact binary systems the light curves are predominantly dependent on the three main geometrical parameters, namely, the mass ratio (q), inclination (i), and the degree of contact (f). Other factors such as gravity darkening, reflections effects, effective temperature, albedos, and limb darkening only have a small influence on the shape of the observed light curve. In the analysis of V633 Vir the gravity darkening was set at 0.32, limb darkening coefficients were interpolated from van Hamme (1993), the bolometric albedos were fixed at 0.5, and simple refection treatment was applied. The maximum magnitude of the stars is not well known, therefore, the photometric data were normalized to the mean magnitude between phases 0.24 and 0.26 in each case. This methodology has previously been applied to the analysis of All Sky Automated Survey and ground-based amateur observations (Wadhwa 2005, 2006).

The mass ratio (q) search grid is illustrated in Figure 1. The minimum error is seen at q = 0.14 and this was accepted as the true mass ratio for the system. The WDWIN 5.6 package also computes other parameters such as relative luminosities, relative radii (pole, back, and side), potential (fillout), temperature of the secondary, and inclination. These are summarized in Table 1. The observed and fitted light curves are illustrated in Figure 2 while the three-dimensional representation of the system (Bradstreet 1983) is shown in Figure 3.

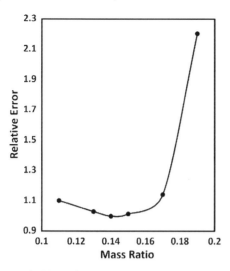

Figure 1. Mass ratio (q) search grid. The relative error has been normalized to the minimum error, which occurs at q = 0.14.

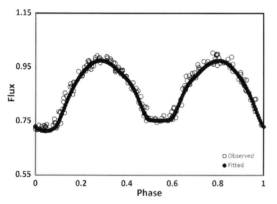

Figure 2. Fitted and observed light curve for V633 Virginis with q = 0.14 and other parameters as noted in Table 1.

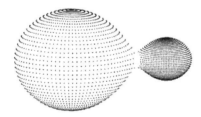

Figure 3. Three-dimensional representation of V633 Vir produced using BINARY MAKER 2.0 (Bradstreet 1983).

Table 1. Light curve solution parameters for V633 Vir.

Parameter	V633 Vir
T_1 (Fixed) K	6700
T_2 K	6464 ± 47
Mass Ratio (q)	0.14 ± 0.01
Potential	2.05 ± 0.01
Fillout (%)	33 ± 2.05
Inclination (°)	80.9 ± 1.7
L_1	10.56 ± 0.06
L_2	1.92
r_1 (mean)	0.565
r_2 (mean)	0.24

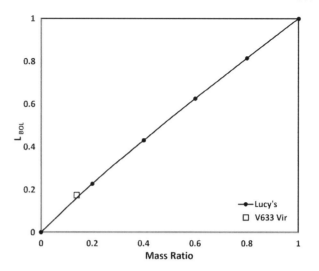

Figure 4. Mass Ratio – Bolometric Luminosity ratio. V633 Vir, being a low mass ratio system, lies close to Lucy's relationship (solid line) as predicted; see text.

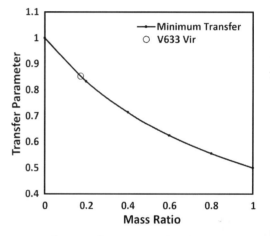

Figure 5. Mass ratio – Transfer parameter. The minimum energy transfer parameter is represented by the solid line. A Type systems of low mass ratio are expected to fall on or near the minimum. V633 Vir conforms with the predicted value; see text.

4. Astrophysical quantities and absolute parameters

Some astrophysical quantities of the newly analyzed contact system were compared with other systems; the mass ratio-luminosity relationship and energy transfer characteristics were reviewed. In addition, as the distance to the system is known, we derived some absolute parameters for the system.

Lucy (1968) found that observable luminosity ratio is proportional to the mass ratio ($L_2/L_1 = q^{0.92}$); later reviews, however, have found the relationship a little more complex and better reflected as a relationship between bolometric luminosity and mass ratio. Csizmadia and Klagyivik (2004) showed that different sub-types of contact binary systems occupied specific regions of such a diagram. Those with mass ratios below 0.2 still closely follow Lucy's relationship, however, A-type stars with mass ratios >0.2 tend to occupy the space below the line representing Lucy's relationship.

As the temperatures and relative luminosities of the stars are known from the light curve solutions it is possible to calculate the bolometric luminosities via

$$(L_2/L_1)_{Bol} = (L_2/L_1)V \times 10^{0.4(BC1-BC2)} \quad (1)$$

where BC1 and BC2 are bolometric corrections interpolated from Flower (1996). The bolometric luminosity and mass ratio relationship is shown in Figure 4, where the solid line represents Lucy's relationship $L_2/L_1 = q^{92}$. As expected, V633 Vir having a mass ratio below 0.2 is close to Lucy's relationship.

As contact binary stars approach contact and both stars fill their roche lobes there is a transfer of energy usually from the primary to the secondary. Csizmadia and Klagyivik (2004) studied this transfer by defining the transfer parameter $\beta = L_{1\,Observed}/L_{1\,ZAMS}$, which given the existing relationship between mass and luminosity can be expressed as a combination of the mass ratio and the temperatures of the component stars. In addition, they determined that the minimum rate of the transfer parameter could be expressed as $\beta_{Env} = 1/(1+(L_2/L_1)_{Bol})$ and that the transfer parameter corresponds to this minimum rate except in the cases of high mass ratio ($q > 0.72$). We calculated the transfer parameter for each of V633 Vir and the results are illustrated in Figure 5, where the solid line represents the minimum transfer parameter. V633 Vir is on or close to the minimum transfer parameter as predicted by the relationship described by Csizmadia and Klagyivik (2004).

The determination of absolute parameters is critically dependent on knowing the actual stellar masses. In the absence of radial velocity data, total mass cannot be unequivocally calculated; however, if the distance to the system is known, Kepler's relationships along with mass-luminosity relationship can be employed to estimate the total mass and other physical parameters. In our case the estimated distance for V633 Vir is available (511 Pc) from the SIMBAD data service.

As the primary eclipse is total, the measured magnitude at mid-eclipse represents the light from the primary star only, as the secondary is totally eclipsed. Given the distance we can estimate the absolute magnitude of the primary star. Knowing the absolute magnitude an estimation of the luminosity relative to the Sun is given by $L_1/L_\odot = 10^{0.4(M-M\odot)}$ (M = absolute magnitude). Knowing the luminosity one can use the mass luminosity relationship to derive the mass of the primary. Individual stellar masses and the mass of the system as a whole can be derived using the mass ratio from the light curve solution. Using Kepler's third law and the derived total mass one can estimate the semi-major axis of the system.

As noted above the WD light curve solution provides an estimation of the relative radii for each star in three formats. The average of each of the reported values for r_1 and r_2 were used in conjunction with the calculated semi-major axis to estimate the true radius of each star. Knowing the actual radius and temperature then allows an estimation of the bolometric magnitude of each star. All results of the actual parameters of each system as summarized in Table 2.

Table 2. Absolute parameters for V633 Vir derived from parallax distance of 511 Pc (SIMBAD). Bolometric magnitude assumes solar temperature of 5772 K and relative radius assumes $R_\odot = 997,500$ KM.

Absolute Parameter	V633 Vir
Distance (Pc)	511
Total Mass ($M_\odot$)	1.596
Mass Primary ($M_\odot$)	1.40
Mass Secondary ($M_\odot$)	0.196
Semi-Major Axis ($R_\odot$)	1.89
Primary Radius / $R_\odot$	1.07
Secondary Radius / $R_\odot$	0.45
Primary M_{Bol}	3.96
Secondary M_{Bol}	0.45

5. Conclusion

A photometric solution of V633 Vir is presented which indicates an A-type system with the secondary star cooler than the primary and extreme low mass ratio of 0.14. There is relatively shallow contact at less than 35%, however thermal contact is quite good with temperature difference between the stars of less than 250 K. The system behaves similarly to others of its type with respect to some astrophysical quantities.

References

Bradstreet, D. H. 1983, BINARY MAKER 2.0, Contact Software, Norristown, PA.

Csizmadia, Sz., and Klagyivik, P. 2004, *Astron. Astrophys.*, **426**, 1001.

Drake, A. J., et al. 2014, *Astrophys. J., Suppl. Ser.*, **213**, 9.

Flower, P. J. 1996, *Astrophys. J.*, **469**, 355.

Jayasinghe, T., et al. 2018, *Mon. Not. Roy. Astron. Soc.*, **486**, 1907 (https://doi.org/10.1093/mnras/stz844).

Lucy, L. B. 1968, *Astrophys. J.*, **151**, 1123.

Nelson, R. H. 2009, WDWINT56A: Astronomy Software by Bob Nelson (https://www.variablestarssouth.org/bob-nelson).

Noll, L. C. 1994–2013, "Stellar Classification Table" (http://www.isthe.com/chongo/tech/astro/HR-temp-mass-table-byhrclass.html).

Ruciński, S. M. 1993, *Publ. Astron. Soc. Pacific*, **105**, 1433.

Terrell, D., and Wilson, R. E. 2005, *Astrophys. Space Sci.*, **296**, 221.

van Hamme, W. 1993, *Astron. J.*, **106**, 2096.

Wadhwa, S. S. 2005, *Astrophys. Space Sci.*, **300**, 289.

Wadhwa, S. S. 2006, *Astrophys. Space Sci.*, **301**, 195.

High Occurrence Optical Spikes and Quasi-Periodic Pulses (QPPs) on X-ray Star 47 Cassiopeiae

Gary Vander Haagen
Stonegate Observatory, 825 Stonegate Road, Ann Arbor, MI 48103; garyvh2@gmail.com

Received May 6, 2019; revised July 23, 2019; accepted September 9, 2019

Abstract The high cadence flare search of X-ray star 47 Cas revealed 46 B-band flares at 0.26 to 0.78 mag peak above the mean. Durations of the very short flares or spikes ranged from 20 to 220 ms and energy levels from 22.7 to 199.5 × 10^{31} ergs. The study collected 2.6 × 10^7 photometric measurements over 32.3 hours from October 19, 2018, through April 9, 2019. The 46 flares identified represent a flare rate of 4.6 flares/hour during the 500 samples/sec sampling period. The analysis showed numerous quasi-periodic pulses (QPPs) accompanying the spikes, with the most predominant frequencies of 2 to 12 Hz.

1. Introduction

47 Cas is a type F0V, high proper motion, high luminosity X-ray star with B-mag 5.590, V-mag 5.268, and parallaxes of 30.16 mas (Wenger *et al.* 2000). Gudel *et al.* (1998) note that 47 Cas has been studied periodically since 1995 for its high X-ray luminosity and for better understanding of stellar evolution in the region between solar type stars beyond F5V and early F-stars. X-ray and microwave data analysis from the same study shows the star is a triple system; 47 Cas A as 1.5 Ms and 47 Cas B as 1.1 Ms with an orbital period (AB) of 1616 days. 47 Cas C was detected in a VLA radio map at 3.6 and 6 cm wavelengths at a distance of 2.5" SE of the binary. No other data have been obtained on the third member. In the latest X-ray study by Pandey and Karmakar (2015), using flare data detected by the XMM-Newton mission, an X-ray flare sequence contained 40 times less energy than the total energy emitted at the flares' peak. A large amount of residual energy was released at other longer wavelengths. Also concluded was that the main source of energy for the flare is the star's strong magnetic field and its release likely occurs within a single coronal loop structure.

At optical wavelengths, photometry conducted by Olsen (1983) consisted of uvby data on A5 to G0 stars brighter than mag 8.3 V. The photometric data are consistent with those published later by Hoffleit and Warren (1995) in the *Bright Star Catalogue*. Neither of these studies did time-series work or looked for flaring events. The AAVSO data archives contain no 47 Cas data. Gudel *et al.* (1998) noted that the Villanova University's Automatic Photometric Telescope (APT) reported some "marginally significant variability over a few hours, and probably some longer-term monthly variations" but no data were published.

The energetic nature of 47 Cas at very short wavelengths and near total lack of photometry made it an intriguing candidate for study. The study boundaries included: 1) collection of high cadence B-band photometric data for detection of possible sub-second or longer flare events, 2) where flares are detected, analysis of the length, peak σ, ΔB-mag, total energy released and, (3) a fast Fourier transform (FFT) analysis to isolate any quasi-periodic pulses (QPPs) or short-term variability.

2. Optical system, data collection, and analysis tools

The optical system consisted of a 43-cm corrected Dall-Kirkham scope, a high-speed silicon photomultiplier (SPM), and a data acquisition system capable of sub-millisecond data collection intervals. The Hamamatsu C13366 series SPM was chosen because it has comparable sensitivity to a standard single channel vacuum photomultiplier yet a more robust mechanical and electrical design with the disadvantage of higher dark counts (Vander Haagen 2012).

The optical system is shown in Figure 1. The incoming beam is split approximately 85/4 with the reflected portion passing through the B-band filter, an f-stop yielding a 57 arc-second field, and onto the SPM detector (Vander Haagen 2017). An integral wide bandwidth pulse amplifier amplifies the SPM signal, producing a 2- to 3-volt pulse of approximately 30 ns for each converted photon. These photon pulses are sent to a PC based data acquisition system, a Measurement Computing (2019) DaqBoard 1000 series, where they are gated and counted based upon the collection rate. A 2-ms gate was used for most of the measurements generating a 500 samples/sec acquisition rate. The balance of the incoming beam passes straight through to a B-band filter and a conventional CCD camera used for initial alignment, guiding, and measurement of both the guide star flux and background flux. Referring to Figure 2, the flux value counts from the target data stream along with GPS 1-second time stamps are recorded in the DAQ Log File by the data acquisition system. The CCD Data and Control stream consists of reference and background flux values plus pixel counts for each guide star sample, typically every 5 seconds. These values are stored in the AG (auto-guider) Tracker Log file.

Upon completion of the night's search the DAQ Log File and the CCD's AG Tracker Log File are merged as shown in Figure 3. The large DAQ Log File with up to1 million data samples is parsed with the much slower occurring auto-guider, AG Tracker Log. Here, every target data point is matched to the time-appropriate Tracker data with the target counts corrected for the SPM dark counts and sky background in a quadrature calculation and each sample GPS time stamped. This parsing operation results in an integrated file or "data set" with all constituents ready for analysis, with the file containing up to 1 million sample lines each containing target, reference, and background data. Files of this size are too large for spreadsheet

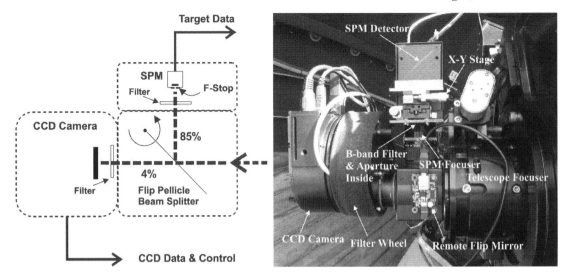

Figure 1. The optical train schematic and photo shows the pellicle beam splitter with both reflected and pass-through beams. The reflected beam passes through the B-band filter, aperture, and onto the silicon photomultiplier (SPM). The pellicle can be flipped to allow 100% light transmission for initial target alignment using the CCD camera. The SPM is mounted on a X–Y stage for precise centering of detector to the centerline of the CCD camera. Guiding is provided with pellicle in position shown. The target SPM photon counts and data from the CCD camera are processed and stored as shown in data acquisition pipeline (Figure 2).

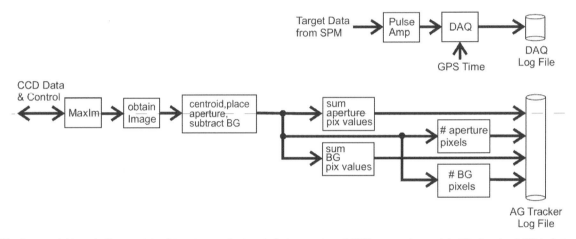

Figure 2. The data acquisition pipeline consists of two streams, the target photon counts and CCD camera image data. The low-level SPM photon counts are amplified and counted by the data acquisition system (DAQ) for the period selected (ex. 2 ms) and the data written to the DAQ Log File. A GPS synced time stamp is added every second to the appropriate data line. The DAQ Log is limited to 1 million data lines per file. The CCD data feeds the camera control program that extracts the guide-star image, centroids the image, and extracts both flux and pixel count for the guide star and background. This occurs at a rate determined by the guide star exposure, typically every 5 seconds. These data are written to the AG Tracker Log File.

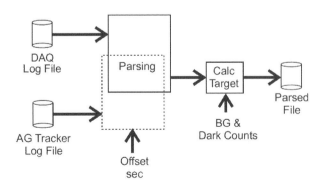

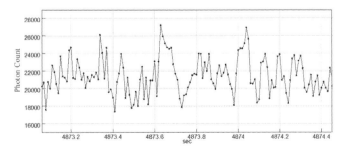

Figure 3. Upon completion of the data collection the DAQ and AG Tracker Log files are merged using a parsing operation. The lower resolution tracker file data are aligned with the DAQ target photon counts, and the target counts are reduced using a quadrature calculation to remove the sky background and SPM dark counts, thereby producing an integrated file of sample by sample target counts, reference star flux, and sky background, all referenced to an accurate UT time stamp in seconds, e.g., 5402 sec equals 1:30:02 UT with resolution to the sample period.

Figure 4. 1/19/2018 flare 1-1.2 is an example of a flare sampled at 10 ms, 100 samples/sec. Note the minimal detail on the rise side of the first large flare at 4873.6 sec reaching 5.8σ and the second flare at 4874.04 sec during the fall period. Such conditions suggest undersampling of the signal.

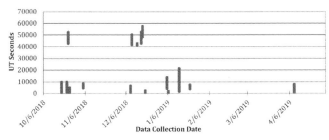

Figure 5. Span of nightly 47 Cas data collection dates. Total data collection time 116.3 Ksec, 32.3 hours.

analysis but are easily analyzed using signal processing software such as SIGVIEW (SignalLab 2019). SIGVIEW 3.2 is capable of quickly handling files up to 10^6 lines of time-based data. Any portion of the data can be reviewed near instantly with a powerful suite of tools: statistical analysis, smoothing, averaging, filtering, resampling, probability distributions, FFTs including spectrograms (FFT segmented over time), and complex calculation capability for correlation and convolution. These data files were reviewed for potential flares using statistical techniques, resampled for better detection and analysis of longer flares, and viewed using digital filters and FFT for detection of possible periodic occurrences.

3. Flare search and data analysis

A criterion was developed to isolate short duration flares in large sample sizes from randomly occurring event sequences (Vander Haagen 2015). A flare must contain a minimum of three consecutive data points, two at or above 3σ and one at or above 5σ. Since the minimum number of photons per gate was always 100 or more, normal distribution statistics were used to compute the standard deviation. Statistics were collected for a minimum 600 seconds prior to the event where possible using digital signal processing software, SIGVIEW 3.2. This process is similar in direction to that followed by flare searches (Byrne et al. 1994). The probability of this sequence being a random event can be represented by Equation (1), where N is the number of photons gated each measurement and σ is for the positive events only.

$$\Pi_{3,3,5\sigma} = P_r(3\sigma)^2 5\sigma = (1.35 \times 10^{-3})^2 \times (2.9 \times 10^{-7}) N$$
$$= 5.2 \times 10^{-13} N \quad (1)$$

With N generally ranging from 4,000 to 5,000 photons per gate at 500 samples/sec the probability of the event sequence being random is approximately 2.5×10^{-9}. This criterion was used for each of the data sets to isolate short duration flares.

Four flares were detected during the 100 samples/sec data collection period October 19, 2018, through January 5, 2019. Figure 4 shows flare 1–1.2 where there is insufficient detail on the rise and fall curves, indicating the flares were sampled at too low a rate. The rate was increased to 500 samples/sec for the data collection runs January 6, 2019, through April 9, 2019. Figure 5 shows the span of nightly data collection times in seconds across the dates. The total data collection time was 32.3 hours or 116.3 Ksec, comprising 2.6×10^7 data points. Table 1 summarizes the 46 flares detected. Four flares were detected at

Table 1. Summary of the 47 Cas flare data over the collection period October 19, 2018, through April 9, 2019.

Date	Flare ID	ΔB mag.	Peak σ	P_b ms	E_b ergs $\times 10^{31}$	Duration ms
10/19/2018	1-1.2	0.27	5.8	231.9	199.5	220
10/19/2018	1-2.2	0.26	5.6	95.5	82.1	80
10/23/2018	1-1.1	0.33	5.1	115.1	99.0	100
11/4/2018	1-1.1	0.43	5.5	120.5	103.6	110
1/14/2019	1-1.1	0.33	5.1	95.6	82.3	48
1/14/2019	2-1.3	0.39	5.4	166.8	143.5	104
1/14/2019	2-2.3	0.37	5.1	139.3	119.8	118
1/14/2019	2-3.3	0.37	5.2	69.3	59.6	62
1/14/2019	3-1.2	0.43	6.0	120.5	103.6	102
1/14/2019	3-2.2	0.37	5.1	75.3	64.8	62
1/14/2019	4-1.1	0.37	5.2	97.8	84.1	80
1/14/2019	5-1.1	0.34	5.4	58.0	49.9	48
1/14/2019	6-1.4	0.30	5.1	32.9	28.3	28
1/14/2019	6-2.4	0.34	5.8	71.6	61.5	64
1/14/2019	6-3.4	0.31	5.2	75.1	64.6	64
1/14/2019	6-4.4	0.30	5.0	30.8	26.5	26
1/14/2019	7-1.2	0.29	5.5	173.9	149.5	156
1/14/2019	7-2.2	0.28	5.3	128.9	110.8	112
1/14/2019	8-1.11	0.30	5.1	115.9	99.7	102
1/14/2019	8-2.11	0.42	7.5	84.2	72.4	72
1/14/2019	8-3.11	0.40	7.2	118.9	102.2	102
1/14/2019	8-4.11	0.32	5.5	53.2	45.8	44
1/14/2019	8-5.11	0.30	5.1	40.0	34.4	38
1/14/2019	8-6.11	0.30	5.0	30.6	26.4	30
1/14/2019	8-7.11	0.30	5.1	40.0	34.4	34
1/14/2019	8-8.11	0.37	6.4	96.5	83.0	80
1/14/2019	8-9.11	0.33	5.6	26.4	22.7	20
1/14/2019	8-10.11	0.31	5.3	66.4	57.1	50
1/14/2019	8-11.11	0.30	5.1	59.6	51.3	54
1/14/2019	9-1.3	0.40	6.9	50.8	43.7	38
1/14/2019	9-2.3	0.36	6.1	91.8	78.9	74
1/14/2019	9-3.3	0.33	5.5	132.7	114.1	108
1/14/2019	10-1.2	0.31	5.2	65.3	56.1	86
1/14/2019	10-2.2	0.31	5.2	91.8	78.9	82
1/22/2019	1-1.2	0.49	6.2	68.1	58.6	54
1/22/2019	1-2.2	0.53	7.0	58.3	50.1	46
1/22/2019	2-1.2	0.52	6.9	82.9	71.3	76
1/22/2019	2-2.2	0.54	7.3	40.7	35.0	32
4/9/2019	1-1.3	0.72	5.1	55.5	47.7	40
4/9/2019	1-2.3	0.72	5.0	61.4	52.8	44
4/9/2019	1-3.3	0.73	5.1	90.7	78.0	80
4/9/2019	2-1.3	0.76	5.1	84.0	72.2	64
4/9/2019	2-2.3	0.77	5.1	85.6	73.6	60
4/9/2019	2-3.3	0.78	5.2	48.3	41.5	36
4/9/2019	3-1.2	0.69	5.1	80.5	69.2	58
4/9/2019	3-2.2	0.71	5.3	108.0	92.9	90

100 samples/sec or 0.18 flares/hr and 42 flares at 500 samples/sec or 4.6 flares/hr. High cadence studies by Vander Haagen (2017, 2015, 2013) on cooler flare stars have shown much lower flaring activity: CR Dra 0.016 flares/hr (Vander Haagen 2017), BY Dra 0.04 flares/hr (Vander Haagen 2015), and AR Lac

0.04 flares/hr, II Peg 0.03 flares/hr, UX Ari 0.04 flares/hr (Vander Haagen 2013), and rarely are any QPPs detected. 47 Cas at 500 samples/sec with 4.6 flares/hr has minimally 100 times the flare rate of those cited.

The median flare duration was 64 ms, ranging from 20 to 220 ms, which is consistent with the studies cited above. The duration was measured between the mean flux value crossings on the rise and fall curve of each flare rounded off to the nearest sample point. The highest flare peak was 7.5σ and largest B-magnitude increase was 0.78.

The energy calculation for each flare starts with conversion of B-band photometric measurements into flux. Equation (2) converts B_0 magnitude into flux (Henden and Kaitchuck 1990):

$$F_b = 10^{-0.4(B_0-q_b)} \text{ watts / (cm}^2 \text{ Angstrom)}, \quad (2)$$

where B_0 is the star's B-mag corrected for atmospheric extinction and q_b is the absolute zero-point constant (Henden and Kaitchuck 1990). Collection of all the radiated star flux within the B-band (BW_b) results in Equation (3) representing the quiescent level for the star:

$$F_b = 4\pi d^2\, 10^{-0.4(B_0-q_b)}\, BW_b \text{ watts.} \quad (3)$$

To determine the actual flare flux, Gershberg (1972) computes the "equivalent duration" of a flare, Equation (4):

$$P_b = \int [(I_f - I_0) / I_0]\, dt \text{ sec.} \quad (4)$$

I_f is the flare flux count over the interval dt and I_0 is the quiescent or mean flux count for each data set. Incorporation of the equivalent duration P_b into Equation (3) and conversion from watt-sec to ergs results in the energy level for the flare, Equation (5):

$$E_b = 4\pi d^2\, 10^{-0.4(B_0-q_b)}\, BW_b\, 10^7\, P_b \text{ ergs.} \quad (5)$$

Adding the physical properties for 47 Cas results in Equation (6): d = 30.16 mas (Wenger *et al.* 2000) or 1.04×10^{20} cm, B_0 = 5.59 (Wenger *et al.* 2000), q_b = –37.86 with the error estimated at 10–20% (Henden and Kaitchuck 1990). The Sloan B-band filter BW_b = 1500 Å (Astrodon 2019). P_b was calculated by numerically integrating in EXCEL each flare's photon count data. The flare energy data along with other key information for each flare are summarized in Table 1.

$$E_b = 8.6 \times 10^{33}\, P_b \text{ ergs} \quad (6)$$

Six representative flares of the 46 are shown in Figures 6–11. All of the flares shown were acquired at 500 samples/sec. The upper frame of each figure is the actual data plotted with the data at 2-ms intervals. The lower frame is the FFT for the data span shown in the flare plot. Several of the sequences have very evident QPPs, as evidenced in Figures 6, 8, 9, and 11. Those QPPs can be identified in the photon plots and in the FFT for each flare sequence. Figure 12 compiles the peak FFT frequencies from the six data groups. Figure 13 slices the same data by period. From these data the greatest QPP activity is in the 12 Hz and lower frequencies, with equivalent periods generally below 0.1 second.

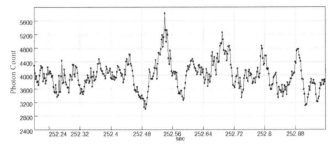

Figure 6a. 1/14/2019, flare 2-3.3 photon count versus seconds: mean 4123, 62 ms, peak 252.54 sec, 5812 photon counts.

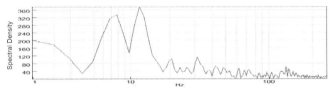

Figure 6b. Relative power spectral density versus frequency; FFT peaks 1.9, 7.8, 11.7. 20.5, and 31.3 Hz.

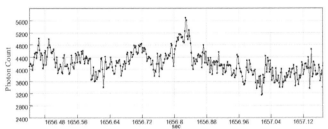

Figure 7a. 1/14/19, flare1-1.1, photon count versus seconds: mean 4183, 48 ms, peak 1656.828 sec, 5692 photon counts.

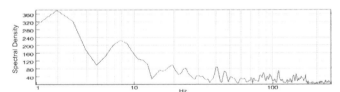

Figure 7b. Relative power spectral density versus frequency; FFT peaks 1.9, 7.8, 20.5, and 40 Hz.

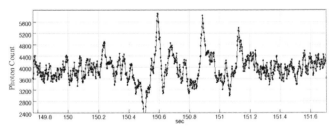

Figure 8a. 1/14/2019, files 8-2.11-4.11, photon count versus seconds: mean 4015, 3 events, peaks 150.59 sec, 72 ms, 5899 photon counts; 150.886 sec, 102 ms, 5817 photon counts; 151.126 sec, 44 ms, 5391 photon counts.

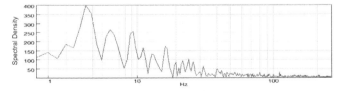

Figure 8b. Relative power spectral density versus frequency; FFT peaks 3.4, 5.9, 9.3, 11.2, 13.2, and 16.6 Hz.

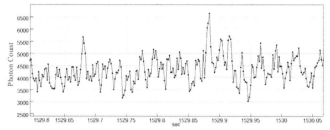

Figure 9a. 1/22/2019, flare 1-2.2, photon count versus seconds: mean 4069, 46 ms, peak 1529.884 sec, 6658 photon counts.

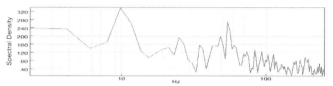

Figure 9b. Relative power spectral density versus frequency; FFT peaks 3.9, 9.8, 25.3, and 54.7 Hz.

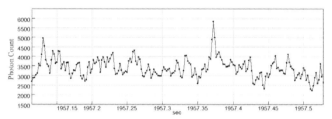

Figure 10a. 4/9/2019, flare 1-3.3, photon count versus seconds: mean 2985, 80 ms, peak 1957.375 sec, 5829 photon counts.

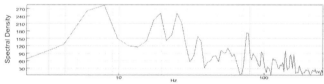

Flare 10b. Relative power spectral density versus frequency; FFT peaks 5.9, 7.8, 19.5, 25.4, 35.2, and 76.2 Hz.

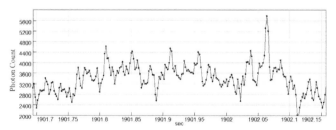

Figure 11a. 4/9/2019, flare 3-2.2, photon count versus seconds: mean 2992, 90 ms, peak 1902.064 sec, 5775 photon counts.

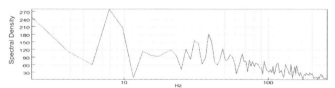

Figure 11b. Relative power spectral density versus frequency; FFT peaks 1.9, 7.8, 9.8, 13.7, 21.5, 31.3, 39.1, and 48.8 Hz.

The search for longer duration flares was undertaken by resampling all the high-speed data into 5-second integration bins and replotted. This improves the S/N and greatly enhances the ability to detect slower outbursts. No flares other than those already noted were identified.

A method of quantifying the activity of flare stars was introduced by Gershberg (1972) and Lacy, Moffett, and Evans (1976). The process develops a cumulative flare frequency distribution diagram of log f of the flare frequency for energy E_1 or greater verses log E. The number (N) of flare energy levels from E_1 to E_{max} is summed and repeated for successive higher values of E. These data are plotted against the flaring frequency, f = N/t, where t is the total monitoring time in hours. The flare data of Table 1 were plotted in the cumulative flare frequency distribution diagram for 47 Cas (Figure 14). The resultant power-law plot significantly reduces random variations and shows a linear relationship until the knee at –0.1 to 0.0 log f. The nonlinearity at higher flare rates is generally due to the system detection threshold at lower energy levels. A similar discontinuity can occur at higher energy levels as the maximum flare energy level is reached for the star. The power-law fit (Lacy, Moffett, and Evans 1976) for flaring activity has the general form of $\log f = \alpha + \beta \log E_b$. Solving using the flare data, $\alpha = 116.43$ and $\beta = -3.548$. The power-law trend line is plotted in Figure 14 along with the cumulative distribution and shows good correlation up to the knee.

4. Summary

X-ray star 47 Cas proved to be a highly energetic optical flare star with a flare rate of 4.6 flares/hr during the 500 s/sec data collection period. This represents a flare rate minimally 100 times that reported on other short duration or spike flare studies. The mechanisms are unknown since no published work has been found on simultaneous X-ray and fast cadence photometry. Of further importance were the numerous QPPs present ranging in frequency from sub-hertz to 25 Hz and above accompanying many of the flare sequences. Quasi-periodic pulses (QPPs) or quasi-periodic oscillations (QPOs) are largely interchangeable terms due to their loose definition. QPPs have been observed in solar and stellar activity and their mechanisms debated. Balona *et al.* 2015 suggest such mechanisms in stellar super-flares as magnetohydromagnetic forces acting on flare loops or acoustic oscillations from high energy particle impulses or quakes but little conclusive evidence, concluding that "new processes need to be found." Warner and Woudt (2008) discuss QPOs in CVs, citing possible mechanisms but with no applicable conclusions. Furthermore, the type of QPPs seen in 47 Cas are different from those cited; they are higher frequency, occur with great rapidity and amplitude, and are part of the actual flare sequence in many cases. A definitive QPP mechanism is currently unknown. Future availability of concurrent data at X-ray, radio, and optical wavelengths may help answer the causal factors for such phenomena along with improved modeling of star systems demonstrating repetitive oscillations or pulses.

The cumulative flare frequency profile followed well the typical relationship for flaring activity.

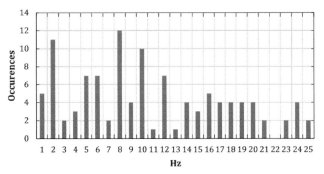

Figure 12. FFT data of occurrences versus frequency compiled from Figures 6–11.

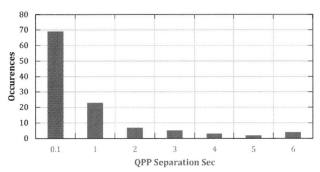

Figure 13. FFT data from Figures 6–11 converted to equivalent periods.

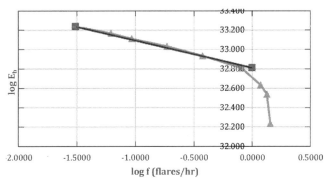

Figure 14. Cumulative B-band flare frequency distribution diagram with trend line, log f = 116.43 – 3.548 log E_b.

5. Acknowledgements

The author expresses his thanks to the referee for probing and questioning in several important areas and Dr. A. Henden for assistance on the 47 Cas quiescent energy calculation. This research has made use of the SIMBAD database, operated at CDS, Strasbourg, France.

References

Astrodon. 2019, Sloan optical filters (https://astrodon.com/products/astrodon-photometrics-sloan-filters).

Balona, L. A., Broomhall, A.-M., Kosovichev, A., Nakariakov, V. M., Pugh, C. E., and Van Doorsselaere, T. 2015, *Mon. Not. Roy. Astron. Soc.*, **450**, 956.

Byrne, P. B., Lanzafame, A. C., Sarro, L. M., and Ryans, R. 1994, *Mon. Not. Roy. Astron. Soc.*, **270**, 427.

Gershberg, R. E. 1972, *Astrophys. Space Sci.*, **19**, 75.

Gudel, M., Guinan, E. F., Etzel, P. B., Mewe, R., Kaastra, J. S., and Skinner, S. L. 1998, in *The Tenth Cambridge Workshop on Cool Stars, Stellar Systems and the Sun*, eds. R. A. Donahue, J. A. Bookbinder, ASP Conf. Ser. 154, Astronomical Society of the Pacific, San Francisco, 1247.

Henden, A. A., and Kaitchuck, R. H. 1990, *Astronomical Photometry*, Willmann-Bell, Richmond, VA.

Hoffleit, D., and Warren, W. H., Jr. 1995, *Bright Star Catalogue*, 5th ed.

Lacy, C. H., Moffett, T. J., and Evans, D. S. 1976, *Astrophys. J., Suppl. Ser.*, **30**, 85.

Measurement Computing. 2019, data acquisition systems (https://www.mccdaq.com/productsearch.aspx?q=daqboard).

Olsen, E. H. 1983, *Astron. Astrophys.*, **54**, 55.

Pandey, J. C., and Karmakar, S. 2015, *Astron. J.*, **149**, 47.

SignalLab. 2019, SIGVIEW 3.2 software for DSP applications (http://www.sigview.com/index.htm).

Vander Haagen, G. A. 2012, in *The Society for Astronomical Sciences 31st Annual Symposium on Telescope Science*, eds. B. D. Warner, R. K. Buchleim, J. L. Foote, D. Mais, Society for Astronomical Sciences, Rancho Cucamonga, CA, 165.

Vander Haagen, G. A. 2013, *J. Amer. Assoc. Var. Star Obs.*, **41**, 320.

Vander Haagen, G. A. 2015, *J. Amer. Assoc. Var. Star Obs.*, **43**, 219.

Vander Haagen, G. A. 2017, *J. Amer. Assoc. Var. Star Obs.*, **45**, 36.

Warner, B., and Woudt, P. A. 2008, in *Cool Discs, Hot Flows: The Varying Faces of Accreting Compact Objects*, AIP Conf. Proc. 1054, AIP Publishing, Melville, NY, 101.

Wenger, M., *et al.* 2000, "The SIMBAD astronomical database" (http://simbad.u-strasbg.fr/simbad) *Astron. Astrophys., Suppl. Ser.*, **143**, 9.

BVR$_c$I$_c$ Photometric Observations and Analyses of the Totally Eclipsing, Solar Type Binary, OR Leonis

Ronald G. Samec
Faculty Research Associate, Pisgah Astronomical Research Institute, 1 PARI Drive, Rosman, NC 28772; ronaldsamec@gmail.com

Daniel B. Caton
Dark Sky Observatory, Physics and Astronomy Department, Appalachian State University, 525 Rivers Street, Boone, NC 28608-2106; catondb@appstate.edu

Danny R. Faulkner
Johnson Observatory, 1414 Bur Oak Court, Hebron, KY 41048; dfaulkner@answersingenesis.org

Received May 11, 2019; revised May 28, 2019; accepted May 29, 2019

Abstract CCD, BVR$_c$I$_c$ light curves of OR Leo were taken on 21, 22, 24 January, 11, 25 February, and 11 March 2018 at Dark Sky Observatory in North Carolina with the 0.81-m reflector of Appalachian State University by D. Caton. OR Leo was discovered by the SAVS survey which classified it as a V = 0.51 amplitude, EW variable. Ten times of minimum light were calculated, five primary and five secondary eclipses, from our present observations. The following quadratic ephemeris was determined from all available times of minimum light. The 15-year (~20,000 orbits) period study shows that the orbital period is increasing at a very significant level of confidence: JD Hel Min I = 2458188.65373 (0.00039)d + 0.2709786(0.0000002) × E + 5.6(0.2) × 10^{-10} × E^2. The mass ratio is found to be somewhat extreme, $M_2/M_1 = 0.1827 \pm 0.0004$ ($M_1/M_2 = 5.5$). The total eclipses assure this determination. Its Roche Lobe fill-out is ~24%. The solution had no need of spots. The temperature difference of the components is about ~60 K, with the more massive component the hotter one, so it is an A-type W UMa binary. The inclination is $81.1 \pm 0.2°$. The secondary eclipse shows a time of constant light with an eclipse duration of 27 minutes.

1. Introduction

In this paper, we continue our analysis of solar type (F-K) eclipsing binaries. These have included all evolutionary configurations, including pre-contact (e.g. Samec *et al.* 2018a), semidetached, Algol (e.g. Caton *et al.* 2017), and V1010 Oph type (e.g. Samec *et al.* 2017a), critical contact (e.g. Samec *et al.* 2017b), overcontact binaries (e.g. Samec *et al.* 2018b), and extreme contact binaries (e.g, Caton *et al.* 2019), which are followed by the violent (Tylenda and Kamiński 2016) red novae stage where the stars coalesce into a single, fast rotating earlier type star. In clusters, these are known as blue stragglers. Here we present the first precision photometry and light curve analysis of another such a binary, the overcontact system OR Leonis.

2. History

CCD photometry at the Star'a Lesn'a Observatory (Pribulla, Vanko, and Hambálek 2009) revealed that two extremely short-period ASAS variables are overcontact binaries. One was J071829-0336.7 (OR Leo), observed during three nights (14 January, 3 February, and 26 March 2009). From the R$_c$ and I$_c$ curves, they estimated a mass ratio, q = 0.15, an inclination, i = 88°, and a fill-out 0.5. The following ephemeris and three times of minimum light were given:

JD Hel Min I = 2454 905.2867(3) + 0.270969(4)d × E. (1)

Their light curves are shown in Figure 1.

VSX gives a magnitude range of R = 13.4–14.0 and the following ephemeris for this binary:

HJD Min I = 2454905.2867 + 0.270969d × E. (2)

The system was observed by the All Sky Automated Survey as ASASN-V J113030.89-010156.9 (Pojmański 2002). It gives a Vmax =13.514, an amplitude of 0.39, and an EW (W UMa) designation, J=12.439, K=12.043, B-V=0.552 (E(B-V)= 0.028), and a distance of 195 pc. The ephemeris given is:

HJD Min I = 2457537.80197 + 0.2709608d × E. (3)

From the ASAS curves we were able to phase the data with Equation (3) and do parabolic fits to the primary and secondary minima to locate times of "low light" within 0.001 phase of each minimum (see Table 3). Finally, the binary was listed in the 81st Name-List of Variable Stars (Kazarovets *et al.* 2015).

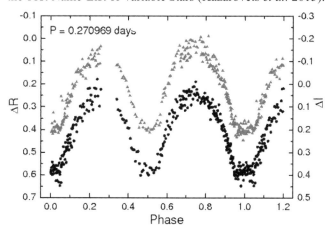

Figure 1. R and I light curves (Pribulla, Vanko, and Hambálek 2009).

Table 1. Information on the stars used in this study.

Star	Name	R.A. (2000) h m s	Dec. (2000)[1] ° ′ ″	V	B–V[2]	J–K[3]
V	OR Leo GSC 04930-00180 2MASS J11303081-0101570 ASAS 113031-0101.9 TYC 04930-00180	11 30 30.8174	–01 01 57.01	13.84	0.55	0.38 ± 0.02[4]
C	GSC 04930-00167	11 30 43.8242	–01 02 49.573	13.74	—	0.38 ± 0.046
K (Check)	GSC 2751-0129 3UC178-119845	11 30 32.1915	–01 05 02.008	14.40	—	0.42 ± 0.0462

[1]UCAC3 (U.S. Naval Obs. 2012), [2]APASS (Henden et al. 2015), [3]2MASS (Skrutskie et al. 2006), [4]spectral type G4.

3. 2018 BVR_cI_c photometry

The 2018 BVR_cI_c light curves were taken with the Dark Sky Observatory (DSO) 0.81-m reflector at Philips Gap, North Carolina, on 21, 22, 24 January, 11, 25 February, and 11 March 2018 with a thermoelectrically-cooled (–35° C) 2K × 2K Apogee Alta camera and Johnson-Cousins BVR_cI_c filters. The observations were taken by D. Caton. Reduction and analyses were done by R. Samec. This system was observed as a part of our professional collaborative studies of interacting binaries at Pisgah Astronomical Research Institute from data taken from DSO observations. Individual observations included 327 images in B, 325 in V, 338 in R_c, and 337 in I_c. The probable error of a single observation was 9 mmag in B and R_c, 8 mmag in V, and 10 mmag in I_c. The nightly comparison – check star (C–K) values stayed constant throughout the observing run with a precision of 80 mmag in B and R_c and 35 mmag in V and I_c. Exposure times varied from 100s in B, 30–40 s in V, and 20–25 s in R_c and I_c. To produce these images, nightly images were calibrated with 25 bias frames, at least five flat frames in each filter, and ten 300-second dark frames.

The coordinates and magnitudes of the variable star, comparison star, and check star are given in Table 1.

The nightly C–K values stayed constant throughout the observing run with a precision of ≈0.01 mag. Our B–V and R_c-I_c comparison-variable magnitude curves show that the variable and comparison stars are near-spectral matches, with Δ(B–V) and $\Delta(R_c-I_c)$ near zero mag.

Figures 2a and 2b show sample observations of B, V, and B–V color curves on the night of 14 September and 15 October 2015.

Our observations are listed in Table 2, with magnitude differences ΔB, ΔV, ΔR_c, and ΔI_c in the sense variable minus comparison star. The finder chart is given in Figure 3 with the variable star (V), comparison star C), and check star (K) shown.

4. Orbital period study

Ten mean times (from BVR_cI_c data) of minimum light were calculated from our present observations, five primary and five secondary eclipses:

HJD I = 2458139.87973 ± 0.00005, 2458142.8597 ± 0.0005, 2458160.741430 ± 0.000003, 2458174.8331 ± 0.0002, 2458188.6529 ± 0.0012,

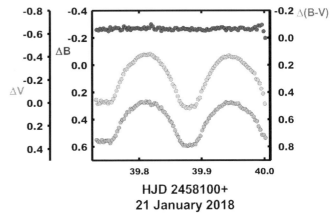

Figure 2a. B, V, and B–V color curves on the night of 21 January 2018.

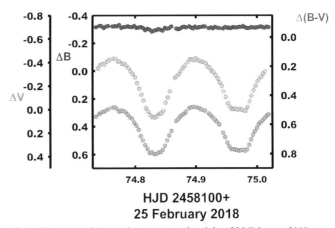

Figure 2b. B, V, and B–V color curves on the night of 25 February 2018.

HJD II = 2458139.7442 ± 0.0012, 2458142.9903 ± 0.0037, 2458160.8800 ± 0.0008, 2458188.78740 ± 0.0011, 2458174.9709 ± 0.0005.

A least squares minimization method (Mikulášek et al. 2014) was used to determine the minima for each curve in B, V, R_c, and I_c. These minima were weighted as 1.0 in the period study. In addition, four times of minimum light were calculated from the data of Pribulla, Vanko, and Hambálek (2009) (weighted 1.0 and a weak timing 0.5). Seven times of "low light" were determined from the ASAS light curve. These are weighted 0.1. A quadratic ephemeris was determined from these data:

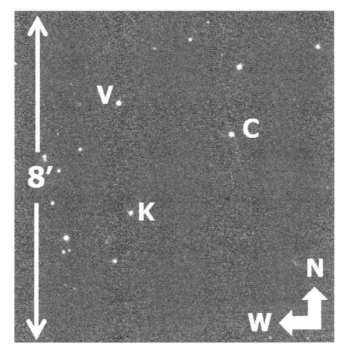

Figure 3. Finder chart, OR Leo (V), Comparison star (C), and check (K).

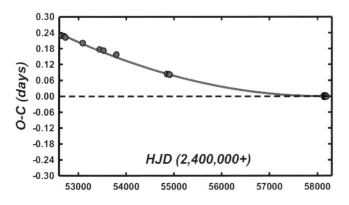

Figure 4. O–C eclipse timing residuals (Equation 3) and quadratic fit of OR Leo from Equation (4).

$$\text{JD Hel Min I} = 2458188.65373 + 0.27097862d \times E + 5.6 \times 10^{-10} \quad (4)$$
$$\pm 0.00058 \quad \pm 0.00000031 \quad \pm 0.2 \times 10^{-10}.$$

A plot of the quadratic residuals is given in Figure 4. The quadratic fit carried a precision of 28 sigma. The O–C quadratic residual calculation results are given in Table 3.

This period study covers a time interval of over 15 years and shows an orbital period that is increasing (at a very significant level of confidence). A possible cause of this effect is mass transfer to the primary component making the mass ratio more extreme. Thus, the primary component is absorbing the secondary component. If this continues, the mass ratio will become more extreme. Ultimately the system would become unstable, resulting in a red novae event and finally a single fast rotating spectrally, earlier star (Tylenda and Kamiński 2016). Alternately, the period change could be a part of a sinusoidal variation due to the presence of a third body.

Using a main sequence mass for the primary, the ephemeris yields a $dP/dt = 1.51 \times 10^{-6}$ d/yr or a mass exchange rate of

$$\frac{dM}{dt} = \frac{\dot{P} M_1 M_2}{3P (M_1 - M_2)} = \frac{3.5 \times 10^{-7} M_\odot}{d} \quad (5)$$

in a conservative scenario (the primary component is the gainer).

5. Light curves and temperature determination

The light curve characteristics of OR Leo are given in Table 4. The B, V, R_c, and I_c curves are shown in Figures 5a and 5b. The curves are of fair precision, averaging slightly above 1% photometric precision. The amplitude of the light curves varies from 0.32 to 0.38 magnitude, I_c to B. The O'Connell effect, an indicator of spot activity, averages less than the noise level, 0.003–0.01 magnitude. But, night-to-night variations and scatter in the light curves demonstrate a high magnetic activity level. The differences in minima are small, 0.005–0.04 magnitude, indicating an overcontact binary in fair thermal contact. A time of constant light occurs at our secondary minima and lasts some 27 minutes due to a total eclipse. The 2MASS, J–K = 0.38 ± 0.02 for the binary, and the APASS (B–V)–E(B–V) = 0.52. These correspond to ~G4V±2.5 spectral type which yields a temperature of 5750±400 K. Fast rotating binary stars of this type are noted for having convective atmospheres, so the binary is of solar type with a convective atmosphere.

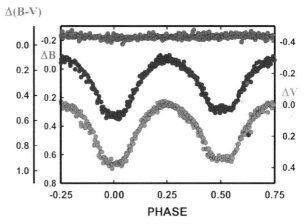

Figure 5a. 2018 B,V light curves and B–V color curves (Variable-Comparison, magnitudes phased with Equation 4.

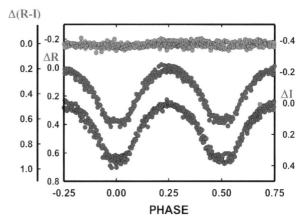

Figure 5b. 2018 R_c, I_c light curves and R_c–I_c color curves (Variable-Comparison, magnitudes phased with Equation 4.

6. Light curve solution

The B, V, R_c, and I_c curves were pre-modeled with BINARY MAKER 3.0 (Bradstreet and Steelman 2002) and fits were determined in BVR_cI_c filter bands which were very stable. The solution was that of an overcontact eclipsing binary. The parameters were then averaged and input into a four-color simultaneous light curve calculation using the Wilson-Devinney program (Wilson and Devinney 1971; Wilson 1990, 1994; Van Hamme and Wilson 1998). The BINARY MAKER parameters included q (mass ratio) = 0.16 ± 0.2, fill-out = 25%, inclination (i) = 84°, and $T_2 - T_1 = 50 \pm 150$ K. The computation was completed in Mode 3 and converged to a solution. Convective parameters g = 0.32, A = 0.5 were used. An eclipse duration of ~27 minutes was determined for our secondary eclipse (phase 0.5) and the light curve solution. The more massive component is the hotter, making the system an A-type W UMa contact binary. In modeling the B,V,R_c,I_c curves simultaneously, the I_c curve was found to be discordant. So we modeled the B,V,R_c curves simultaneously and the I_c curve separately. We feel that there is physical reason for this, such as asymmetrically-placed circumbinary dust or gas or an IR thin disk affecting the I_c curves much more than the visual ones. Our CCD is back-illuminated so red noise could affect the results, but this has not been seen previously on many dozens of cool binaries. We tried third light but that did not solve any fitting issues. More observations are needed to sort this out. Only the simultaneous BVR_cI_c results will be commented on here. The solutions are given in Table 5. The B,V,R_c,I_c normalized fluxes overlaid by our 2018 solution of OR Leo are given as Figures 6a and 6b. Figures 7a–7d display quarter phases of the Roche lobe surface.

7. Discussion

OR Leo is an overcontact W UMa binary with a Roche lobe fill-out of 24%. Since the eclipses were total, the mass ratio, q, is well determined. The system has a rather extreme mass ratio of ~0.18 (M_2/M_1), and the component temperature difference is small, only ~60 K. No spots were needed in the modeling. The inclination is ~81 degrees, which results in total eclipses. Its photometric spectral type indicates a surface temperature of ~5750 K for the primary component, making it a solar type binary. Such a main sequence star would have a mass of ~0.92 $M_\odot$ and the secondary (from the mass ratio) would have a mass of ~0.17 $M_\odot$, making it very much undersized for its temperature (~5680 K), but the contact produces a secondary atmosphere with much the same temperature as that of the primary component. Such a main sequence star would be of type M6V. So its true core temperature is completely masked. The W UMa is of A-type, which may mean that it is a mature, very old W UMa binary. However, the fill-out is somewhat low for such a system.

8. Conclusion

The period study of this overcontact W UMa binary has a 15-year time duration. The period is found to be increasing at about the 27 sigma level. This can happen due to mass transfer

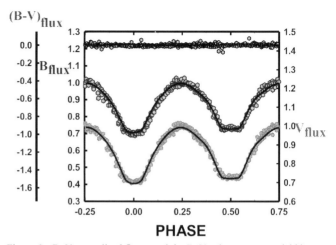

Figure 6a. B, V normalized fluxes and the B–V color curves overlaid by our 2018 solution of OR Leo.

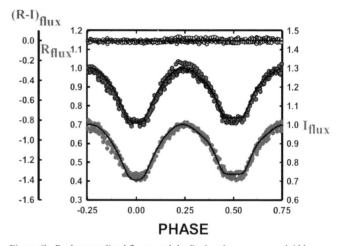

Figure 6b. R_c, I_c normalized fluxes and the R_c–I_c color curves overlaid by our 2018 solution of OR Leo. The I_c curve solution is that shown in Table 5, right hand column.

Figure 7a. OR Leo, geometrical representation at phase 0.0.

Figure 7b. OR Leo, geometrical representation at phase 0.25.

Figure 7c. OR Leo, geometrical representation at phase 0.50.

Figure 7d. OR Leo, geometrical representation at phase 0.75.

to the primary component. We would expect this magnetic braking due to the solar type of the system. If this is occurring, it is moderating the changing of the period, thus making the mass transfer rate actually higher. The mass exchange, if continuous, has decreased the mass of the secondary star, making the mass ratio rather extreme. We would expect that this will eventually cause the system to coalesce following a red novae event (Tylenda and Kamiński 2016). After some system mass loss, one would theorize that the binary will become a rather normal, fast rotating, single ~G0V type field star. As stated in section 4, the observed period increase could be a part of a light-time effect caused by an additional and undetected third body in the system. Finally, radial velocity curves are needed to obtain absolute (not relative) system parameters. More photometric monitoring is needed to determine the nature of the I_c observations. Deeper IR observations could help resolve this problem of the discordant I_c curve.

9. Acknowledgements

Dr. Samec wishes to thank Dr. Bob Hill for his continued observing help and kind friendship through the years.

References

Bradstreet, D. H., and Steelman, D. P. 2002, *Bull. Amer. Astron. Soc*, **34**, 1224.

Caton, D., Gentry, D. R., Samec, R. G., Chamberlain, H., Robb, R., Faulkner, D. R., and Hill, R. 2019, *Publ. Astron. Soc. Pacific*, **131**, 4203.

Caton, D. B., Samec, R. G., Robb, R., Faulkner, D. R., van Hamme, W., Clark, J. D., and Shebs, T. 2017, *Publ. Astron. Soc. Pacific*, **129**, 4202.

Henden, A. A., *et al.* 2015, AAVSO Photometric All-Sky Survey, data release 9 (http://www.aavso.org/apass).

Kazarovets, E. V., Samus, N. N., Durlevich, O. V. Kireeva, N. N., and Pastukhova, E. N. 2015, *Inf. Bull. Var. Stars*, No. 6151, 1.

Mikulášek, Z., Chrastina, M., Liška, J., Zejda, M., Janík, J., Zhu, L.-Y., and Qian, S.-B. 2014, *Contrib. Astron. Obs. Skalnaté Pleso*, **43**, 382.

Pojmański, G. 2002, *Acta Astron.*, **52**, 397.

Pribulla, T., Vanko, M., and Hambalek, L. 2009, *Inf. Bull. Var. Stars*, No. 5886, 1.

Samec, R. G., Caton, D. B., and Faulkner, D. R. 2018a, *J. Amer. Assoc. Var. Star Obs.*, **46**, 57.

Samec, R. G., Caton, D. B., Faulkner, D. R., and Hill, R. 2018b, *J. Amer. Assoc. Var. Star Obs.*, **46**, 106.

Samec, R. G., Norris, C. L., Hill, B. L., van Hamme, W., and Faulkner, D. R. 2017a, *J. Amer. Assoc. Var. Star Obs.*, **45**, 3.

Samec, R. G., Olsen, A., Caton, D. B., Faulkner, D. R., and Hill, R. L. 2017b, *J. Amer. Assoc. Var. Star Obs.*, **45**, 173.

Skrutskie, M. F., *et al.* 2006, *Astron. J.*, **131**, 1163.

Tylenda, R., and Kamiński, T. 2016, *Astron. Astrophys.*, **592A**, 134.

U.S. Naval Observatory. 2012, UCAC-3 (http://www.usno.navy.mil/USNO/astrometry/optical-IR-prod/ucac).

van Hamme, W. V., and Wilson, R. E. 1998, *Bull. Amer. Astron. Soc.*, **30**, 1402.

Wilson, R. E. 1990, *Astrophys. J.*, **356**, 613.

Wilson, R. E. 1994, *Publ. Astron. Soc. Pacific*, **106**, 921.

Wilson, R. E., and Devinney, E. J. 1971, *Astrophys. J.*, **166**, 605.

Table 2. OR Leo observations, ΔB, ΔV, ΔR$_c$, and ΔI$_c$, variable star minus comparison star.

ΔB	BHJD 2458100+	ΔB	BHJD 2458100+	ΔB	BHJD 2458100+	ΔB	BHJD 2458100+	ΔB	BHJD 2458100+
0.257	39.7316	0.042	39.9165	0.161	42.8849	0.163	60.9020	0.277	74.9721
0.286	39.7343	0.023	39.9193	0.143	42.8876	0.112	60.9092	0.281	74.9753
0.283	39.7371	0.008	39.9220	0.107	42.8903	−0.018	74.7390	0.287	74.9786
0.262	39.7398	−0.009	39.9247	0.091	42.8931	−0.042	74.7426	0.289	74.9818
0.273	39.7425	−0.024	39.9274	0.060	42.8958	−0.051	74.7462	0.262	74.9851
0.275	39.7452	−0.037	39.9301	0.041	42.8985	−0.046	74.7498	0.221	74.9883
0.266	39.7479	−0.048	39.9329	0.033	42.9012	−0.066	74.7534	0.180	74.9916
0.283	39.7507	−0.058	39.9356	0.013	42.9039	−0.077	74.7570	0.142	74.9948
0.272	39.7534	−0.053	39.9383	0.001	42.9066	−0.088	74.7641	0.110	74.9981
0.275	39.7561	−0.059	39.9410	−0.025	42.9094	−0.088	74.7677	0.077	75.0014
0.251	39.7588	−0.067	39.9438	−0.030	42.9121	−0.069	74.7713	0.054	75.0046
0.225	39.7615	−0.059	39.9465	−0.036	42.9148	−0.066	74.7749	0.041	75.0078
0.190	39.7642	−0.056	39.9492	−0.055	42.9175	−0.050	74.7785	0.019	75.0111
0.162	39.7670	−0.055	39.9519	−0.038	42.9202	−0.050	74.7820	0.003	75.0143
0.127	39.7697	−0.055	39.9546	−0.041	42.9229	−0.038	74.7856	−0.008	75.0176
0.104	39.7724	−0.046	39.9574	−0.061	42.9257	−0.021	74.7892	−0.045	88.5686
0.069	39.7751	−0.042	39.9601	−0.058	42.9284	0.004	74.7928	−0.053	88.5744
0.038	39.7778	−0.032	39.9628	−0.065	42.9311	0.022	74.7964	−0.034	88.5801
0.030	39.7806	−0.016	39.9655	−0.043	42.9365	0.046	74.8000	−0.043	88.5859
0.010	39.7833	−0.009	39.9683	−0.038	42.9393	0.088	74.8036	−0.075	88.5917
−0.001	39.7860	0.004	39.9710	−0.034	42.9420	0.127	74.8071	−0.057	88.6032
−0.009	39.7887	0.019	39.9737	−0.002	42.9447	0.172	74.8107	0.016	88.6090
−0.018	39.7914	0.027	39.9764	−0.019	42.9474	0.214	74.8143	0.042	88.6147
−0.035	39.7942	0.041	39.9792	0.006	42.9501	0.268	74.8182	0.090	88.6205
−0.050	39.7969	0.054	39.9819	0.019	42.9528	0.292	74.8217	0.142	88.6262
−0.058	39.7996	0.064	39.9846	0.025	42.9556	0.319	74.8253	0.238	88.6320
−0.065	39.8023	0.106	39.9873	0.046	42.9583	0.334	74.8289	0.288	88.6378
−0.070	39.8050	0.112	39.9901	0.057	42.9610	0.336	74.8325	0.321	88.6493
−0.074	39.8078	0.138	39.9928	0.074	42.9637	0.328	74.8361	0.342	88.6551
−0.072	39.8105	0.176	39.9955	0.106	42.9664	0.310	74.8396	0.296	88.6666
−0.073	39.8132	0.212	39.9982	0.128	42.9692	0.307	74.8432	0.259	88.6723
−0.077	39.8159	0.286	40.0010	0.169	42.9719	0.289	74.8468	0.163	88.6781
−0.080	39.8186	0.046	40.7307	0.184	42.9746	0.231	74.8503	0.100	88.6839
−0.070	39.8214	0.024	40.7334	0.217	42.9773	0.193	74.8539	0.060	88.6896
−0.048	39.8268	0.011	40.7361	0.245	42.9800	0.149	74.8575	0.031	88.6954
−0.046	39.8295	0.001	40.7388	0.281	42.9827	0.051	74.8658	−0.030	88.7011
−0.027	39.8323	−0.014	40.7415	0.292	42.9855	0.037	74.8690	−0.049	88.7069
−0.007	39.8350	−0.018	40.7443	0.268	42.9882	0.023	74.8723	−0.096	88.7127
0.020	39.8404	−0.041	40.7470	0.277	42.9909	−0.020	74.8788	−0.070	88.7184
0.037	39.8431	−0.042	40.7497	0.257	42.9936	−0.037	74.8820	−0.096	88.7242
0.060	39.8458	−0.048	40.7524	0.256	42.9963	−0.062	74.8873	−0.061	88.7300
0.087	39.8486	−0.059	40.7551	0.055	60.6394	−0.084	74.8906	−0.001	88.7415
0.107	39.8513	−0.055	40.7578	0.092	60.6440	−0.081	74.8938	0.004	88.7472
0.144	39.8540	−0.053	40.7605	0.032	60.6472	−0.078	74.8971	0.045	88.7530
0.162	39.8567	−0.060	40.7632	0.015	60.6532	−0.088	74.9003	0.107	88.7588
0.193	39.8594	−0.062	40.7660	−0.023	60.6599	−0.075	74.9035	0.180	88.7645
0.232	39.8622	−0.042	40.7687	−0.018	60.6668	−0.072	74.9068	0.175	88.7703
0.266	39.8649	−0.028	40.7714	−0.062	60.6736	−0.065	74.9100	0.278	88.7761
0.286	39.8676	−0.030	40.7741	−0.048	60.6805	−0.063	74.9132	0.284	88.7996
0.301	39.8703	−0.019	40.7768	−0.037	60.6874	−0.053	74.9165	0.235	88.8054
0.306	39.8730	−0.018	40.7795	0.040	60.6942	−0.030	74.9197	0.159	88.8112
0.317	39.8757	−0.006	40.7822	0.058	60.7011	−0.022	74.9230	0.100	88.8169
0.312	39.8785	0.019	40.7850	0.091	60.7080	−0.001	74.9262	0.052	88.8227
0.319	39.8812	0.046	40.7877	0.135	60.7148	0.009	74.9294	0.011	88.8285
0.302	39.8839	0.024	40.7904	0.227	60.7220	0.032	74.9327	−0.002	88.8342
0.303	39.8866	0.287	42.8459	0.306	60.7289	0.056	74.9363	−0.021	88.8400
0.309	39.8893	0.335	42.8550	0.294	60.7358	0.084	74.9396	−0.042	88.8458
0.291	39.8921	0.311	42.8577	−0.050	60.8324	0.109	74.9428	−0.061	88.8515
0.254	39.8948	0.324	42.8604	0.006	60.8392	0.149	74.9461	−0.066	88.8573
0.243	39.8975	0.323	42.8659	0.052	60.8461	0.190	74.9493	−0.081	88.8630
0.196	39.9002	0.322	42.8686	0.119	60.8529	0.233	74.9526	−0.023	88.8688
0.170	39.9030	0.321	42.8713	0.198	60.8598	0.257	74.9558	0.003	88.8746
0.143	39.9057	0.293	42.8740	0.270	60.8667	0.275	74.9591	0.039	88.8861
0.112	39.9084	0.265	42.8767	0.287	60.8804	0.275	74.9623		
0.092	39.9111	0.228	42.8795	0.283	60.8876	0.276	74.9656		
0.065	39.9138	0.207	42.8822	0.241	60.8948	0.281	74.9688		

Table continued on following pages

Table 2. OR Leo observations, ΔB, ΔV, ΔR_c, and ΔI_c, variable star minus comparison star, cont.

ΔV	VHJD 2458100+	ΔV	VHJD 2458100+	ΔV	VHJD 2458100+	ΔV	VHJD 2458100+	ΔV	VHJD 2458100+
0.325	39.7326	0.114	39.9148	0.214	42.8858	0.025	74.7450	0.341	74.9840
0.339	39.7353	0.102	39.9175	0.168	42.8885	0.020	74.7486	0.302	74.9872
0.342	39.7380	0.087	39.9202	0.153	42.8913	0.010	74.7521	0.256	74.9905
0.325	39.7407	0.065	39.9229	0.130	42.8940	−0.001	74.7558	0.218	74.9938
0.342	39.7434	0.058	39.9256	0.113	42.8967	−0.002	74.7594	0.187	74.9970
0.332	39.7461	0.038	39.9284	0.091	42.8994	−0.014	74.7629	0.159	75.0003
0.331	39.7489	0.034	39.9311	0.075	42.9021	−0.018	74.7665	0.127	75.0035
0.328	39.7516	0.024	39.9338	0.061	42.9049	−0.003	74.7701	0.109	75.0068
0.351	39.7543	0.012	39.9365	0.046	42.9076	0.020	74.7737	0.094	75.0100
0.325	39.7570	−0.002	39.9392	0.019	42.9103	0.011	74.7773	0.077	75.0133
0.308	39.7597	0.007	39.9420	0.022	42.9130	0.015	74.7808	0.055	75.0165
0.288	39.7625	0.005	39.9447	0.019	42.9157	0.025	74.7844	0.046	75.0197
0.250	39.7652	−0.002	39.9474	0.017	42.9184	0.045	74.7880	0.054	88.5641
0.220	39.7679	0.007	39.9501	−0.003	42.9212	0.069	74.7916	0.026	88.5705
0.173	39.7706	0.013	39.9529	−0.003	42.9266	0.079	74.7952	−0.011	88.5763
0.149	39.7733	0.011	39.9556	0.005	42.9293	0.113	74.7987	0.016	88.5821
0.142	39.7761	0.009	39.9583	0.001	42.9320	0.140	74.8023	0.025	88.5878
0.109	39.7788	0.033	39.9610	0.010	42.9347	0.174	74.8059	0.033	88.5936
0.082	39.7815	0.047	39.9637	0.014	42.9375	0.228	74.8095	0.025	88.5994
0.081	39.7842	0.057	39.9665	0.019	42.9402	0.263	74.8131	0.074	88.6051
0.053	39.7869	0.057	39.9692	0.027	42.9429	0.303	74.8167	0.148	88.6167
0.057	39.7897	0.070	39.9719	0.031	42.9456	0.336	74.8205	0.179	88.6224
0.044	39.7924	0.086	39.9746	0.053	42.9483	0.371	74.8241	0.259	88.6282
0.029	39.7951	0.088	39.9774	0.058	42.9511	0.366	74.8277	0.326	88.6339
0.019	39.7978	0.104	39.9801	0.076	42.9538	0.380	74.8313	0.372	88.6397
0.006	39.8005	0.125	39.9828	0.084	42.9565	0.379	74.8349	0.384	88.6570
−0.003	39.8033	0.156	39.9856	0.108	42.9592	0.366	74.8384	0.390	88.6628
−0.004	39.8060	0.165	39.9883	0.116	42.9619	0.362	74.8420	0.360	88.6685
0.003	39.8087	0.209	39.9910	0.132	42.9646	0.345	74.8456	0.291	88.6743
−0.010	39.8114	0.238	39.9937	0.155	42.9674	0.309	74.8491	0.193	88.6801
−0.007	39.8169	0.282	39.9965	0.186	42.9701	0.264	74.8527	0.153	88.6858
0.021	39.8196	0.283	39.9992	0.215	42.9728	0.222	74.8563	0.133	88.6916
0.017	39.8223	0.308	40.0019	0.243	42.9755	0.185	74.8599	0.088	88.6973
0.006	39.8250	0.100	40.7316	0.285	42.9782	0.100	74.8680	0.054	88.7031
0.012	39.8278	0.091	40.7343	0.314	42.9810	0.084	74.8712	0.048	88.7089
0.034	39.8305	0.064	40.7370	0.345	42.9837	0.062	74.8744	0.017	88.7146
0.054	39.8332	0.046	40.7398	0.338	42.9864	0.050	74.8777	−0.009	88.7204
0.044	39.8359	0.039	40.7425	0.345	42.9891	0.039	74.8809	−0.008	88.7262
0.077	39.8386	0.038	40.7452	0.358	43.0027	0.022	74.8841	0.019	88.7319
0.086	39.8413	0.020	40.7479	0.217	60.6337	−0.009	74.8895	0.027	88.7357
0.108	39.8441	0.020	40.7506	0.144	60.6386	−0.015	74.8927	0.068	88.7377
0.127	39.8468	0.004	40.7533	0.167	60.6409	−0.020	74.8960	0.077	88.7434
0.159	39.8495	0.007	40.7560	0.144	60.6433	−0.019	74.8992	0.108	88.7492
0.174	39.8522	−0.001	40.7587	0.062	60.6510	−0.014	74.9024	0.145	88.7550
0.205	39.8549	0.009	40.7615	0.016	60.6642	−0.009	74.9057	0.196	88.7607
0.242	39.8577	0.001	40.7642	0.011	60.6711	0.001	74.9089	0.238	88.7665
0.277	39.8604	0.013	40.7669	0.007	60.6780	0.006	74.9121	0.321	88.7722
0.302	39.8631	0.025	40.7696	0.006	60.6849	0.014	74.9154	0.349	88.7780
0.334	39.8658	0.015	40.7723	0.046	60.6917	0.038	74.9219	0.358	88.7838
0.354	39.8685	0.051	40.7750	0.062	60.6986	0.052	74.9251	0.296	88.8074
0.373	39.8712	0.028	40.7778	0.135	60.7055	0.070	74.9283	0.208	88.8131
0.372	39.8740	0.083	40.7805	0.149	60.7123	0.091	74.9316	0.125	88.8246
0.378	39.8767	0.082	40.7832	0.257	60.7192	0.107	74.9348	0.084	88.8304
0.376	39.8794	0.095	40.7859	0.340	60.7264	0.130	74.9385	0.066	88.8362
0.367	39.8821	0.102	40.7913	0.344	60.7332	0.200	74.9450	0.045	88.8419
0.372	39.8849	0.344	42.8469	0.392	60.7401	0.237	74.9482	0.049	88.8477
0.364	39.8876	0.364	42.8505	0.372	60.7538	0.283	74.9515	0.022	88.8534
0.363	39.8903	0.375	42.8614	0.105	60.8367	0.316	74.9547	0.014	88.8592
0.332	39.8930	0.382	42.8641	0.127	60.8436	0.340	74.9580	0.012	88.8650
0.314	39.8957	0.370	42.8668	0.178	60.8504	0.339	74.9612	0.036	88.8707
0.290	39.8984	0.376	42.8695	0.284	60.8573	0.343	74.9645	0.071	88.8765
0.242	39.9012	0.343	42.8722	0.332	60.8642	0.347	74.9677		
0.213	39.9039	0.324	42.8750	0.285	60.8993	0.352	74.9710		
0.183	39.9066	0.292	42.8777	0.190	60.9066	0.349	74.9742		
0.167	39.9093	0.264	42.8804	0.062	74.7358	0.344	74.9775		
0.138	39.9120	0.249	42.8831	0.040	74.7414	0.354	74.9807		

Table continued on following pages

Table 2. OR Leo observations, ΔB, ΔV, ΔR$_c$, and ΔI$_c$, variable star minus comparison star, cont.

ΔR	RHJD 2458100+	ΔR	RHJD 2458100+	ΔR	RHJD 2458100+	ΔR	RHJD 2458100+	ΔR	RHJD 2458100+
0.139	40.729	0.021	42.922	0.258	60.856	−0.007	74.889	0.097	88.606
0.111	40.732	0.013	42.924	0.334	60.863	−0.021	74.892	0.128	88.612
0.076	40.735	0.024	42.927	0.380	60.870	−0.015	74.895	0.171	88.618
0.096	40.738	0.028	42.930	0.389	60.877	−0.010	74.899	0.206	88.624
0.070	40.740	0.025	42.933	0.378	60.884	−0.015	74.902	0.281	88.629
0.052	40.743	0.034	42.935	0.383	60.891	−0.008	74.905	0.331	88.635
0.044	40.746	0.041	42.938	0.328	60.898	−0.008	74.908	0.379	88.641
0.040	40.748	0.054	42.941	0.192	60.905	−0.002	74.912	0.417	88.647
0.031	40.751	0.053	42.943	0.071	74.741	0.006	74.915	0.417	88.658
0.029	40.754	0.057	42.946	0.054	74.744	0.018	74.918	0.404	88.664
0.035	40.757	0.075	42.949	0.034	74.748	0.035	74.921	0.370	88.670
0.025	40.759	0.081	42.952	0.035	74.751	0.050	74.924	0.282	88.675
0.033	40.762	0.093	42.954	0.019	74.755	0.072	74.928	0.218	88.681
0.027	40.765	0.108	42.957	0.021	74.759	0.075	74.931	0.174	88.687
0.027	40.767	0.116	42.960	0.004	74.762	0.106	74.934	0.146	88.693
0.037	40.770	0.140	42.962	−0.008	74.766	0.131	74.938	0.100	88.698
0.054	40.773	0.161	42.965	0.008	74.769	0.162	74.941	0.069	88.704
0.050	40.776	0.190	42.968	0.022	74.773	0.189	74.944	0.050	88.710
0.098	40.778	0.199	42.971	0.042	74.777	0.233	74.948	0.026	88.716
0.076	40.781	0.248	42.973	0.044	74.780	0.274	74.951	−0.003	88.722
0.102	40.786	0.288	42.976	0.052	74.784	0.311	74.954	0.023	88.727
0.129	40.789	0.307	42.979	0.070	74.787	0.340	74.957	0.062	88.733
0.325	42.845	0.341	42.981	0.104	74.791	0.350	74.961	0.078	88.739
0.357	42.848	0.336	42.984	0.122	74.794	0.342	74.964	0.125	88.750
0.394	42.851	0.341	42.987	0.135	74.798	0.358	74.967	0.162	88.756
0.369	42.854	0.368	42.990	0.162	74.802	0.356	74.970	0.226	88.762
0.382	42.862	0.377	42.992	0.201	74.805	0.359	74.974	0.279	88.768
0.385	42.865	0.389	42.995	0.232	74.809	0.358	74.977	0.372	88.773
0.377	42.867	0.391	42.998	0.262	74.812	0.368	74.980	0.386	88.779
0.381	42.870	0.338	43.001	0.301	74.816	0.352	74.983	0.388	88.785
0.359	42.873	0.143	60.640	0.353	74.820	0.320	74.987	0.384	88.791
0.329	42.875	0.143	60.643	0.381	74.823	0.284	74.990	0.358	88.803
0.308	42.878	0.161	60.645	0.377	74.827	0.235	74.993	0.300	88.808
0.267	42.881	0.112	60.650	0.360	74.831	0.208	74.996	0.225	88.814
0.236	42.884	0.081	60.656	0.377	74.834	0.171	75.000	0.214	88.820
0.207	42.886	0.032	60.663	0.378	74.838	0.147	75.003	0.146	88.826
0.187	42.889	0.035	60.670	0.389	74.841	0.124	75.006	0.108	88.832
0.164	42.892	0.041	60.677	0.362	74.845	0.105	75.009	0.079	88.837
0.142	42.895	0.037	60.684	0.334	74.848	0.089	75.013	0.039	88.843
0.122	42.897	0.079	60.691	0.294	74.852	0.072	75.016	0.045	88.849
0.112	42.900	0.091	60.697	0.246	74.856	0.073	75.019	0.046	88.855
0.077	42.903	0.153	60.704	0.187	74.859	0.074	88.566	0.036	88.860
0.070	42.905	0.173	60.711	0.110	74.867	0.051	88.572	0.044	88.866
0.063	42.908	0.249	60.718	0.088	74.871	0.031	88.577	0.087	88.872
0.052	42.911	0.399	60.753	0.074	74.874	0.052	88.583	0.101	88.878
0.035	42.914	0.092	60.836	0.058	74.877	0.041	88.589	0.155	88.883
0.033	42.916	0.129	60.842	0.037	74.880	0.029	88.595		
0.030	42.919	0.188	60.849	0.016	74.884	0.063	88.601		

Table continued on next page

Table 2. OR Leo observations, ΔB, ΔV, ΔR_c, and ΔI_c, variable star minus comparison star, cont.

ΔI	IHJD 2458100+	ΔI	IHJD 2458100+	ΔI	IHJD 2458100+	ΔI	IHJD 2458100+	ΔI	IHJD 2458100+
0.111	40.730	0.022	42.925	0.253	60.855	0.020	74.888	0.071	88.607
0.088	40.733	0.006	42.928	0.309	60.862	0.018	74.892	0.156	88.613
0.076	40.735	−0.003	42.930	0.334	60.869	0.019	74.895	0.181	88.619
0.074	40.738	0.019	42.933	0.350	60.876	0.007	74.898	0.217	88.624
0.044	40.741	0.019	42.936	0.388	60.883	0.009	74.901	0.268	88.630
0.039	40.743	0.016	42.938	0.378	60.890	0.014	74.905	0.348	88.636
0.039	40.746	0.035	42.941	0.307	60.897	0.016	74.908	0.390	88.653
0.028	40.749	0.036	42.944	0.241	60.904	0.018	74.911	0.379	88.659
0.008	40.754	0.058	42.947	0.195	60.912	0.023	74.914	0.380	88.665
0.013	40.760	0.075	42.949	0.090	74.740	0.048	74.917	0.324	88.671
0.009	40.762	0.068	42.952	0.052	74.744	0.047	74.921	0.234	88.676
0.013	40.765	0.094	42.955	0.061	74.747	0.074	74.924	0.172	88.682
0.011	40.768	0.097	42.957	0.052	74.751	0.076	74.927	0.146	88.688
0.039	40.771	0.115	42.960	0.017	74.754	0.108	74.930	0.090	88.699
0.037	40.773	0.132	42.963	0.020	74.758	0.116	74.934	0.062	88.705
0.044	40.776	0.165	42.966	0.014	74.762	0.157	74.937	0.039	88.711
0.047	40.779	0.183	42.968	−0.001	74.765	0.194	74.941	0.010	88.717
0.073	40.781	0.213	42.971	−0.004	74.769	0.224	74.944	0.016	88.722
0.103	40.784	0.228	42.974	0.023	74.772	0.250	74.947	0.023	88.728
0.115	40.787	0.286	42.976	0.037	74.776	0.285	74.950	0.046	88.734
0.079	40.790	0.319	42.979	0.044	74.780	0.326	74.954	0.093	88.745
0.328	42.845	0.333	42.982	0.044	74.783	0.337	74.957	0.128	88.751
0.350	42.849	0.337	42.985	0.060	74.787	0.350	74.960	0.159	88.757
0.366	42.854	0.337	42.987	0.089	74.790	0.361	74.963	0.234	88.763
0.376	42.857	0.374	42.990	0.116	74.794	0.379	74.967	0.291	88.769
0.373	42.862	0.337	42.993	0.130	74.797	0.358	74.970	0.351	88.774
0.390	42.865	0.344	42.995	0.146	74.801	0.383	74.973	0.363	88.780
0.370	42.868	0.388	42.998	0.186	74.805	0.374	74.976	0.359	88.786
0.340	42.870	0.378	43.001	0.221	74.808	0.371	74.980	0.379	88.792
0.361	42.873	0.157	60.638	0.249	74.812	0.365	74.983	0.352	88.798
0.321	42.876	0.179	60.640	0.288	74.815	0.343	74.986	0.341	88.804
0.288	42.879	0.154	60.642	0.348	74.819	0.283	74.989	0.274	88.809
0.250	42.881	0.138	60.645	0.351	74.823	0.266	74.993	0.201	88.815
0.236	42.884	0.122	60.649	0.369	74.826	0.224	74.996	0.156	88.821
0.208	42.887	0.086	60.655	0.340	74.830	0.173	74.999	0.120	88.827
0.175	42.889	0.045	60.662	0.355	74.834	0.174	75.002	0.105	88.832
0.149	42.892	0.048	60.669	0.360	74.837	0.152	75.006	0.050	88.838
0.134	42.895	0.030	60.676	0.349	74.841	0.122	75.009	0.059	88.844
0.104	42.898	0.030	60.683	0.339	74.844	0.109	75.012	0.048	88.850
0.090	42.900	0.055	60.690	0.321	74.848	0.107	75.015	0.028	88.856
0.083	42.903	0.084	60.697	0.284	74.851	0.093	75.019	0.043	88.861
0.046	42.906	0.124	60.703	0.249	74.855	0.071	88.567	0.047	88.867
0.056	42.909	0.164	60.710	0.196	74.859	0.053	88.573	0.063	88.873
0.046	42.911	0.205	60.717	0.114	74.867	0.034	88.578	0.114	88.879
0.026	42.914	0.277	60.724	0.088	74.870	0.046	88.584	0.115	88.884
0.013	42.917	0.307	60.731	0.095	74.873	0.050	88.590		
0.023	42.919	0.081	60.835	0.073	74.877	0.027	88.596		
0.016	42.922	0.172	60.848	0.052	74.880	0.048	88.601		

Table 3. O–C residuals for OR Leo.

	Epoch 2400000+	Cycles	Linear Residuals	Quadratic Residuals	Weight	Reference
1	52652.7891	−20430.0	0.0637	−0.0048	0.1	ASAS (Pojmański 2002)
2	52706.7113	−20231.0	0.0628	−0.0028	0.1	ASAS (Pojmański 2002)
3	52733.8048	−20131.0	0.0593	−0.0049	0.1	ASAS (Pojmański 2002)
4	53092.6950	−18806.5	0.0492	0.0030	0.1	ASAS (Pojmański 2002)
5	53446.7043	−17500.0	0.0356	0.0052	0.1	ASAS (Pojmański 2002)
6	53525.5551	−17209.0	0.0340	0.0069	0.1	ASAS (Pojmański 2002)
7	53797.7374	−16204.5	0.0265	0.0099	0.1	ASAS (Pojmański 2002)
8	54857.5970	−12293.0	−0.0147	−0.0010	0.5	Pribulla et al. (2009)
9	54905.4228	−12116.5	−0.0151	−0.0005	1.0	Pribulla et al. (2009)
10	54905.5589	−12116.0	−0.0145	0.0001	1.0	Pribulla et al. (2009)
11	54917.4808	−12072.0	−0.0153	−0.0005	1.0	Pribulla et al. (2009)
12	58139.7442	−180.5	0.0033	0.0021	1.0	Present observations
13	58139.8797	−180.0	0.0034	0.0021	1.0	Present observations
14	58142.8597	−169.0	0.0027	0.0013	1.0	Present observations
15	58142.9903	−168.5	−0.0022	−0.0036	1.0	Present observations
16	58160.7414	−103.0	0.0003	−0.0015	1.0	Present observations
17	58160.8800	−102.5	0.0035	0.0016	1.0	Present observations
18	58174.8331	−51.0	0.0016	−0.0007	1.0	Present observations
19	58174.9709	−50.5	0.0039	0.0016	1.0	Present observations
20	58188.6529	0.0	0.0019	−0.0008	1.0	Present observations
21	58188.7874	0.5	0.0009	−0.0018	1.0	Present observations

Table 4. Light curve characteristics for OR Leo.

Filter	Phase 0.00	Magnitude* Min. I	Phase 0.25	Magnitude* Max. II
B		0.377 ± 0.008		0.016 ± 0.017
V		0.359 ± 0.020		0.007 ± 0.010
R		0.319 ± 0.017		−0.062 ± 0.011
I		0.359 ± 0.020		0.007 ± 0.010

Filter	Phase 0.50	Magnitude Min. II	Phase 0.75	Magnitude Max. I
B		0.365 ± 0.015		0.019 ± 0.013
V		0.354 ± 0.023		0.010 ± 0.003
R		0.275 ± 0.009		−0.072 ± 0.012
I		0.354 ± 0.023		0.010 ± 0.020

Filter	Min. I – Max. II	Max. I – Max. II	Min. I – Min. II
B	0.3607 ± 0.0253	0.0029 ± 0.0305	0.0118 ± 0.0236
V	0.3525 ± 0.0303	0.0029 ± 0.0134	0.0052 ± 0.0429
R	0.3817 ± 0.0282	0.0098 ± 0.0230	0.0440 ± 0.0263
I	0.3525 ± 0.0303	0.0029 ± 0.0299	0.0052 ± 0.0430

*Magnitude is the variable star – comparison star magnitude.

Table 5. B,V,Rc,Ic, solution parameters.

Parameters	B,V, R_c Values	I_c Values
$\lambda B, \lambda V, \lambda R, \lambda I$ (nm)	440, 550, 640	790
$x_{bol1,2}, y_{bol1,2}$	0.570, 0.269, 0.570, 0.269	0.570, 0.269, 0.570, 0.269
$x_{1I,2I}, y_{1I,2I}$	—	0.839, 0.145, 0.839, 0.145
$x_{1R,2R}, y_{1R,2R}$	0.762, 0.232, 0.762, 0.232	—
$x_{1V,2V}, y_{1V,2V}$	0.691, 0.251, 0.691, 0.251	—
$x_{1B,2B}, y_{1B,2B}$	0.607, 0.246, 0.607, 0.246	—
g_1, g_2	0.32	0.32
A_1, A_2	0.5	0.5
Inclination (°)	81.1 ± 0.1	79.3 ± 0.3
T_1, T_2 (K)	5750*, 5683 ± 4	5750*, 5890 ± 12
Epoch	2458188.6529 ± 0.0015	2458188.6523 ± 0.0003
Period	0.270961 ± 0.000001	0.270958 ± 0.000002
$\Omega_1 = \Omega_2$	2.1613 ± 0.0015	2.078 ± 0.003
$q(m_2/m_1)^*$	0.1827 ± 0.0004	0.1505 ± 0.0006
$F_1 = F_2$ (%)	24 ± 1	28 ± 1
$L_1/(L_1+L_2)_I$	—	0.832 ± 0.009
$L_1/(L_1+L_2)_R$	0.826 ± 0.007	—
$L_1/(L_1+L_2)_V$	0.827 ± 0.010	—
$L_1/(L_1+L_2)_B$	0.829 ± 0.009	—

*± 400 K.

VR$_c$I$_c$ Photometric Study of the Totally Eclipsing Pre-W UMa Binary, V616 Camelopardalis: Is it Detached?

Ronald G. Samec
Faculty Research Associate, Pisgah Astronomical Research Institute, 1 PARI Drive, Rosman, NC 28772; ronaldsamec@gmail.com

Daniel B. Caton
Davis R. Gentry
Dark Sky Observatory, Department of Physics and Astronomy, Appalachian State University, 525 Rivers Street, Boone, NC 28608-2106; catondb@appstate.edu, gentrydr@appstate.edu

Danny R. Faulkner
Johnson Observatory, 1414 Bur Oak Court, Hebron, KY 41048

Received May 16, 2019; revised July 8, 26, August 9, 2019; accepted August 16, 2019

Abstract V616 Cam is a F3V±3 type (T~6750±400 K) eclipsing binary. It was observed on March 5, 6, 9, and 30, 2017, at Dark Sky Observatory in North Carolina with the 0.81-m reflector of Appalachian State University. Five times of minimum light were determined from our present observations, which include three primary eclipses and two secondary eclipses. In addition, two other timings were given, one in VSX, and one in Shaw's list of near contact binaries. The following quadratic ephemeris was determined from the available times of minimum light: JD Hel Min I = 2457817.8367 ± 0.0016 d + (0.52835050 ± 0.00000108) × E – (0.00000000238 ± 0.000000000009) × E^2. The rapid period decrease may indicate that the binary is undergoing magnetic braking and is approaching its contact configuration. The possibility of a third body is discussed, but no third light was determined in the solution. VR$_c$I$_c$ simultaneous Wilson-Devinney program solutions preferred a near semi-detached solution (the primary component near filling its critical lobe and the secondary slightly underfilling, ~V1010 Oph type). Mode 2, 4, and 5 solutions were determined to arrive at this result. The noted solution gives slightly better sum of square residuals. This solution gives a mass ratio of ~0.36 and a component temperature difference of ~2090 K. A BINARYMAKER-fitted dark spot altered slightly but was not eliminated in the WD synthetic light curve computations. A 16 ± 2° radius spot is on the larger component above the equator with a T-factor of 0.95. A total eclipse of 38 minutes occurs at phase 0.5.

1. Introduction

In the course of our studies of solar type binaries we have discovered a number of pre-contact systems (Samec *et al.* 2017) on their way to becoming W UMa contact systems. We have designated them as pre-contact W UMa binaries since their orbital period studies have shown them to be systems that are likely undergoing magnetic braking (angular momentum loss, AML). This could lead them to contact and ultimately to coalescence and to the formation of fast rotating earlier spectral type single star following a red novae event (Tylenda and Kamiński 2016). Streaming plasmas moving on stiff, rotating radial patterns away from the binary out to the Alfvén radius (~50 stellar radii) cause this phenomena. Here we present the first precision photometry and light curve analysis of another such candidate, the near-contact system V616 Cam. The first report of these observations was given as a poster paper at the American Astronomical Society Meeting #233 (Samec *et al.* 2019).

2. History

The light curve of NSVS 103152 (V616 Cam) was listed in the near contact binaries list of Shaw and Hou (2007). This list gives the position, magnitude (V = 13.3934), and the ephemeris:

JD Hel MinI = 2451419.956 d + 0.52839d × E. (1)

Figure 1 displays Shaw's light curve.

The AAVSO International Variable Star Index (VSX; Watson *et al.* 2006–2014) gives r = 13.393—? magnitude and an ephemeris of:

HJD = 2440419.95624 d + 0.52838555 × E. (2)

The Two Micron All Sky Survey (2MASS; Skrutskie *et al.* 2006) gives a J–K of 0.220±0.044 mag and the AAVSO Photometric All-Sky Survey, data release 9 (APASS-DR9; Henden *et al.* 2015) gives a B–V = 0.32. Hoffman *et al.* (2008) and VSX give epoch T$_{min}$ = 2401536.696006, and a maximum V = 13.29 (catalogue data), and list it as a W UMa binary.

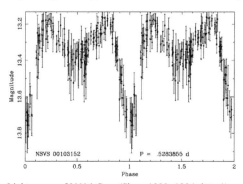

Figure 1. Light curve of V616 Cam (Shaw 1990, 1994; http://www.physast.uga.edu/~jss/ncb/LC/lc10421183.pdf).

This system was observed as a part of our student/ professional collaborative studies of interacting binaries using data taken from Dark Sky Observatory (DSO) observations. The observations were taken by R. Samec, D. Caton, and D. Faulkner. Reduction and analyses were done by R. Samec. The GAIA DR2 determined distance is 1211±23 pc (Bailer-Jones 2015). The light curve shown (Figure 2) is available in the ASAS-SN (All-Sky Automated Survey for Supernovae, https://asas-sn.osu.edu/) Variable Stars Data Base (The ASAS-SN Catalog of Variable Stars: II, Shappee *et al.* 2014 and Kochanek *et al.* 2017) with a J–K = 0.22, an ephemeris:

JD Hel Min I = 2457757.06327 d + 0.5283593 × E,   (3)

and catalog name: ASASSN-V J090553.27+820344.9. They also give a B–V = 0.322 and an E(B–V) of 0.028, making the corrected B–V ~ 0.294.

3. 2017 VR$_c$I$_c$ photometry

The observations were taken with the Dark Sky Observatory (DSO) 0.81-m reflector at Philips Gap, North Carolina, on 5, 6, 9, and 30 March 2017 with a thermoelectrically cooled (–35° C) 2K × 2K Apogee Alta camera and Johnson-Cousins VR$_c$I$_c$ filters. The Individual observations included 280 in V, 287 in R$_c$, and 285 in I$_c$. The standard error of a single observation was 10 mmag V, 16 mmag in R$_c$, and 13 mmag in I$_c$. The nightly check-comparison star (C–K) values stayed constant throughout the observing run with a precision of 1%. Exposure times varied from 25s in V to 15s in R$_c$ and I$_c$. Two sample sets of observations from March 5 and 6 are given as Figures 3 and 4. The coordinates and magnitudes of the variable star, comparison star, and check star are given in Table 1.

The finder chart is given as Figure 5 with the variable star (V), comparison star (C), and check star (K) shown. Our observations are listed in Table 2, with magnitude differences ΔV, ΔR$_c$, and ΔI$_c$ in the sense variable minus comparison star.

4. Orbital period study

Five times of minimum light were calculated, three primary and two secondary eclipses, from our present observations. Two times of minimum light are listed in the literature (Hubscher and Lehmann 2012). VSX also gives a time of minimum light.

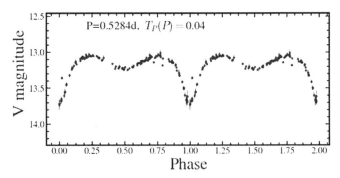

Figure 2. V light curve of V616 Cam (ASASSN-V J090553.27+820344.9) (https://asas-sn.osu.edu/).

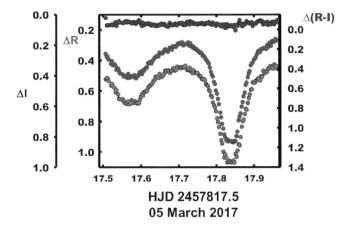

Figure 3. R$_c$, I$_c$, and R$_c$–I$_c$ color curves on the night of 5 May 2017.

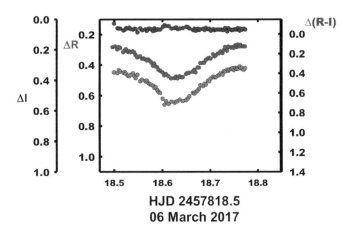

Figure 4. R$_c$, I$_c$, and R$_c$–I$_c$ color curves on the night of 6 March 2017.

Table 1. Photometric targets.

Star		Name[1]	R.A. (2000) h m s	Dec. (2000) ° ′ ″	V[1]	J–K[2]
Variable	V	V616 Cam GSC 4547 0771 3UC345-013290	09 05 52.600	+82 03 44.4	13.19	0.22 ± 0.02
Comparison	C	GSC 4547 0773 3UC345-013321	09 07 26.9480	+82 03 48.200	12.92	0.260 ± 0.035
Check	K	GSC4547 1067 3UC345-013313	09 06 57.1873	+82 00 28.91	13.34	0.420 ± 0.046

[1] UCAC3 (U.S. Naval Obs. 2012). [2] 2MASS (Skrutskie *et al.* 2006)

Shaw and Hou's (2007) list of near contact binaries gives an ephemeris. All of these times of minimum light are given in Table 3. Linear and quadratic ephemerides were determined from these data:

$$\text{JD Hel Min I} = 2457817.8526 \text{ d} + 0.5283783 \\ \pm 0.0127 \quad \pm 0.0000025 \quad (4)$$

$$\text{JD Hel Min I} = 2457817.8367 \text{ d} + 0.52835050 \times E - 0.00000000238 \times E^2 \\ \pm 0.0016 \quad \pm 0.00000108 \quad \pm 0.00000000009. \quad (5)$$

The period study covers a time interval of some 17.5 years and shows a period that is decreasing (at a high precision level; the errors shown here are standard errors). The rapid period decrease may indicate that the binary is undergoing magnetic braking and is approaching its contact configuration. The main problem at this point is the small number of times of minimum light. However, the first "minima" in Table 3 is an epoch from a period study, the second "point" is actually two data points (Hubscher and Lehmann 2012), and the last "one" is actually five minima. The quadratic ephemeris yields a $\dot{P}$ = –3.304 ± 0.014 × 10^{-6} d/yr or a mass exchange rate of

$$\frac{dM}{dt} = \frac{\dot{P} M_1 M_2}{3P(M_1 - M_2)} = \frac{1.62 \pm 0.54 \times 10^{-6} M_\odot}{d} \quad (6)$$

in a conservative scenario. Since the possibility of a third component must be considered, the apparent quadratic curve could be part of a sinusoid. In fact, a large percentage of short-period systems have third components (Tokovinin *et al.* 2006). Further eclipse timings are needed to confirm or disaffirm this scenario.

A plot of the quadratic residuals is given in Figure 6. The quadratic fit carried a precision of 27 sigma. Again, this result should not be taken with high credibility due to the relatively small number of data points. The O–C quadratic residual calculations are given in Table 3.

If the resulting trend is continuous, ultimately the system would become unstable, resulting in a red novae event, and finally coalesce into a single fast rotating spectrally, earlier star (Tylenda and Kamiński 2016). Alternately, the period change could a part of a sinusoidal variation due to the presence of a third body.

5. Light curves

The VR$_c$I$_c$ phased light curves calculated from Equation 2 are displayed in Figures 7a and 7b. The light curve averages at quarter phases and characteristics are given in Table 4. The amplitude of the light curve varies from 0.69 to 0.64 mag in V to I$_c$. The O'Connell effect, as a possible indicator of spot activity (O'Connell 1951; i.e., Guinan *et al.* 1991), is appreciable, averaging 0.06 mag. The differences in minima are large, 0.41–0.46 mag from I$_c$ to V, probably indicating noncontact light curves. The amplitudes of the light curves are 0.64 to 0.69 mag, I$_c$ to V, indicating a fairly large inclination for near contact curves. The V–I$_c$ and R–I$_c$ color curves fall at phase 0.0, which is characteristic of a contact binary, however the color curves rise slightly at phase 0.5, which indicates that the secondary component is under filling its Roche Lobe. Thus, the shape of the curves indicates a near-semidetached binary coming into contact.

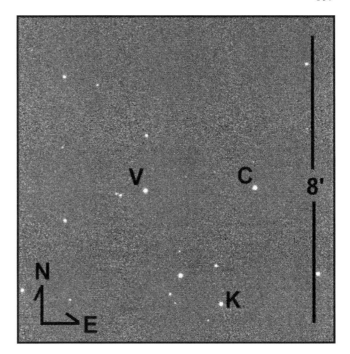

Figure 5. Finder chart: V616 Cam (V), comparison star (C), and check star (K).

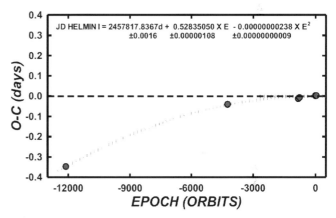

Figure 6. The period study of 17.5 years indicates continuous period decrease for V616 Cam.

6. Temperature

2MASS gives a J–K = 0.22 for the binary. The APASS-DR9 gives B–V = 0.32 (E(B–V) given earlier). These correspond to a F3V±3 eclipsing binary (Mamajek 2019). This yields a temperature of 6750 ± 400 K. Fast rotating binary stars of this type are known for having convective atmospheres, so spots are expected and indeed, one major spot is found.

We have modeled a number of short period F-type binaries with magnetic spot activity. These include V500 Peg (Caton *et al.* 2017), FF Vul (Samec *et al.* 2016a), V500 And (Samec *et al.* 2016b), and GSC 3208 1986 (Samec *et al.* 2015), to name a few.

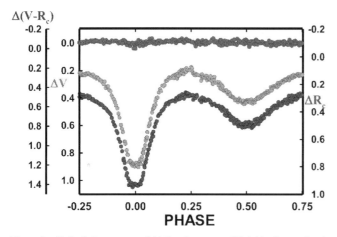

Figure 7a. V, R_c light curves and V–R color curves (Variable-Comparison), magnitudes phased with Equation 2.

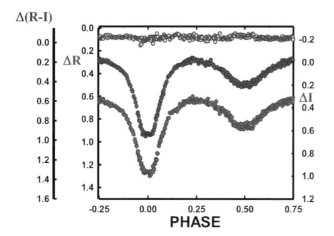

Figure 7b. R_c,I_c light curves and R_c–I_c color curves (Variable-Comparison), magnitudes phased with Equation 2.

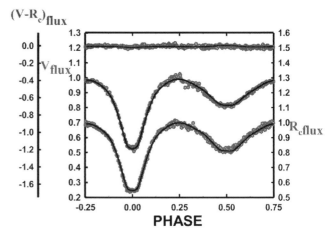

Figure 8a. V, R_c normalized fluxes and the V–R color curves overlaid by the detached solution for V616 Cam.

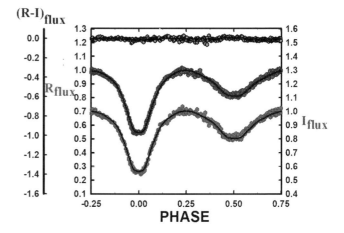

Figure 8b. R_c, I_c normalized fluxes and the R–I color curves overlaid by the detached solution of NSV 103152.

7. Light curve solution

The V, R_c, and I_c curves were pre-modeled with BINARY MAKER 3.0 (Bradstreet and Steelman 2002, but uses black body atmospheres) and fits were determined in the three filter bands. The result of the average best fit is that of a critical contact eclipsing binary with a fill-out of 100.5%. The fill-out of a contact binary is

$$F = \frac{(\Omega_1 - \Omega_{ph})}{(\Omega_1 - \Omega_2)}, \quad (7)$$

where the photometric potential, Ω_{ph}, is between the inner critical potential, Ω_1, and the outer Ω_2, $\Omega_2 \leq \Omega_{ph} \leq \Omega_1$. Other results included q (mass ratio) = 0.37, inclination (i) = 86°, and $\Delta T = 1918$ K. The fill-out would indicate in themselves that the Roche lobes of both components were near or slightly over contact. However, the large difference in component temperatures would show that the components are not in contact. The parameters were then averaged and input into a three-color simultaneous light curve calculation using the WILSON-DEVINNEY program (WD; Wilson and Devinney 1971; Wilson 1990, 1994; Van Hamme 1998). Convective parameters g = 0.32, A = 0.5 were used. Third light was tried and it gave nonphysical (negative) values. The solution was started in Mode 2 to allow the computation itself to determine the actual configuration. The curves converged to a solution in the detached mode, however, the primary component fill-out was 99.99%. The secondary was under-filling at 99.8%. We define the fill-out for a semi-detached or detached binary as simply the critical surface potential, (Ω_1), divided by the actual surface potential, (Ω_{ph}), and it may be expressed in percentages. A Mode 4 solution was computed (with the primary component critically filling its Roche lobe, and the secondary under-filling, the V1010 Oph (Shaw 1994) type binary configuration). The V1010 Oph type binary is the first contact phase of the near contact binary, i.e., the more massive binary is the first to fill its Roche lobe. The solution was completed and the sum of square residuals were very similar with the detached one but slightly worse. The fill-out of the secondary component was 99.5%. Thus, the binary is very near this configuration and very near contact. The difference in the component temperatures, some 2090 K, precludes contact. Additionally, as suggested by the referee, a Mode 5 (Algol, secondary component filling its critical lobe and the primary star underfilling) solution was run. Again, the sum of square residuals for the detached solution was slightly better than either the V1010 Oph (Mode 4) or the Algol (Mode 5). The solutions follow in Tables 5a and 5b. The normalized V and the

Figure 9a. V616 Cam, geometrical representation at phase 0.00.

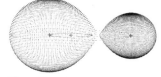

Figure 9b. V616 Cam, geometrical representation at phase 0.25.

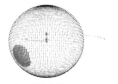

Figure 9c. V616 Cam, geometrical representation at phase 0.50.

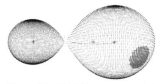

Figure 9d. V616 Cam, geometrical representation at phase 0.75.

$R_c I_c$ light curves with the detached solution curves overlain are displayed in Figures 8a and 8b. The Roche lobe surfaces are given in Figures 9a–9d. The limb darkening coefficients are two dimensional coefficients logarithmic (x and y) provided in the WD program (Wilson 1990).

8. Discussion

V616 Cam is apparently between a precontact and critical contact W UMa binary configuration with the primary component very near or at its critical contact. This configuration can result when a binary is first evolving into contact. Its spectral type indicates a surface temperature of ~6750 K for the primary component. The secondary component has a temperature of ~4657 K (K3.5V), which means that it is under-massive as compared to a single main sequence star ($M_1 = 1.4 M_\odot$, $M_2 = 0.73 M_\odot$). The mass ratio is ~0.36 rather than the expected 0.52. This would be expected for a system undergoing magnetic decay or where the primary component is receiving mass from the secondary and the mass ratio is becoming more extreme. The fill-out of both components is between 99% and critical contact by potential with the primary nearer one. The inclination is 85.7°, which causes a total eclipse of 38 minutes to occur at phase 0.5. The primary component has an iterated cool spot region of 16 ± 2° with a mean T-factor of ~0.95 (T~6410 K). This was not unexpected for solar type binaries. Again, there is certainly a possibility of a third component. However, no third light was determined in the solutions. The apparent quadratic curve could be part of a sinusoid. Further eclipse timings are needed to confirm or disaffirm this scenario.

9. Conclusion

The period study of this pre-contact W UMa binary has a 17.5-year time duration. The period is found to be strongly decreasing at a high level of significance. This is expected for a massive solar type binary undergoing magnetic braking. The presence of a cool magnetic spot would confirm this scenario. If this is indeed the case, the system will slowly coalesce over time as it loses angular momentum due to ion winds moving radially outward on stiff magnetic field lines rotating with the binary (out to the Alfvén radius). If it is not already in contact, the system will soon become a W UMa contact binary and, ultimately, one would expect the binary to become a rather normal, fast rotating, single A2V type field star after a red novae coalescence event (with a ~5% mass loss, Tylenda and Kamiński 2016). Finally, radial velocity curves are needed to obtain absolute (not relative) system parameters.

10. Acknowledgements

Dr. Samec wishes to thank Appalachian State University for allowing us use of the DSO 32-inch F/8 R-C reflector for our many projects.

References

Bailer-Jones, C. A. L. 2015, *Publ. Astron. Soc. Pacific*, **127**, 994.

Bradstreet, D. H., and Steelman, D. P. 2002, *Bull. Amer. Astron. Soc.*, 34, 1224.

Caton, D. B., Samec, R. G., Robb, R., Faulkner, D. R., Van Hamme, W., Clark, J. D., and Shebs, T. 2017, *Publ. Astron. Soc. Pacific*, **129**, 064202.

Guinan, E. F., Bradstreet, D. H., Etzel, P. B., Ibanoglu, C., Ready, C. J., Steelman, D. P., and Thrash, T. A. 1991, *Bull Amer. Astron. Soc.*, **23**, 1412.

Henden, A. A., et al. 2015, AAVSO Photometric All-Sky Survey, data release 9 (http://www.aavso.org/apass).

Hoffman D. I., Harrison T. E., Coughlin J. L., McNamara B. J., Holtzman J. A., Taylor G. E., and Vestrand W. T. 2008, *Astron. J.*, **136**, 1067.

Hubscher, J., and Lehmann, P. B. 2012, *Inf. Bull. Var. Stars*, No. 6026, 1.

Kochanek, C. S., et al. 2017, *Publ. Astron. Soc. Pacific*, **129**, 104502.

Mamajek, E. 2019, "A Modern Mean Dwarf Stellar Color and Effective Temperature Sequence" Version 2019.3.22 (http://www.pas.rochester.edu/~emamajek/EEM_dwarf_UBVIJHK_colors_Teff.txt).

O'Connell, D. H. K. 1951, *Publ. Riverside Coll. Obs.*, **2**, 85.

Samec, R. G., Caton, D., and Faulkner, D. R. 2019, AAS Meeting #233, id.348.02.

Samec, R. G., Chamberlain, H., Caton, D. B., Faulkner, D. R., Clark, J. D., and Shebs, T. 2016a, *J. Amer. Assoc. Var. Star Obs.*, **44**, 108.

Samec, R. G., Kring, J. D., Robb, R., Van Hamme, W., and Faulkner, D. R. 2015, *Astron. J.*, **149**, 90.

Samec, R. G., Nyaude, R., Caton, D., and van Hamme, W. 2016b, *Astron. J.*, **152**, 199.

Samec, R. G., Olsen, A., Caton, D., and Faulkner, D. R. 2017. *J. Amer. Assoc. Var. Star Obs.*, **45**, 148.

Shappee, B. J., et al. 2014, *Astrophys. J.*, **788**, 48.

Shaw, J. S. 1990, in *Proceedings of the NATO Advanced Study Institute on Active Close Binaries*, ed., C. Ibanoglu, Kluwer, Dordrecht, 241.

Shaw, J. S. 1994, *Mem. Soc. Astron. Ital.*, **65**, 95.

Shaw, J. S., and Hou, D. 2007, "Near-Contact Binaries of the Northern Sky Variability Survey" (http://www.physast.uga.edu/~jss/ncb/).

Skrutskie, M. F., et al. 2006, *Astron. J.*, **131**, 1163.

Tokovinin, A., Thomas, S., Sterzik, M., and Udry, S. 2006, *Astron. Astrophys.*, **450**, 681.

Tylenda, R., and Kamiński, T. 2016, *Astron. Astrophys.*, **592A**, 134.

U.S. Naval Observatory. 2012, UCAC-3 (http://www.usno.navy.mil/USNO/astrometry/optical-IR-prod/ucac).

Van Hamme, W. V., and Wilson, R. E. 1998, *Bull. Amer. Astron. Soc.*, **30**, 1402.

Watson, C., Henden, A. A., and Price, C. A. 2006–2014, AAVSO International Variable Star Index VSX (Watson+, 2006–2014; http://www.aavso.org/vsx).

Wilson, R. E. 1990, *Astrophys. J.*, **356**, 613.

Wilson, R. E. 1994, *Publ. Astron. Soc. Pacific*, **106**, 921.

Wilson, R. E., and Devinney, E. J. 1971, *Astrophys. J.*, **166**, 605.

Wozniak, P. R., *et al.* 2004, *Astron. J.*, **127**, 2436.

Table 2. V616 Cam observations, ΔV, ΔR_c, and ΔI_c, variable star minus comparison star.

ΔV	HJD 2457800+	ΔV	HJD 2457800+	ΔV	HJD 2457800+	ΔV	HJD 2457800+	ΔV	HJD 2457800+
0.287	17.5086	0.259	17.678	0.879	17.831	0.246	18.527	0.247	18.731
0.300	17.5121	0.240	17.681	0.879	17.834	0.252	18.531	0.253	18.734
0.307	17.5144	0.243	17.684	0.878	17.837	0.256	18.534	0.248	18.738
0.316	17.5166	0.243	17.686	0.878	17.839	0.271	18.537	0.254	18.740
0.345	17.5232	0.243	17.689	0.891	17.842	0.263	18.542	0.232	18.743
0.348	17.5254	0.238	17.692	0.879	17.845	0.271	18.544	0.242	18.747
0.361	17.5276	0.228	17.694	0.867	17.848	0.275	18.547	0.236	18.749
0.360	17.5327	0.226	17.698	0.845	17.850	0.297	18.551	0.227	18.752
0.388	17.5352	0.235	17.700	0.851	17.853	0.247	18.554	0.240	18.755
0.392	17.5377	0.229	17.702	0.798	17.856	0.289	18.557	0.237	18.758
0.397	17.5418	0.231	17.706	0.775	17.858	0.284	18.561	0.237	18.760
0.404	17.5443	0.224	17.708	0.745	17.861	0.308	18.564	0.219	18.764
0.408	17.5468	0.234	17.711	0.694	17.864	0.332	18.567	0.220	18.766
0.413	17.5542	0.235	17.714	0.650	17.866	0.336	18.570	0.222	18.768
0.406	17.5567	0.230	17.717	0.606	17.870	0.342	18.574	0.231	18.773
0.418	17.5592	0.235	17.719	0.585	17.872	0.349	18.577	0.720	21.513
0.435	17.5635	0.222	17.722	0.553	17.874	0.353	18.580	0.812	21.518
0.430	17.5659	0.227	17.725	0.508	17.878	0.347	18.583	0.859	21.524
0.446	17.5684	0.220	17.727	0.477	17.880	0.372	18.586	0.882	21.527
0.435	17.5720	0.229	17.730	0.456	17.883	0.372	18.589	0.894	21.530
0.418	17.5745	0.232	17.733	0.424	17.886	0.374	18.592	0.900	21.536
0.415	17.5770	0.238	17.735	0.403	17.888	0.388	18.595	0.874	21.539
0.428	17.5801	0.231	17.738	0.399	17.891	0.393	18.598	0.906	21.544
0.434	17.5826	0.250	17.741	0.361	17.894	0.418	18.604	0.883	21.547
0.441	17.5883	0.246	17.743	0.338	17.896	0.416	18.607	0.882	21.550
0.430	17.5908	0.259	17.746	0.352	17.899	0.431	18.611	0.830	21.553
0.433	17.5933	0.251	17.749	0.311	17.902	0.424	18.618	0.800	21.556
0.412	17.5965	0.258	17.751	0.316	17.905	0.420	18.623	0.768	21.559
0.399	17.5990	0.273	17.755	0.306	17.907	0.433	18.628	0.711	21.562
0.398	17.6014	0.274	17.757	0.275	17.913	0.431	18.631	0.687	21.564
0.391	17.6046	0.284	17.760	0.265	17.915	0.426	18.635	0.599	21.570
0.402	17.6071	0.299	17.763	0.251	17.919	0.417	18.639	0.556	21.573
0.393	17.6096	0.281	17.766	0.256	17.921	0.412	18.642	0.483	21.579
0.384	17.6127	0.329	17.768	0.253	17.924	0.423	18.646	0.471	21.582
0.376	17.6152	0.317	17.771	0.241	17.927	0.412	18.653	0.451	21.585
0.367	17.6176	0.316	17.774	0.256	17.929	0.406	18.656	0.259	21.636
0.362	17.6209	0.344	17.776	0.232	17.932	0.407	18.659	0.242	21.639
0.358	17.6234	0.353	17.780	0.233	17.935	0.381	18.666	0.230	21.644
0.361	17.6258	0.383	17.782	0.225	17.937	0.381	18.669	0.215	21.647
0.352	17.6291	0.399	17.785	0.239	17.940	0.370	18.672	0.227	21.652
0.321	17.6316	0.413	17.788	0.208	17.943	0.369	18.676	0.193	21.655
0.329	17.6341	0.432	17.790	0.222	17.945	0.353	18.679	0.212	21.658
0.313	17.6378	0.456	17.793	0.196	17.948	0.352	18.682	0.199	21.666
0.313	17.6403	0.477	17.796	0.201	17.951	0.346	18.688	0.186	21.671
0.314	17.6427	0.515	17.798	0.199	17.954	0.345	18.691	0.223	21.674
0.324	17.6458	0.543	17.801	0.201	17.956	0.323	18.694	0.232	21.677
0.291	17.6482	0.598	17.804	0.183	17.960	0.319	18.698	0.224	21.679
0.310	17.6507	0.626	17.806	0.196	17.962	0.308	18.701	0.254	21.696
0.285	17.6552	0.650	17.809	0.176	18.499	0.315	18.704	0.239	21.704
0.282	17.6577	0.719	17.812	0.210	18.502	0.296	18.709	0.241	21.707
0.288	17.6602	0.735	17.815	0.223	18.505	0.293	18.712	0.254	21.709
0.273	17.6633	0.781	17.817	0.235	18.509	0.278	18.715	0.263	21.712
0.263	17.6658	0.816	17.821	0.230	18.512	0.272	18.719	0.242	21.715
0.267	17.6683	0.867	17.823	0.232	18.515	0.275	18.722	0.271	21.718
0.267	17.6732	0.862	17.826	0.238	18.522	0.272	18.725	0.298	21.720
0.271	17.6757	0.866	17.829	0.229	18.524	0.256	18.729	0.267	21.723

Table continued on following pages

Table 2. V616 Cam observations, ΔV, ΔR$_c$, and ΔI$_c$, variable star minus comparison star, cont.

ΔR	HJD 2457800+	ΔR	HJD 2457800+	ΔR	HJD 2457800+	ΔR	HJD 2457800+	ΔR	HJD 2457800+
0.357	17.507	0.317	17.676	0.933	17.832	0.315	18.528	0.282	18.738
0.378	17.510	0.326	17.679	0.930	17.835	0.311	18.531	0.284	18.741
0.376	17.512	0.330	17.682	0.930	17.837	0.317	18.534	0.289	18.745
0.395	17.515	0.307	17.684	0.929	17.840	0.312	18.539	0.285	18.747
0.423	17.521	0.312	17.687	0.938	17.843	0.319	18.542	0.290	18.750
0.405	17.524	0.310	17.690	0.938	17.845	0.336	18.545	0.273	18.753
0.431	17.526	0.313	17.692	0.929	17.848	0.352	18.548	0.285	18.756
0.426	17.531	0.302	17.695	0.929	17.851	0.344	18.554	0.268	18.758
0.436	17.533	0.291	17.698	0.897	17.853	0.360	18.558	0.280	18.762
0.462	17.536	0.288	17.700	0.868	17.856	0.359	18.561	0.270	18.764
0.453	17.540	0.286	17.704	0.794	17.859	0.365	18.565	0.280	18.766
0.469	17.542	0.294	17.706	0.770	17.862	0.372	18.568	0.280	18.770
0.487	17.545	0.282	17.709	0.748	17.864	0.365	18.571	0.280	18.773
0.492	17.552	0.302	17.712	0.704	17.867	0.374	18.574	0.754	21.509
0.506	17.555	0.295	17.714	0.674	17.870	0.384	18.577	0.756	21.510
0.505	17.557	0.295	17.717	0.637	17.872	0.406	18.581	0.812	21.513
0.502	17.561	0.294	17.720	0.612	17.876	0.392	18.583	0.867	21.516
0.516	17.564	0.289	17.723	0.563	17.878	0.410	18.586	0.935	21.522
0.489	17.566	0.283	17.725	0.547	17.880	0.417	18.590	0.935	21.524
0.514	17.570	0.304	17.728	0.512	17.884	0.433	18.593	0.951	21.530
0.510	17.572	0.289	17.731	0.488	17.886	0.439	18.595	0.939	21.533
0.504	17.575	0.303	17.733	0.471	17.889	0.449	18.601	0.925	21.536
0.508	17.578	0.315	17.736	0.433	17.892	0.469	18.604	0.940	21.542
0.509	17.581	0.310	17.739	0.421	17.894	0.461	18.608	0.924	21.547
0.497	17.583	0.323	17.741	0.416	17.897	0.475	18.613	0.905	21.550
0.519	17.586	0.327	17.744	0.396	17.900	0.489	18.618	0.893	21.553
0.511	17.589	0.326	17.747	0.390	17.903	0.487	18.625	0.846	21.556
0.504	17.591	0.335	17.749	0.381	17.905	0.477	18.629	0.807	21.559
0.493	17.594	0.340	17.753	0.362	17.908	0.489	18.632	0.772	21.562
0.476	17.597	0.351	17.755	0.344	17.911	0.475	18.636	0.721	21.565
0.483	17.599	0.347	17.758	0.338	17.913	0.481	18.640	0.677	21.568
0.491	17.603	0.356	17.761	0.347	17.916	0.481	18.643	0.648	21.571
0.465	17.605	0.372	17.764	0.328	17.919	0.460	18.650	0.607	21.573
0.469	17.607	0.372	17.766	0.327	17.921	0.454	18.653	0.588	21.576
0.456	17.611	0.381	17.769	0.323	17.925	0.452	18.656	0.580	21.579
0.449	17.613	0.393	17.772	0.320	17.927	0.446	18.664	0.311	21.634
0.447	17.616	0.401	17.774	0.322	17.930	0.434	18.667	0.311	21.636
0.419	17.619	0.412	17.778	0.327	17.933	0.449	18.670	0.297	21.639
0.420	17.621	0.421	17.780	0.301	17.935	0.418	18.674	0.310	21.642
0.411	17.624	0.447	17.782	0.315	17.938	0.418	18.677	0.289	21.647
0.421	17.627	0.475	17.786	0.290	17.941	0.389	18.680	0.307	21.653
0.405	17.629	0.496	17.788	0.284	17.943	0.388	18.685	0.293	21.655
0.399	17.632	0.507	17.790	0.305	17.946	0.386	18.688	0.268	21.658
0.384	17.636	0.542	17.794	0.274	17.949	0.368	18.691	0.271	21.661
0.379	17.638	0.556	17.796	0.280	17.952	0.359	18.695	0.290	21.664
0.373	17.641	0.598	17.799	0.280	17.954	0.355	18.698	0.275	21.666
0.372	17.644	0.620	17.802	0.261	17.958	0.360	18.701	0.284	21.672
0.374	17.646	0.654	17.804	0.271	17.960	0.331	18.706	0.314	21.674
0.349	17.649	0.688	17.807	0.267	17.962	0.345	18.709	0.302	21.680
0.355	17.653	0.743	17.810	0.285	18.497	0.339	18.712	0.280	21.683
0.355	17.656	0.774	17.813	0.280	18.503	0.328	18.716	0.209	21.685
0.351	17.658	0.817	17.815	0.299	18.507	0.324	18.719	0.331	21.694
0.335	17.661	0.848	17.818	0.283	18.510	0.313	18.722	0.310	21.696
0.328	17.664	0.887	17.821	0.290	18.513	0.324	18.726	0.302	21.699
0.337	17.666	0.904	17.823	0.317	18.519	0.304	18.729	0.321	21.702
0.344	17.671	0.926	17.827	0.307	18.522	0.284	18.732	0.321	21.704
0.327	17.674	0.919	17.829	0.298	18.525	0.308	18.736	0.280	21.707

Table continued on next page

Table 2. V616 Cam observations, ΔV, ΔR$_c$, and ΔI$_c$, variable star minus comparison star, cont.

ΔI	HJD 2457800+	ΔI	HJD 2457800+	ΔI	HJD 2457800+	ΔI	HJD 2457800+	ΔI	HJD 2457800+
0.426	17.507	0.383	17.676	0.966	17.832	0.371	18.529	0.333	18.736
0.417	17.511	0.370	17.680	0.960	17.835	0.375	18.532	0.340	18.739
0.420	17.513	0.366	17.682	0.964	17.837	0.365	18.534	0.331	18.741
0.436	17.515	0.353	17.684	0.970	17.840	0.356	18.539	0.327	18.745
0.466	17.522	0.353	17.688	0.970	17.843	0.374	18.542	0.323	18.748
0.455	17.524	0.367	17.690	0.958	17.846	0.386	18.545	0.335	18.750
0.471	17.526	0.364	17.692	0.971	17.848	0.365	18.549	0.319	18.754
0.482	17.531	0.363	17.692	0.914	17.851	0.346	18.551	0.328	18.756
0.510	17.533	0.341	17.696	0.933	17.854	0.407	18.554	0.326	18.758
0.496	17.536	0.351	17.701	0.893	17.856	0.397	18.559	0.312	18.762
0.524	17.540	0.344	17.704	0.851	17.859	0.454	18.561	0.324	18.764
0.553	17.542	0.351	17.707	0.859	17.862	0.408	18.565	0.321	18.767
0.555	17.545	0.349	17.709	0.803	17.864	0.403	18.568	0.333	18.771
0.556	17.552	0.343	17.712	0.736	17.868	0.419	18.572	0.323	18.774
0.570	17.555	0.344	17.715	0.700	17.870	0.426	18.575	0.799	21.509
0.580	17.557	0.336	17.717	0.689	17.873	0.430	18.577	0.861	21.513
0.587	17.562	0.356	17.720	0.653	17.876	0.435	18.581	0.842	21.516
0.582	17.564	0.363	17.723	0.619	17.878	0.451	18.584	0.888	21.519
0.570	17.567	0.360	17.725	0.625	17.881	0.445	18.587	0.968	21.525
0.588	17.570	0.349	17.729	0.580	17.884	0.466	18.590	0.974	21.528
0.581	17.573	0.371	17.731	0.548	17.886	0.478	18.593	0.969	21.531
0.562	17.575	0.364	17.733	0.525	17.889	0.484	18.596	0.999	21.542
0.569	17.578	0.386	17.737	0.494	17.892	0.522	18.601	0.986	21.545
0.581	17.581	0.366	17.739	0.491	17.895	0.562	18.605	0.964	21.551
0.569	17.583	0.378	17.741	0.446	17.897	0.539	18.608	0.943	21.554
0.573	17.587	0.384	17.745	0.449	17.900	0.557	18.614	0.884	21.556
0.583	17.589	0.383	17.747	0.446	17.903	0.541	18.619	0.854	21.559
0.579	17.591	0.383	17.750	0.434	17.905	0.547	18.626	0.801	21.562
0.582	17.595	0.397	17.753	0.431	17.909	0.541	18.629	0.711	21.568
0.559	17.597	0.404	17.756	0.421	17.911	0.538	18.632	0.679	21.571
0.539	17.600	0.406	17.758	0.420	17.913	0.528	18.637	0.651	21.574
0.522	17.603	0.423	17.761	0.434	17.917	0.566	18.640	0.640	21.577
0.531	17.605	0.425	17.764	0.406	17.919	0.537	18.643	0.622	21.579
0.509	17.608	0.424	17.766	0.410	17.922	0.531	18.650	0.586	21.582
0.507	17.611	0.431	17.770	0.369	17.925	0.520	18.654	0.344	21.634
0.500	17.613	0.455	17.772	0.388	17.927	0.506	18.657	0.360	21.637
0.502	17.616	0.465	17.775	0.383	17.930	0.483	18.664	0.347	21.645
0.479	17.619	0.471	17.778	0.375	17.933	0.488	18.667	0.334	21.648
0.470	17.622	0.497	17.780	0.376	17.935	0.479	18.670	0.319	21.650
0.467	17.624	0.504	17.783	0.355	17.938	0.469	18.674	0.336	21.656
0.457	17.627	0.533	17.786	0.353	17.941	0.456	18.677	0.299	21.658
0.452	17.630	0.541	17.788	0.338	17.944	0.433	18.680	0.325	21.661
0.437	17.632	0.577	17.791	0.352	17.946	0.440	18.686	0.337	21.664
0.430	17.636	0.597	17.794	0.347	17.949	0.424	18.689	0.333	21.667
0.410	17.638	0.608	17.796	0.354	17.952	0.430	18.691	0.320	21.669
0.410	17.641	0.634	17.799	0.327	17.954	0.404	18.696	0.346	21.675
0.405	17.644	0.680	17.802	0.337	17.958	0.407	18.699	0.339	21.678
0.416	17.646	0.705	17.805	0.340	17.960	0.390	18.702	0.327	21.680
0.389	17.649	0.737	17.807	0.357	17.963	0.377	18.707	0.333	21.683
0.410	17.653	0.782	17.811	0.355	18.497	0.382	18.710	0.341	21.694
0.400	17.656	0.824	17.813	0.350	18.500	0.381	18.713	0.321	21.699
0.403	17.658	0.847	17.816	0.354	18.507	0.371	18.717	0.360	21.702
0.387	17.662	0.900	17.819	0.335	18.510	0.367	18.720	0.368	21.705
0.368	17.664	0.938	17.821	0.344	18.513	0.355	18.723	0.406	21.718
0.371	17.666	0.918	17.824	0.352	18.519	0.365	18.727	0.401	21.721
0.366	17.671	0.941	17.827	0.347	18.522	0.342	18.729	0.395	21.724
0.372	17.674	0.970	17.829	0.359	18.525	0.344	18.732	0.416	21.727

Table 3. O–C residual calculations of V616 Cam.

	Cycle	Epochs 2400000.0+	Error	Linear Residuals	Quadratic Residuals	Weight	Reference
1	−12108.5	51419.9562	—	−0.0280	−0.0002	1.0	Shaw (1990, 1994)
2	−12108.0	51420.2204	—	−0.0281	−0.0002	0.5	VSX (Watson et al. 2014)
3	−4238.5	55578.3813	0.0210	0.0600	0.0009	1.0	Hubscher and Lehmann (2012)
4	−4238	55578.6463	0.0041	0.0608	0.0017	1.0	Hubscher and Lehmann (2012)
5	−828.5	57380.0848	0.0027	−0.0064	−0.0119	0.5	This work; NSVS (Wozniak et al. 2004)
6	−770.0	57410.9985	0.0003	−0.0028	−0.0069	0.5	This work; NSVS (Wozniak et al. 2004)
7	−0.5	57817.5742	0.0013	−0.0142	0.0017	1.0	This work
8	0.0	57817.8376	0.0001	−0.0150	0.0009	1.0	This work
9	1.5	57818.6310	0.0002	−0.0142	0.0018	1.0	This work
10	7.0	57821.5364	0.0004	−0.0149	0.0013	1.0	This work
11	47.0	57842.6710	0.0002	−0.0158	0.0015	1.0	This work

Table 4. Light curve characteristics for V616 Cam.

Filter	Phase 0.25	Magnitude Min. I	± σ	Phase 0.75	Magnitude Max. I	± σ
V		0.886	± 0.011		0.193	± 0.014
R_c		0.935	± 0.007		0.275	± 0.009
I_c		0.971	± 0.011		0.335	± 0.018

Filter	Phase 0.5	Magnitude Min. II	± σ	Phase 0.0	Magnitude Max. II	± σ
V		0.428	± 0.009		0.283	± 0.010
R_c		0.497	± 0.014		0.385	± 0.005
I_c		0.563	± 0.020		0.335	± 0.014

Filter	Min. I – Max. I	± σ	Max. II – Max. I	± σ	Min. I – Min. II	± σ
V	0.693	± 0.025	0.090	± 0.024	0.458	± 0.020
R_c	0.660	± 0.016	0.090	± 0.014	0.439	± 0.022
I_c	0.636	± 0.029	0.000	± 0.032	0.408	± 0.031

Filter	Max. II – Max. I	± σ	Min. II – Max. I	± σ
V	0.090	± 0.024	0.235	± 0.023
R_c	0.109	± 0.014	0.221	± 0.023
I_c	0.000	± 0.0325	0.227	± 0.038

Table 5a. VR_cI_c synthetic curve solution input parameters for V616 Cam.

Parameters	Values
$\lambda_V, \lambda_R, \lambda_I$ (nm)	550, 640, 790
$x_{bol1,2}, y_{bol1,2}$ (bolometric limb darkening)	0.638, 0.637, 0.248, 0.148
$x_{1I,2I}, y_{1I,2I}$ (limb darkening)	0.539, 0.656, 0.281, 0.160
$x_{1R,2}R, y_{1R,2R}$ (limb darkening)	0.624, 0.744, 0.291, 0.128
$x_{1V,2V}, y_{1V,2V}$ (limb darkening)	0.698, 0.799, 0.282, 0.054
g_1, g_2 (gravity darkening)	0.32
A_1, A_2 (albedo)	0.5

Table 5b. V, R_c, I_c Wilson-Devinney program solution parameters.

Parameters	Mode 2 Detached Solution	Mode 4 Semi-detached (V1010 Oph)	Mode 5 Semi-detached (Algol)
Inclination (°)	85.0 ± 0.2	86.75 ± 0.23	86.75 ± 0.30
T_1, T_2 (K)	6750*, 4662 ± 5	6750*, 4703 ± 6	6750*, 4734 ± 6
Ω_1 potential	2.5857 ± 0.0022	2.5889	2.583 ± 0.002,
Ω_2 potential	2.5871 ± 0.0021	2.602 ± 0.002	2.572
$q (m_2/m_1)$	0.3554 ± 0.0006	0.3572 ± 0.0006	0.3489 ± 0.0012
Fill-outs: F_1 (%),	99.99	100	99.6
F_2 (%)	99.78	99.5	100
$L_1/(L_1+L_2)_I$	0.9076 ± 0.0007	0.9060 ± 0.0007	0.9021 ± 0.0007
$L_1/(L_1+L_2)_R$	0.9259 ± 0.0010	0.9242 ± 0.0010	0.9205 ± 0.0010
$L_1/(L_1+L_2)_V$	0.9451 ± 0.0006	0.9432 ± 0.0006	0.9399 ± 0.0006
JDo (days)	2457817.8373 ± 0.0001	2457817.8373 ± .0001	2457817.8373 ± 0.0001
Period (days)	0.52847 ± 0.00003	0.528468 ± 0.00003	0.528466 ± 0.000035
r_1, r_2 (pole)	0.442 ± 0.001, 0.273 ± 0.004	0.4419 ± 0.0009, 0.2715 ± 0.0035	0.4417 ± 0.0009, 0.2723 ± 0.0005
r_1, r_2 (point)	0.6 ± 0.1, 0.38 ± 0.08	—, 0.3626 ± 0.0265	0.574 ± 0.010, —
r_1, r_2 (side)	0.473 ± 0.002, 0.285 ± 0.004	0.473 ± 0.001, 0.283 ± 0.004	0.472 ± 0.001, 0.2836 ± 0.0006
r_1, r_2 (back)	0.4995 ± 0.0025, 0.317 ± 0.007	0.499 ± 0.001, 0.313 ± 0.007	0.498 ± 0.002, 0.3164 ± 0.0006
Spot Parameters	Star 1, Cool Spot		
Colatitude (°)	110 ± 5.	107 ± 2	107 ± 4
Longitude (°)	134.5 ± 3.4	128.5 ± 3.1	135 ± 1
Spot radius (°)	18.4 ± 0.8	14.0 ± 0.4	14.7 ± 0.5
Tfact	0.959 ± 0.003	0.940 ± 0.004	0.940 ± 0.004
$\Sigma(res)^2$	0.262956	0.265574	0.270075

*The 6750 K primary temperature carries a ±400 K uncertainty.

The Long-term Period Changes and Evolution of V1, a W Virginis Star in the Globular Cluster M12

Pradip Karmakar
Department of Mathematics, Madhyamgram High School (H.S.), Madhyamgram, Sodepur Road, Kolkata 700129, India; pradipkarmakar39@gmail.com

Horace A. Smith
Department of Physics and Astronomy, Michigan State University, East Lansing, MI 48824; smith@pa.msu.edu

Nathan De Lee
Department of Physics, Geology, and Engineering Technology, Northern Kentucky University, Highland Heights, KY 41099; deleenm@nku.edu

Received May 29, 2019; revised June 19, July 8, 2019; accepted July 10, 2019

Abstract We present new *B*, *V*, and *I* band photometry of the W Virginis-type variable star V1 in the globular cluster M12. Observations made from 1916 through 2018 show that during this interval the period of V1 has not shown a constant rate of period change, as might be expected were evolution alone responsible for the period changes. It has, however, shown period changes that appear to have a more abrupt character, probably both increases and decreases.

1. Introduction

Type II Cepheids are believed to be evolved low-mass pulsating variable stars, located within the instability strip at a level brighter than the RR Lyrae stars (e.g. Percy 2007; Catelan and Smith 2015). Type II Cepheids are often divided into two subgroups, depending upon their period. Those with periods shorter than 4 to 8 days are often termed BL Her stars, whereas those of longer period are often denoted W Virginis stars. Metal-poor BL Her variables are believed to be stars that have evolved from the blue horizontal branch, and which are now passing through the instability strip heading toward the asymptotic red giant branch (see, for example, Neilson *et al.* 2016; Osborn *et al.* 2019). The evolution of W Virginis stars poses greater difficulties. W Virginis stars have been sometimes considered to be stars undergoing thermal pulse instabilities that cause them to loop to the blue from the asymptotic giant branch. As they loop to the blue, they enter the instability strip at luminosities brighter than those of BL Her variables. However, some recent theoretical calculations indicate that thermal pulses are not sufficient to create such blue loops (Bono *et al.* 2016), opening once more the question of the nature of such stars. Studies of the long-term period changes of W Viriginis stars have the potential to shed light upon this perplexing problem. The periods of pulsating stars are often known to greater accuracy than any other property, and changes in pulsation period may reveal the direction and speed of stellar evolution through the instability strip. If W Virginis stars are undergoing loops into the instability strip, the pulsation equation tells is that we should expect to detect some stars with increasing periods (those on the redward portion of the loop) and some with decreasing periods (those on the blueward part of the loop). The rates of period change should be consistent with the theoretical predictions of the durations of blue loops.

V1 in the globular cluster M12 (NGC 6218) was discovered by Sawyer (1938a), and, with a period of 15.5 days, it is classified as a W Virginis star. Clement *et al.* (1988) used photographic observations to study its period between 1916 and 1985. They concluded that during this interval the pulsation period of V1 underwent both increases and decreases. In this paper we examine the light curve and period of V1 using observations obtained between 2002 and 2018, with the goal of detecting any long-term period change that might be attributed to evolution.

2. The observations and data set

We obtained new *B*, *V*, and Cousins *I* band images of M12 between 2006 and 2011 using the 0.6-m telescope of the Michigan State University campus observatory with an Apogee Alta U47 CCD camera (0.6 arc-second pixel, 10×10 arcmin field of view). Bias and dark images were subtracted in the conventional way, and twilight images were used as flat field images. Exposures were about 1 minute long, varying somewhat with sky conditions.

As noted by Sawyer (1938b) and Clement *et al.* (1988), photometry of V1 is made difficult by the presence of a neighboring star with a blue photographic magnitude near 14.0. Taking advantage of excellent seeing, Klochkova *et al.* (2003) were able to analyze the spectra both of V1 and its close companion, although they found the companion to be less than an arc second from V1. They found the companion star to have an effective temperature of 4200 K, making it considerably cooler than V1. As the seeing at the Michigan State University observatory is typically 3 or 4 arc seconds, the companion is deeply imbedded within the image of the brighter V1. The companion is also blended with V1 in the ASAS-SN data, which we discuss below.

We considered which of three methods of reduction might be best under this circumstance: aperture photometry, profile fitting photometry, and image differencing photometry. We decided against aperture photometry because the crowded nature of the M12 field means that V1 has other neighbors, somewhat more distant than the close companion, but near enough to fall within any large aperture or sky annulus. Although profile fitting photometric routines, such as DAOPHOT/ALLSTAR (Stetson 1987, 1994), can mitigate the effect of blends, in this case the blended companion was too deeply within the profile of V1 for that mitigation to be successful. Nor did image differencing (Alard and Lupton 1998) succeed in separating V1 from its close companion. In the end, we decided to proceed with profile fitting photometry using the DAOPHOT and ALLSTAR routines as in Rabidoux et al. (2010). Because of the near neighbor we will, however, emphasize period determination for V1 rather than details of the light curve in this paper.

Instrumental magnitudes obtained from DAOPHOT were transformed to the standard system as in Rabidoux et al. (2010), applying color terms as in equations 1, 2, and 3 of that paper. We used seven uncrowded local standards with APASS magnitudes (data release 10; Henden et al. 2018) to set the magnitude zero-points for B and V. APASS provides Sloan i' magnitudes, whereas we used a Cousins I-band filter. Because of this, and because of the unknown effect of the close companion on the I photometry, we could not use the APASS stars to calibrate our I-band photometry. Peter Stetson has created local standard stars in and near M12, which can be found at the website: http://www.cadc-ccda.hia-iha.nrc-cnrc.gc.ca/en/community/STETSON/standards/. Although most of the Stetson standards are too faint and too crowded for our use, a few are sufficiently bright and uncrowded to provide a check on our calibration. Because the Stetson standards include Cousins I-band photometry, we used three uncrowded Stetson local standards to set the I calibration. However, the circumstance that the star blended with V1 is cooler than V1 means that it will cause our I photometry to be too bright by an even greater amount than in B or V. Thus, particular caution attaches to any use of the I light curve. The Michigan State University (MSU) CCD photometry for V1 is listed in Table 1.

ASAS-SN observations (Shappee et al. 2014; Kochanek et al. 2017) of V1 were downloaded from the Sky Patrol option on the ASAS-SN webpage, using the position for V1 from the Clement et al. (2001) catalogue of variable stars in globular clusters. Only V-band observations are available for most of the 2012–2018 time period, though g-band data are more recently available. Because the g data gave period results identical to those from V, we include only our analyses for the V periods in Table 2. Few ASAS-SN observations of V1 for 2012 are available, but the number increases in later years. The large ASAS-SN pixels mean, however, that the 1,090 V data points that we found useful show the effects of blending even more seriously than is the case for the MSU CCD observations. Thus, the ASAS-SN observations are valuable mainly for the investigation of period changes, but they are very important for that purpose.

A search was made for additional observations of V1 in the Harvard DASCH photometry (http://dasch.rc.fas.harvard.edu/project.php), but no useful observations of V1 were found.

3. Period determinations for V1

We have used two period-finding routines to search for periodicities in the V1 data, PERIOD04 (Lenz and Berger 2005) and a date-compensated discrete Fourier transform, as implemented in PERANSO 2.0 (Vanmunster 2006). Period searches were carried out, with the results shown in Table 2. The searches were carried out for the MSU B and V data, for infrared K-band observations of V1 by Matsunaga et al. (2006), and for ASAS-SN V observations. The column headed N(obs) indicates the number of observations.

The primary periods found by PERIOD04 and PERANSO 2.0 agree well. For PERIOD04, the listed uncertainties derive from the least squares fitting routine. For the PERANSO 2.0 results, uncertainties depend upon the noise in the amplitude spectrum, which we estimated independently of the default values in the PERANSO routine. Clement et al. (1988) derived the periods at the beginning of their Table IV from their phase shift diagram for V1, but gave no uncertainties for the derived periods. The N(obs) values in Table 2 for Clement et al. (1988) do not double-count the photographic observations but, of course, since Clement et al. (1988) derived periods from phase shifts, they actually use observations from more than one time interval in deriving periods. As a check on the Clement et al. (1988) periods, and to verify that our period determinations are consistent with theirs, we reanalyzed the Clement et al. (1988) data using PERIOD04 and PERANSO 2.0. However, only three of the data subsets used by Clement et al. (1988) contained enough observations and were sufficiently free of cycle-count uncertainties to permit the application of our techniques. We have included our results directly beneath the Clement et al. (1988) periods in Table 2. The two approaches generally agree to within the uncertainties. The referee was concerned by the large 1916–1938 interval for which Clement et al. (1988) derived their initial period. We therefore selected a subset of those data covering just the 1931–1938 interval for an additional period search, with the result shown in Table 2.

W Virginis itself is known to show multiple periods (Templeton and Henden 2007). The MSU CCD observations are too sparse to effectively search for a second, weaker, period but the 2013–2018 ASAS-SN V-band observations were prewhitened in PERIOD04 by removing the main frequency and four higher harmonics. A period search was conducted on the residuals from this prewhitening. Although a weak signal was found for a period of 15.296 days (full amplitude of 0.02 mag compared to 0.18 mag for the 15.544-day period), we do not regard this detection as significant. The supposed secondary period and the 15.544-day period would beat with an interval about as long as the time interval of the entire ASAS-SN dataset. To illustrate the ASAS-SN period search results, we show the PERIOD04 amplitude spectrum for the ASAS-SN 2013–2014 V-data in Figure 1. The peaks on either side of the main peak are one cycle per year aliases of the main peak. We find no significant secondary period.

The MSU B, V, and Cousins I phased light curves for V1 are shown in Figures 2–4, with approximate fits to guide the eye. Fits were made by applying the routine from Ková́cs and Kupi (2007). The ASAS-SN 2013–2018 V light curve for V1 is shown in Figure 5.

Table 1. The MSU CCD Photometry.

B observations			V observations			I observations		
HJD	B(V1)	Error	HJD	V(V1)	Error	HJD	I(V1)	Error
2453892.6375	12.730	0.03	2453892.6402	11.318	0.02	2455714.6071	10.391	0.04
2453895.6427	13.051	0.03	2453895.6457	11.609	0.02	2455714.6346	10.398	0.04
2453903.6287	12.248	0.02	2453906.6322	11.174	0.02	2453892.6354	10.240	0.04
2453906.6298	12.461	0.02	2453907.6292	11.237	0.02	2453895.6395	10.462	0.04
2453907.6268	12.642	0.03	2453910.6360	11.580	0.03	2453895.6409	10.467	0.04
2453910.6383	13.055	0.04	2453935.6369	11.050	0.02	2453899.6313	10.486	0.04
2453935.6394	12.279	0.03	2453936.6407	11.092	0.02	2455726.6587	10.560	0.04
2453936.6420	12.406	0.03	2454360.5477	11.672	0.03	2455727.6430	10.600	0.04
2453937.6245	12.480	0.03	2455706.6150	11.583	0.02	2453903.6274	10.177	0.04
2453943.6165	12.911	0.03	2455714.6104	11.306	0.02	2453935.6435	10.190	0.04
2454357.5254	12.941	0.04	2455714.6373	11.284	0.02	2453906.6273	10.072	0.04
2454360.5476	13.016	0.04	2455714.6392	11.251	0.02	2453906.6284	10.101	0.04
2454649.6132	12.476	0.02	2455718.6231	11.339	0.02	2455739.6460	10.532	0.04
2454653.6415	12.980	0.03	2455726.6506	11.648	0.02	2455741.6244	10.590	0.04
2455706.6114	13.013	0.03	2455726.6548	11.636	0.02	2453937.6269	10.199	0.05
2455714.6086	12.460	0.03	2455727.6387	11.524	0.02	2453943.6192	10.614	0.04
2455714.6356	12.529	0.03	2455739.6406	11.803	0.03	2453943.6205	10.555	0.04
2455718.6213	12.626	0.03	2455741.6154	11.666	0.02	2455749.6379	10.191	0.04
2455726.6568	12.881	0.03	2455749.6324	11.244	0.02	2454357.5276	10.067	0.04
2455727.6409	12.684	0.03				2454360.5495	10.415	0.04
2455739.6441	13.148	0.04				2453910.6357	10.316	0.04
2455741.6225	12.886	0.03				2454649.6072	10.113	0.04
2455749.6360	12.506	0.02				2455706.6040	10.386	0.04

Table 2. Period determinations for V1 for the years 1916–2018.

Years	period	Uncertainty	N(obs)	Reference
1916–1938	15.50	—	56	Clement *et al.* 1988
1916–1938	15.51	0.01	56	PERANSO result for Clement *et al.* 1988 data
1916–1938	15.47	0.01	56	PERIOD04 result for Clement *et al.* 1988 data
1931–1938	15.50	0.02	43	PERANSO result for Clement *et al.* 1988 data
1931–1938	15.50	0.01	43	PERIOD04 result for Clement *et al.* 1988 data
1938–1946	15.55	—	20	Clement *et al.* 1988
1938–1946	15.54	0.02	20	PERANSO result for Clement *et al.* 1988 data
1938–1946	15.53	0.02	20	PERIOD04 result for Clement *et al.* 1988 data
1946–1962	15.51	—	19	Clement *et al.* 1988
1962–1970	15.54	—	19	Clement *et al.* 1988
1970–1975	15.51	—	40	Clement *et al.* 1988
1970–1975	15.51	0.03	40	PERANSO result for Clement *et al.* 1988 data
1970–1975	15.52	0.02	40	PERIOD04 result for Clement *et al.* 1988 data
1975–1985	15.57 (or 15.49)	—	14	Clement *et al.* 1988
2002–2005	15.47	0.03	19	PERANSO result for Matsugana *et al.* 2006 data
2002–2005	15.47	0.01	19	PERIOD04 result for Matsunaga *et al.* 2006 data
2006–2011	15.486	0.02	23	PERANSO result for MSU B data
2006–2011	15.484	0.006	23	PERIOD04 result for MSU B data
2006–2011	15.487	0.02	19	PERANSO result for MSU V data
2006–2011	15.489	0.009	19	PERIOD04 result for MSU V data
2013–2018	15.544	0.008	1059	PERANSO result for ASAS-SN data
2013–2018	15.542	0.001	1059	PERIOD04 result for ASAS-SN data
2012	15.623	0.30	14	PERANSO result for ASAS-SN data
2012	15.623	0.12	14	PERIOD04 result for ASAS-SN data
2013	15.597	0.19	63	PERANSO result for ASAS-SN data
2013	15.597	0.12	63	PERIOD04 result for ASAS-SN data
2014	15.606	0.016	140	PERANSO result for ASAS-SN data
2014	15.606	0.009	140	PERIOD04 result for ASAS-SN data
2015	15.555	0.015	225	PERANSO result for ASAS-SN data
2015	15.555	0.008	225	PERIOD04 result for ASAS-SN data
2016	15.575	0.014	218	PERANSO result for ASAS-SN data
2016	15.575	0.008	218	PERIOD04 result for ASAS-SN data
2017	15.504	0.013	267	PERANSO result for ASAS-SN data
2017	15.495	0.008	267	PERIOD04 result for ASAS-SN data
2018	15.566	0.016	163	PERANSO result for ASAS-SN data
2018	15.565	0.011	163	PERIOD04 result for ASAS-SN data

Clement *et al.* (1988) divided their photographic photometry of V1 into discrete groups, comparing the light curves for those groups to the light curve they obtained from observations obtained in 1970 to derive the phase-shift diagram shown in their Figure 2. A mean period of 15.527 days was used in their analysis. It was from their phase shift diagram that they determined the periods in their Table IV. However, they ran into an ambiguity in interpreting the phase-shift diagram of observations obtained *in 1985*, which showed a very large jump in phase. Should that point be plotted at a phase shift of +0.48 or –0.52? They chose to just plot the +0.48 value in Figure 2, but they noted that, depending upon which choice of phase shift was made, the resultant period could be either 15.57 or 15.49 days. Because the 1975–1985 observations are composed of two sets of observations separated by a long gap in time, a direct period determination for this interval does not resolve the ambiguity, but fits both of the Clement *et al.* (1988) alternatives.

We attempted to add our MSU *B* and *V* observations, as well as the annual ASAS-SN observations between 2014 and 2018 to the Clement *et al.* (1988) Figure 2 diagram (ASAS-SN data for 2012 and 2013 being fewer in number). We determined a phase shift for the recent observations by comparing them to the Clement *et al.* (1988) 1970 light curve, following the procedure they adopted. The 1970 curve has significant gaps, however, introducing some uncertainty in the size of the shift. In deriving these shifts, we scaled our light curves in amplitude and applied zero-point shifts to better match the Clement *et al.* (1988) photographic observations. Infrared data were not included in this comparison because the time of maximum in *I* and longer wavelength bands can show significant shifts with respect to *B* or *V* (see, for example, Osborn *et al.* 2019). Our phase shifts are shown in Table 3, where the reference times are the approximate mid-points of each set of observations. In interpreting our results, we quickly ran into a problem with deciding upon the correct cycle-count between observed epochs.

In Figure 6, we plot the phase shifts from Clement *et al.* (1988) as filled points, with the exception of the point for their 1985 data, which is plotted twice, once with at +0.48 (filled point) and again at –0.52 (cross). We then plotted two alternatives for the more recent data, with the crosses shifted one cycle lower in phase.

How do we tell which, if either, alternative in Figure 6 correctly plots the phase differences between the 1970 light curve and the later observations? At first, one might think that the crosses indicate that all except the 1985 point could be fit by a period near 15.527 days. However, that is not the case. A period near 15.527 days fits neither the MSU *B* and *V* light curves nor the ASAS-SN data, as was found in our period searches. It is apparent that gaps in time coverage and the jumps in phase shown in Figure 6 are too large for us to determine recent periods from the phase diagram alone. The data are continuous enough, however, for the restricted time interval covered by the ASAS-SN data for the phase shifts for those observations alone to yield a period of 15.543 days, consistent with the direct period determinations in Table 2.

4. Conclusions

What can we conclude from both the phase diagram and the direct period determinations? As shown in Figure 7, an increase in period to 15.54 days in the 2013–2018 time interval from a period near *15.48 days in 2002–2011 seems certain from the direct period determinations*. This would thus appear to be a relatively abrupt period increase. In their Table IV, Clement *et al.* (1988) used their phase-shift diagram to determine periods for V1 between 15.50 and 15.57 days (or 15.49 and 15.55 days, depending upon the interpretation of the 1975–1985 data). Can we establish the reality of the Clement *et al.* (1988) period jumps independently of their phase-shift diagram? The answer is yes but to a limited degree. For the 1938–1946 time interval, our PERANSO and PERIOD04 period determinations of 15.53 and 15.54 days are in reasonable agreement with the Clement *et al.* (1988) value of 15.55 days. For the 1970–1975 data, our periods of 15.51 or 15.52 days agree with the Clement *et al.* (1988) value of 15.51 days. It remains true, however, that most of the periods in Table IV of Clement *et al.* (1988) depend mainly upon the interpretation of the phase-diagram in their Figure 2. It is nonetheless not possible to revise the cycle-counting used in creating their Figure 2 without introducing alternative jumps in period. We therefore conclude that the period of V1 was not constant nor did it change at a constant rate between 1916 and 2018. We consider the Clement *et al.* (1988) interpretation that V1 has undergone both increases and decreases in period to be likely.

The referee asked whether the simpler interpretation of a period slightly smaller than 15.527 days might approximately explain all of the phase shift points after about 1930, except for that of the 1985 observations, assuming the lower crosses to be correct for the post-2000 observations. The answer would be yes, if the phase shift points were all that needed to be considered. However, as noted above, we see problems with such an interpretation. A period of 15.527 days or smaller produces a light curve for the ASAS-SN observations with much more scatter than the 15.542 day period. The direct period determinations in Table 2 for the post-2000 observations indicate periods near either 15.48 days (for the Matsunaga et al. 2006 and MSU CCD data) or 15.54 days (for the ASAS-SN 2013–2018 data). The average is 15.51 days, but the average period would not produce the best light curve for any of the post-2000 datasets. Moreover, with a gap of some two decades between the Clement *et al.* (1988) 1985 data and the MSU CCD observations, periods of 15.48 or 15.54 days would lead to different cycle counts across the gap. We thus prefer not to rely upon the phase-shift diagram in deriving periods for this interval, placing greater trust in the period determinations listed in Table 2.

Instead of finding that V1 in M12 underwent a constant rate of increase or decrease in period over the past century, as might have been expected from stellar evolution theory, our preferred interpretation is that its period changed in a more variable way. It fluctuated in period between about 15.47 and 15.57 days. If there is any long-term evolutionary period change happening in V1, it is effectively masked by more abrupt jumps in period.

Not all W Virginis variables show period fluctuations similar to those of V1. For example, Templeton and Henden

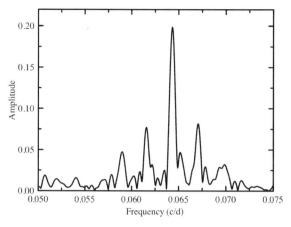

Figure 1. The amplitude spectrum from the PERIOD04 period search of the ASAS-SN data.

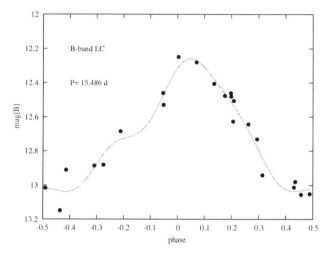

Figure 2. The light curve of V1 from the MSU *B* data, phased with a period of 15.486 days.

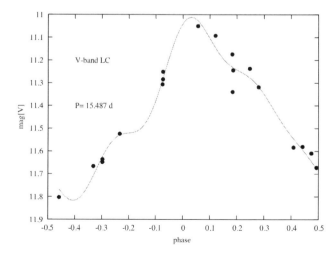

Figure 3. The light curve of V1 from the MSU *V* data, phased with a period of 15.487 days.

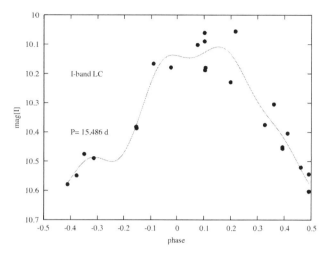

Figure 4. The light curve of V1 from the MSU *I* data, phased with a period of 15.486 days.

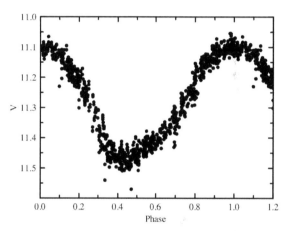

Figure 5. The light curve of V1 from the 2013–2018 ASAS-SN *V* observations, phased with a period of 15.542 days.

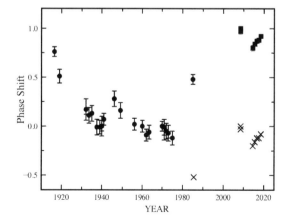

Figure 6. Recent observations are added to the phase-shift diagram of Clement *et al.* (1988). This cannot be done unambiguously, and alternative phase shifts are presented as squares and crosses. The two alternatives for the 1985 Clement *et al.* data are plotted as a circle and a cross.

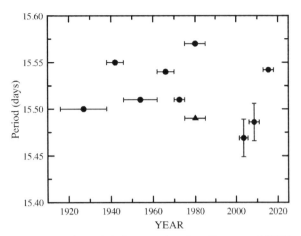

Figure 7. The changing period of V1 over time, from Clement *et al.* (1988) and from this paper. The triangle is an alternative period from Clement *et al.* (1988) for their 1975–1985 data. The Clement *et al.* (1988) periods are derived from phase shifts, and do not have associated error bars, but see Table 2.

Table 3. Phase Shift determinations for V1 for the years 2008–2018.

Epoch	Phase Shift	Uncertainty	Dataset
2008.5	–0.03	0.05	MSU V
2008.5	0.00	0.05	MSU B
2014.5	0.80	0.04	ASAS-SN
2015.5	0.84	0.04	ASAS-SN
2016.5	0.87	0.04	ASAS-SN
2017.5	0.88	0.04	ASAS-SN
2018.5	0.92	0.04	ASAS-SN

(2007) found a long-term period decrease for W Virginis itself. RV Tauri variables have been observed to show random changes of period (e.g. Percy and Coffey 2005), which can be superposed upon long-term period changes. However, RV Tauri behavior is typically seen at longer periods than the 15.5 days of V1. It may be noted that the long-term period changes of relatively few W Virginis variables have been studied. The relatively small number of W Virginis variables with long-term period studies, and the variety of period change behavior observed among those stars that have been studied (Rabidoux *et al.* 2010; Neilson *et al.* 2016) are an encouragement to further study the periods of variables such as V1.

5. Acknowledgements

We thank Charles Kuehn, who obtained three of the CCD images used in this study. We thank Christine Clement for her help and advice. We thank the referee for constructive advice that led us to make significant revisions in this paper.

References

Alard, C., and Lupton, R. H. 1998, *Astrophys. J.*, **503**, 325.
Bono, G. *et al.* 2016, *Mem. Soc. Astron. Ital.*, **87**, 358.
Catelan, M., and Smith, H. A. 2015, *Pulsating Stars*, Wiley-VCH, Weinheim, Germany, 183.
Clement, C. M., Hogg, H. S., and Yee, A. 1988, *Astron. J.*, **96**, 1642.
Clement, C. M., *et al.* 2001, *Astron. J.*, **122**, 2587.
Henden, A., Levine, S.,; Terrell, D., Welch, D. L., Munari, U., and Kloppenborg, B. K. 2018, *Amer. Astron. Soc.*, AAS Meeting No. 232, i.d. 223.06.
Klochkova, V. G., Panchuk, V. E., Tavolganskaya, N. S., and Kovtyukh, V. V. 2003, *Astron. Lett.*, **29**, 748.
Kochanek, C. S., *et al.* 2017, *Publ. Astron. Soc. Pacific*, **129**, 104502.
Kovács, G., and Kupi, G. 2007, *Astron. Astrophys.*, **462**, 1007.
Lenz, P., and Breger, M. 2005, *Commun. Asteroseismology*, **146**, 53.
Matsunaga, N., *et al.* 2006, *Mon. Not. Roy. Astron. Soc.*, **370**, 1979.
Neilson, H. R., Percy, J. R., and Smith, H. A. 2016, *J. Amer. Assoc. Var. Star Obs.*, **44**, 179.
Osborn, W., *et al.* 2019, *Acta Astron.*, in press.
Percy, J. R. 2007, *Understanding Variable Stars*, Cambridge University Press, Cambridge.
Percy, J. R., and Coffey, J. 2005, *J. Amer. Assoc. Var. Star Obs.*, **33**, 193.
Rabidoux, K., *et al.* 2010, *Astron. J.*, **139**, 2300.
Sawyer, H. B. 1938a, *Publ. Dom. Astrophys. Obs.*, **7**, 121.
Sawyer, H. B. 1938b, *Publ. David Dunlap Obs.*, **1**, 57.
Shappee, B. J., *et al.* 2014, *Astrophys. J.*, **788**, 48.
Stetson, P. B. 1987, Publ. Astron. Soc. Pacific, **99**, 191.
Stetson, P. B. 1994, in *Astronomy with the CFHT Adaptive Optics Bonnette*, ed. R. Arsenault, Canada-France-Hawaii Telescope Corp., Kamuela, HI, 72.
Templeton, M. R., and Henden, A. A. 2007, *Astron. J.*, **134**, 1999.
Vanmunster, T. 2006, light curve and period analysis software PERANSO 2.0, CBA Belgium (http://www.cbabelgium.com/peranso).

A New Candidate δ Scuti Star in Centaurus: HD 121191

Roy A. Axelsen
P. O. Box 706, Kenmore Qld 4069, Australia; reaxelsen@gmail.com

Received June 7, 2019; revised June 20, 2019; accepted July 15, 2019

Abstract During a study of the δ Scuti star V1393 Centauri by digital single lens reflex photometry, it was found that two of the chosen comparison stars were variable. This paper reports the subsequent investigation of one of them, the class A star HD 121191, which revealed that it is a candidate δ Scuti star with a period of 0.046282 d and an amplitude of 0.048 magnitude in V.

1. Introduction

δ Scuti stars are pulsating variables with short periods and mostly low amplitudes. They comprise about one third of A5-F2 III-V stars (Percy 2007). Amateur astronomers performing time series photometry occasionally discover previously unrecognized δ Scuti-like periods (Moriarty et al. 2013). This paper reports one such discovery. While performing time series DSLR photometry on the δ Scuti star V1393 Cen, the author found that two of the chosen comparison stars were variable. As no AAVSO finder charts were available for V1393 Cen at the time of the observations, the author simply chose what appeared to be appropriate comparison stars, after checking in SIMBAD (Wenger et al. 2000) and the *General Catalogue of Variable Stars* (GCVS; Samus et al. 2017) that they were not reported as variable. However, after check star light curves showed obvious variability, observations were concentrated on two new variables. This paper reports the results of the study of one of them, HD 121191, an A5 IV/V star (Melis et al. 2013) in Centaurus with coordinates (J2000) R.A. $13^h\ 57^m\ 56.44^s$ Dec. $-53°\ 42'\ 15.34"$ (ICRS coordinates, SIMBAD astronomical database).

2. Methods

Time series photometry was performed on five consecutive nights from 26 to 31 May 2019. Images were taken with a Canon EOS 500D digital single lens reflex (DSLR) camera through an 80mm f/7.5 refractor on an equatorial mount. Exposures of 180 seconds were taken at ISO 400, with a 5-second gap between consecutive exposures. Autoguiding was performed by the software PHD2 GUIDING, with an Orion StarShoot Autoguider imaging through an 80 mm f/5 refractor.

Images were converted to the FITS format and pre-processed in IRIS (Buil 1999–2018) using dark, bias, and flat frames, and images from the blue and green channels were extracted. The latter images were imported into ASTROIMAGEJ (Collins et al. 2017) for aligning and aperture photometry. Comparison and check stars were HD 120858 and HD 121277, respectively. Values of V and B–V for the comparison star were taken to be 8.706 and 1.356. For the check star, the corresponding values were taken to be 9.162 and 1.234. HD 121191 (the new variable) has a magnitude in V of approximately 8.17 and a B–V color index of approximately 0.24. These V and B–V values were taken from the planetarium program GUIDE 9.0 (Project Pluto 1996–2016). The author has found that V and B-V values in GUIDE 9.0 differ very little from those of standard stars from the E regions (Menzies et al. 1989), and are thus considered adequate for the purposes of this study.

Flux values from ASTROIMAGEJ were imported into an EXCEL spreadsheet. Instrumental magnitudes and transformed magnitudes in B and V were calculated, using transformation coefficients derived from images of standard stars from the E regions (Menzies et al. 1989). Atmospheric extinction coefficients were not used. B magnitudes were used only for the determination of the transformation coefficients, and for the calculation of transformed V magnitudes of the variable. As the precision of B magnitudes determined by DSLR photometry for the variable star is less than that for V magnitudes, the former are not reported in this paper.

The data were analyzed in VSTAR (Benn 2013) and in PERIOD04 (Lenz and Breger 2005).

3. Results

The transformed V magnitude of HD 121191 was determined for 653 time points over the five nights of observation. The light curve for one night is shown in Figure 1. The average B–V color index of HD 121191 over the five nights was calculated to be 0.233, very close to the B–V value of 0.243 as published in GUIDE 9.0.

A DCDFT (Date Compensated Discrete Fourier Transform) applied to the data in VSTAR revealed a frequency of 21.603 c d^{-1} corresponding to a period of 0.046291 d. The power spectrum from this analysis (Figure 2) shows this frequency and four other nearby labelled frequencies, with the latter representing one cycle per day aliasing due to the periodicity inherent in the nightly observation sets. ANOVA applied to binned means of the VSTAR residuals (0.04 phase step per bin) revealed an F-value of 0.6180 on 25 and 627 degrees of freedom with a p value of 0.9278. These results indicate that no further periods could be found within these data. A phase plot based on the period 0.046291 d is shown in Figure 3.

The mathematical output of the model created in VSTAR is:

$$f(t:real):real\ \{8.182428 + 0.023603 * \cos(2*PI*21.602471*(t-2458632.1)) + 0.004152 * \sin(2*PI*21.602471*(t-2458632.1))\} \quad (1)$$

Fourier analysis was also performed in the period analysis software PERIOD04 and revealed a frequency of 21.607 (±002) c d^{-1}, corresponding to a period of 0.046282 (± 0.000004) d. This period is 0.000009 d shorter than that found in VSTAR, but is so close that it is considered to be the same period. The signal-

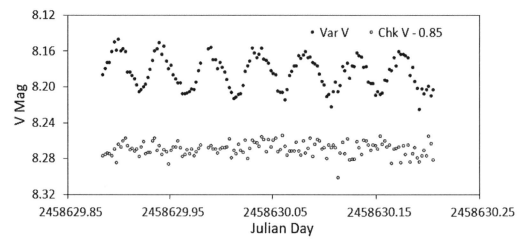

Figure 1. The light curves of the variable star HD 121191 and the check star HD 121277 obtained by DSLR photometry during one night. Almost seven complete cycles of the variable star are evident. Var V = Variable star V magnitude. ChkV – 0.85 = Check star V magnitude minus 0.85.

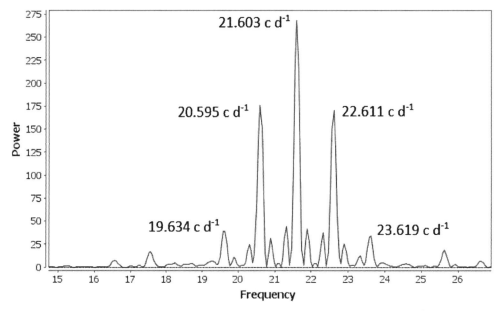

Figure 2. Power spectrum of the Date Compensated Discrete Fourier Transform from vstar. The frequency 21.603 c d^{-1} is accompanied by four other nearby labelled frequencies representing one cycle per day aliasing.

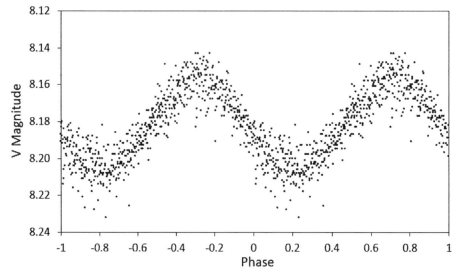

Figure 3. Phase plot of HD 121191 over two complete cycles, based on a period of 0.046291 d, created from data captured during five nights of DSLR photometry.

to-noise ratio for a box size of 2 is 3.01, somewhat low for an acceptable result, which should be at least 4.

4. Discussion

The purpose of this paper is to report an example of δ Scuti-like variability which has apparently not been recognized previously. The basis for this statement is that identity and position searches (i.e., R.A. and Dec.) in SIMBAD, and position searches in the *General Catalogue of Variable Stars* and through the SAO/NASA ADS Astronomy Query Form (http://adsabs.harvard.edu/abstract_service.html) failed to find any specific reference to the target star HD 121191.

The facts that HD 121191 is an A type star (Melis *et al.* 2013) exhibiting stable variability with a period of just over one hour, and that the amplitude is just under 0.05 magnitude in V, suggest that the variability is of δ Scuti type.

The circumstances of the discovery of the variability emphasize the well-known necessity of employing comparison and check stars in differential photometry, and flag the need to be confident of the absence of variability in comparison stars used in ensemble photometry. As shown herein, it is not sufficient to note the absence of candidate comparison stars from catalogues of variable stars, and from bibliographic databases which should contain reports of stellar variability. Although the averaging process in ensemble photometry reduces the final error, it is patently not acceptable to employ as a comparison star for photometry one which exhibits variability with an amplitude of just under 50 millimagnitudes.

Finally, it should be noted that the precision of DSLR time series photometry of an eighth-magnitude star using equipment employed in the present study is not optimal for the investigation of low amplitude δ Scuti-like variability. The superior precision and faster cadences achievable using CCD photometry would be preferable. However, it is clear that even the lesser precision of photometry using a DSLR camera is sufficient for valid discovery information to be documented.

5. Conclusion

The A type star HD 121191 is proposed as a new candidate δ Scuti variable, with a period of 0.046282 d and an amplitude of 0.048 magnitude in V. The discovery of this variability was made serendipitously during DSLR photometric studies of another known δ Scuti variable.

6. Acknowledgements

This research has made use of the SIMBAD database, operated at CDS, Strasbourg, France.

References

Benn, D. 2013, vstar data analysis software (http://www.aavso.org/vstar-overview).

Buil, C. 1999–2018, iris astronomical images processing software (http://www.astrosurf.com/buil/iris-software.html).

Collins, K. A., Kielkopf, J. F., Stassun, K. G., and Hessman, F. V. 2017, *Astron. J.*, **153**, 77.

Lenz, P., and Breger, M. 2005, *Commun. Asteroseismology*, **146**, 53.

Melis, C., Zuckerman, B., Rhee, J. H., Song, I., Murphy, S. J., and Bessell, M. S. 2013, *Astrophys. J.*, **778**, 12.

Menzies, J. W., Cousins, A. W. J., Banfield, R. M., and Laing, J. D. 1989, *S. Afr. Astron. Obs. Circ.*, **13**, 1.

Moriarty, D. J. W., Bohlsen, T., Heathcote, B., Richards, T., and Streamer, M. 2013, *J. Amer. Assoc. Var. Star Obs.*, **41**, 182.

Percy, J. R. 2007, *Understanding Variable Stars*, Cambridge University Press, Toronto, 183.

Project Pluto. 1996–2016, guide 9.0 planetarium software (www.projectpluto.com).

Samus, N. N., Kazarovets, E. V., Durlevich, O. V., Kireeva, N. N., and Pastukhova, E. N. 2017, *Astron. Rep.*, **61**, 80.

Wenger, M., *et al.* 2000, *Astron. Astrophys., Suppl. Ser.*, **143**, 9.

Observations and Analysis of a Detached Eclipsing Binary, V385 Camelopardalis

Ronald G. Samec
Faculty Research Associate, Pisgah Astronomical Research Institute, 1 PARI Drive, Rosman, NC 28772; ronaldsamec@gmail.com

Daniel B. Caton
Dark Sky Observatory, Physics and Astronomy Department, Appalachian State University, 525 Rivers Street, Boone, NC 28608-2106; catondb@appstate.edu

Danny R. Faulkner
Johnson Observatory, 1414 Bur Oak Court, Hebron, KY 41048; dfaulkner@answersingenesis.org

Received July 16, 2019; revised September 4, 17, 2019; accepted September 17, 2019

Abstract V385 Cam is found to be a G7±2 type (T~5500 K) pre-contact eclipsing binary. It was observed on December 15, 16, 17, and 18, 2017, at Dark Sky Observatory in North Carolina with the 0.81-m reflector of Appalachian State University. Three times of minimum light were determined from our present observations, which include two primary eclipses and one secondary eclipse. Twelve other minima were determined or found in the literature. This allowed an 18.5-year period study and a possible quadratic ephemeris was found. The resulting weak period decrease may indicate that the solar type binary is undergoing magnetic braking. A BVR_cI_c simultaneous Wilson-Devinney Program (WD) solution gives a detached solution. This model has a primary component with a 96% fill-out and a secondary component with 91% fill-out. The solution has a mass ratio of 0.390±0.001, and a component temperature difference of ~1000 K. The large ΔT in the components verifies that the binary is not yet in contact. Two hot spots were used in the modeling, a polar spot and a smaller southern spot with a 120° colatitude. The binary star inclination is ~84.1±0.2°, resulting in a total eclipse (secondary component) of 27.2 minutes in duration.

1. Introduction

In this paper, we continue our study of pre-contact W UMa binaries. Among those recently studied include AE Cas (Chamberlain *et al.* 2019), NSVS 103152 (Samec *et al.* 2018), TYC 1488-693-1 (Samec *et al.* 2018), NSVS 10083189 (Samec *et al.* 2017a), and GQ Cancri (Samec *et al.* 2017b), etc. We have designated these as pre-contact W UMa binaries since they are in detached or semidetached configurations and their orbital period studies have shown them to be systems that may be undergoing magnetic braking (angular momentum loss, AML). They could be evolving into contact and finally to binary coalescence and the formation of a fast rotating, earlier spectral type single star after a red novae event (Tylenda and Kamiński 2016). Streaming plasmas moving on stiff, rotating radial patterns away from the binary out to the Alfven radius (~50 stellar radii) could cause this. Here we present the first precision photometry and light curve analysis of another such candidate, the pre-contact system V385 Cam. The first report of these observations was given as a poster paper at the American Astronomical Society Meeting #234 (Samec *et al.* 2019).

2. History

The variable was listed as a near contact binary by Shaw (1994) who gives the light curve (see Figure 1), magnitude (V = 13.78), and ephemeris:

$$HJD = 2451396.798279\,d + 0.61521068 \times E, \quad (1)$$

and position of the binary designated also as NSVS 00580289.

Khruslov (2006) included it in a list of New Short-Period Eclipsing Binaries in Camelopardalis (Figure 2). He gave the position, magnitude range (V = 13.65–14.20), finding chart, light curve, the designation GSC 4515-00038, and the ephemeris:

$$HJD = 2451515.845\,d + 0.61525 \times E. \quad (2)$$

The binary was listed in the "80th Name-List of Variable Stars, Part I" (Kazarovets *et al.* 2011).

This system was observed as a part of our student/professional collaborative studies of interacting binaries using data taken from Dark Sky Observatory (DSO) observations. The observations were taken by R. Samec, D. Caton, and D. Faulkner. Reduction and analyses were done by R. Samec. The GAIA DR2 determined distance is 854±13 pc (Bailer-Jones 2015). The patrol light curve shown in Figure 3 is taken from the ASAS-SN, All-Sky Automated Survey for Supernovae Variable Stars Data Base (Shappee *et al.* 2014; Kochanek *et al.* 2017; https://asas-sn.osu.edu). In addition, V385 Cam has a J-K=0.22, an ephemeris:

$$JDHelMIN\ I = 2457386.76033 + 0.6152709 \times E, \quad (3)$$

and catalog name: ASASSN-V J051947.00+773613.3.

3. BVR_cI_c photometry

Our 2017 BVR_cI_c light curves were taken by D. Caton on December 15, 16, 17, and 18, 2017, at Dark Sky Observatory

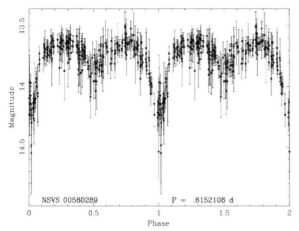

Figure 1. Light curves of V385 Cam (Shaw 1994; http://www.physast.uga.edu/~jss/ncb/).

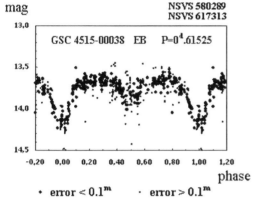

Figure 2. Light curve from Khruslov *et al.* (2006).

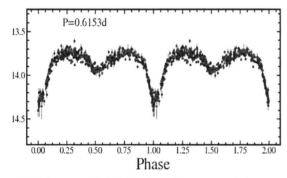

Figure 3. V light curves of V385 Cam (https://asas-sn.osu.edu/).

in North Carolina with the 0.81-m reflector of Appalachian State University at Philips Gap, North Carolina, with a thermoelectrically cooled (–40°C) 2KX2K Apogee Alta with standard BVR_cI_c filters.

Our observations included 347 images in the B filter, 369 images in the V filter, 319 images in the R_c filter, and 306 images in the I_c filter. The probable error of a single observation was 8 mmag in B, 7 mmag in V and I, and 6 mmag in R. Exposure times varied from 40 to 80s in B, 25–60s in V, and 15–40s in R_c and I_c, depending on the count needed to obtain 1% photometry. Photometric targets are given in Table 1.

The nightly C–K magnitudes stayed constant throughout the observing run with a precision of about 1% (see, for instance, Figure 4.). The observed curves from December 16, 2017, are shown in Figure 5.

The V–C differential color indices, $\Delta(B-V)$ and $\Delta(R_c-I_c)$, are seen to be very close to zero. The finder chart is given as Figure 6 with the variable star (V), comparison star (C), and check star (K) shown. Our observations are listed in Table 2, with magnitude differences ΔB, ΔV, ΔR_c, and ΔI_c in the sense variable minus comparison star.

4. Orbital period study

Three times of minimum light were calculated, 2 primary and 1 secondary eclipses, from our present observations:

HJD I = 2457372.62072 ± 0.0003 d, 2457374.46776 ± 0.00043 d,
HJD II = 2457372.92658 ± 0.0011 d,

A least squares minimization method (Mikulášek *et al.* 2014) was used to determine the minima for each curve, in B,V,R_c, and I_c. Eight times of low light were taken from a light curve phased from data (ASAS J051947,00+773613.3) from the All-Sky Automated Survey (ASAS; Pojmański 1997). This plot was used to get times of minima from parabola fits within ± 0.01 of phases 0.0 and 0.5 (https://asas-sn.osu.edu/database/light_curves/340255). The ASAS ephemeris provided another minimum. Another time of minimum light is given by Diethelm (2013). A linear ephemeris and quadratic ephemerides were determined from these data.

All of these times of minimum light are given in Table 3.

Both linear and quadratic ephemerides were determined from these data:

Table 1. Information on the stars used in this study.

Star	Name	R.A. (2000) h m s	Dec. (2000)[1] ° ′ ″	V[2]	J–K[3]
Variable (V)	V385 Cam GSC 4515-00038 2MASS J05194709+7736136 ASASSN-V J051947.00+773613.3 Gaia DR2 552348118709696256	05 19 47.0833	+77 36 13.637	13.78	0.452 ± 0.046
Comparison (C)	GSC 4515 626 3UC 336-015704	05 19 49.3367	+77 33 38.989	13.334	0.34
Check (K)	GSC 4515 1148	05 19 13.0678	+77 36 18.970	11.97	0.593 ± 0.041

[1] *UCAC3 (U.S. Naval Obs. 2012),* [2] *APASS (Henden et al. 2015),* [3] *2MASS (Cutri et al. 2003).*

$$\text{JDHelMinI} = 2457372.61883 + 0.61527147\,d \times E$$
$$\pm 0.00057 \pm 0.00000014 \quad (4)$$

$$\text{JDHelMinI} = 2457372.61875\,d + 0.61527021 \times E - 0.00000000014 \times E^2$$
$$\pm 0.00055 \pm 0.00000068 \pm 0.00000000007 \quad (5)$$

The residual plots for the linear and the quadratic ephemerides are shown in Figures 7a and 7b, respectively. The r.m.s. of the quadratic ephemeris is somewhat better. The higher quality points at the end of the plot cause one to admit of such a possibility. The period study covers a period of some 18.5 years and shows the possibility that the period may be decreasing (at about the ~1.5 sigma level, with the errors shown here are probable errors). This could be due to mass flow to the secondary component or magnetic braking from the solar components.

The quadratic ephemeris yields a $\dot{P} = -1.66 \times 10^{-7}$ d/yr, or a mass exchange rate of

$$\frac{dM}{dt} = \frac{\dot{P}M_1 M_2}{3P(M_1 - M_2)} = \frac{-5.28 \times 10^{-8} M_\odot}{d}. \quad (6)$$

in a conservative scenario. The possibility of a third component should be at least considered; the quadratic curve could be part of a sinusoid. In fact, a large number of short-period systems have third components (Tokovinin *et al.* 2006). Further eclipse timings are needed to confirm or disaffirm this whole scenario.

The O–C quadratic residual calculations are given in Table 3.

5. Light curves and light curve solution

The BVR_cI_c phased light curves calculated from Equation 3 are displayed in Figures 8a and 8b. The curves are of good precision, averaging somewhat better than 1% photometric precision. The amplitude of the light curve varies from 0.64 to 0.46 mag in B to I. The O'Connell effect, an indicator of spot activity, averages several times the noise level, 0.02–0.04 mag. The differences in minima are large, 0.22–0.45 mags, indicating noncontact light curves. The B–V color curves fall ever so very slightly at phase 0.0, however, the color curves rise at phase 0.5, which indicates that the primary component is underfilling its Roche Lobe. This probably indicates that the primary component is not filling its Roche Lobe so we have a detached or semidetached Algol-type binary.

The 2MASS J–K = 0.452 ± 0.046 for the binary (Cutri *et al.* 2003). This corresponds to G7.5 ± 2.5 V eclipsing binary, which yields a temperature of 5500 ± 250 K. Fast rotating binary stars of this type are noted for having convective atmospheres, so spots are expected. Most likely, the spots are magnetic in nature.

The B, V, R_c, and I_c curves were pre-modeled with BINARY MAKER 3.0 (Bradstreet and Steelman 2002) and fits were determined in all filter bands. The result of the best fit was that of a semidetached eclipsing binary with the secondary component filling its critical Roche lobe (classical Algol configuration). The parameters were then averaged and input into a 4-color simultaneous light curve calculation using the 2016 Wilson-Devinney program (WD; Wilson and Devinney 1971; Wilson

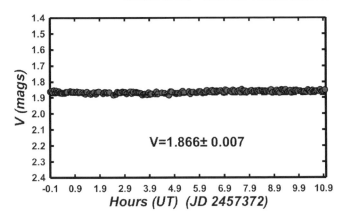

Figure 4. C–K, V magnitudes are shown on the night of JD = 2457372.

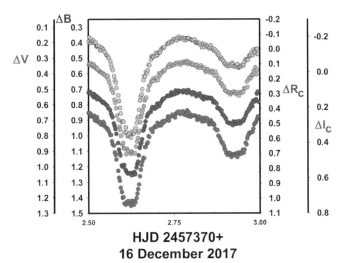

Figure 5. B, V, R_c, I_c curves on the night of December 16, 2017.

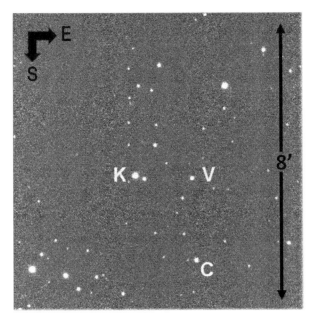

Figure 6. Finder Chart, V385 Cam (V), comparison star (C), and check (K).

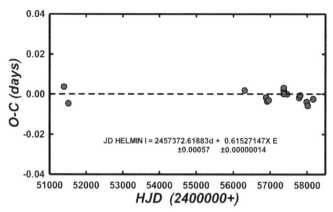

Figure 7a. The period study of 18.5 years with a linear ephemeris.

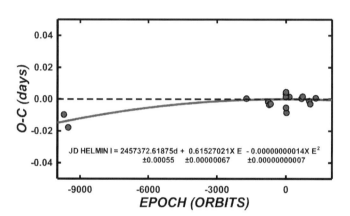

Figure 7b. The period study of 18.5 years, indicating the possibility of a continuous period decrease

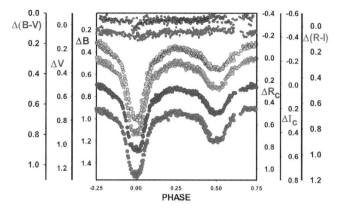

Figure 8a. B, V, R_c, I_c light curves and color curves B–V and R–I ; magnitudes phased with Equation 2.

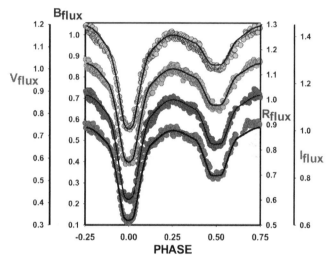

Figure 8b. B,V,R_c,I_c normalized flux overlaid by the detached solution for V385 Cam.

1979, 1990, 1994, 2008, 2012; Van Hamme 1998; Van Hamme and Wilson 2007; Wilson *et al.* 2010; Wilson and Van Hamme 2014). The solution was computed in Mode 2 (detached) with the initial spot parameters and converged to a solution. As the referee noted, there was still an apparent O'Connell effect in the resulting solution at about the 0.75 phase. A small equatorial spot was entered about that phase and allowed to iterate along with all the other solution parameters (see Figure 9d). The model converged to a solution with the two spots, the second one at 120° colatitude with a T-factor of 1.2, and therefore is a magnetic hot spot (facula or plage). The reason for mode 2 was to allow the computation to determine the configuration. If the solution had gone detached in one or the other components the mode would have been changed to mode 4 or 5, dependent on the component that completely filled its respective Roche lobe. The computation converged in the detached mode without doing either. Convective parameters, g = 0.32, A = 0.5 were used. The spots did converge with some changes as expected.

The solution follows as Table 5. The solution was detached with the primary component nearer to critical contact potential-wise (96% fill-out) than the secondary component (91% fill-out). The binary star inclination is ~85.1±0.1°, resulting in a total eclipse (secondary component) of 27.2 minutes in duration. The difference of component temperatures is about 980 K, confirming that the components are not in contact. The modeled period was 0.61498 d.

6. Discussion

V385 Cam is a precontact W UMa binary in a detached but near contact configuration. This configuration could result, evolution-wise, in the well-known V1010 Ophiuchi configuration where the primary more massive component fills its Roche lobe first.

AML (angular momentum loss) could be in play here as indicated by its spectral type and hot spots. Its spectral type is that of a solar type with a surface temperature of 5500 K for the primary component. The secondary component has a temperature of ~4520 K (K4.5V). The mass ratio is 0.39, with an amplitude of 0.64–0.46 mag in B to I, respectively. The inclination is 85° which results in a total eclipse at phase 0.5.

7. Conclusion

The period study of this precontact W U Ma binary has an 18.5-year time duration. The period is found to be weakly decreasing at about the 1.5-sigma level. Decreasing periods

Figure 9a. V385 Cam, geometrical representation of Roche-lobe surfaces at phase 0.00.

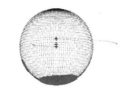

Figure 9c. V385 Cam, geometrical representation of Roche-lobe surfaces at phase 0.50.

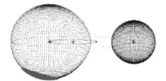

Figure 9b. V385 Cam, geometrical representation of Roche-lobe surfaces at phase 0.25.

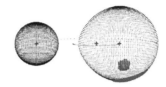

Figure 9d. V385 Cam, geometrical representation of Roche-lobe surfaces at phase 0.75.

are not unusual for a solar type binary since they can undergo magnetic braking. Conservative mass transfer from the primary component to the secondary component could also produce the same effect. The system may coalesce over time as it loses angular momentum due to ion winds moving radially outward on stiff magnetic field lines rotating with the binary (out to the Alfvén radius). The system could soon become an A-type W UMa contact binary and ultimately, one might expect that the binary will become a rather normal, fast rotating, single ~G2V type field star after a red novae coalescence event (Tylenda and Kamiński 2016). Radial velocity curves are needed to obtain absolute (not relative) system parameters.

8. Acknowledgements

Dr. Samec would like to think his former students for working with him over the years. Some have become professional astronomers and physicists. One of these former students, Miss Heather Chamberlain, presented a poster paper at the 2019 American Astronomical Society meeting (Chamberlain *et al.* 2019).

References

Bailer-Jones, C. A. L., *et al.* 2013, *Astron. Astrophys.*, **559A**, 74.
Bradstreet, D. H., and Steelman, D. P. 2002, *Bull. Amer. Astron. Soc.*, **34**, 1224.
Chamberlain, H., Samec, R., Caton, D. B., and Faulkner, D. R. 2019, *Bull. Amer. Astron. Soc.*, **51**, 4.
Cutri, R. M., *et al.* 2003, The IRSA 2MASS All-Sky Point Source Catalog, NASA/IPAC Infrared Science Archive (http://irsa.ipac.caltech.edu/applications/Gator/).
Diethelm, R. 2013, *Inf. Bull. Var. Stars*, No. 6063, 1.
Henden, A. A., *et al.* 2015, AAVSO Photometric All-Sky Survey, data release 9, (https://www.aavso.org/apass).
Kazarovets, E. V., Samus, N. N., Durlevich, O. V., Kireeva, N. N., and Pastukhova, E. N. 2011, *Inf. Bull. Var. Stars*, No. 5969, 1.
Khruslov, A. V. 2006, *Perem. Zvezdy Prilozh.*, **6**, 11.
Kochanek, C. S., *et al.* 2017, *Publ. Astron. Soc. Pacific*, **129**, 104502.
Mikulášek, Z., Chrastina, M., Liška, J., Zejda, M., Janík, J., Zhu, L.-Y., and Qian, S.-B. 2014, *Contrib. Astron. Obs. Skalnate Pleso*, **43**, 382.
Pojmański, G. 1997, *Acta Astron.*, **47**, 467.
Samec, R. G., Caton, D., and Faulkner, D. R. 2019, AAS Meeting #233, id.348.02.
Samec, R. G., Olsen, A., Caton, D., and Faulkner, D. R. 2017b. *J. Amer. Assoc. Var. Star Obs.*, **45**, 148.
Samec, R. G., Caton, D. B., Faulkner, D. R., and Hill, R. 2018, *J. Amer. Assoc. Var. Star Obs.*, **46**, 106.
Samec, R. G., Olsen, A., Caton, D. B. Faulkner, D. R., and Hill, R. L. 2017a, *J. Amer. Assoc. Var. Star Obs.*, **45**, 173.
Shappee, B. J., *et al.* 2014, *Astrophys. J.*, **788**, 48.
Shaw, J. S. 1994, *Mem. Soc. Astron. Ital.*, **65**, 95.
Tokovinin, A., Thomas, S., Sterzik, M., and Udry, S. 2006, *Astron. Astrophys.*, **450**, 681.
Tylenda, R., and Kamiński, T. 2016, *Astron. Astrophys.*, **592A**, 134.
U.S. Naval Observatory. 2012, UCAC-3 (http://www.usno.navy.mil/USNO/astrometry/optical-IR-prod/ucac).
Van Hamme, W. V., and Wilson, R. E. 1998, *Bull. Amer. Astron. Assoc.*, **30**, 1402.
Van Hamme, W., and Wilson, R. E. 2007, *Astrophys. J.*, **661**, 1129.
Wilson, R. E. 1979, *Astrophys. J.*, **234**, 1054.
Wilson, R. E. 1990, *Astrophys. J.*, **356**, 613.
Wilson, R. E. 1994, *Publ. Astron. Soc. Pacific*, **106**, 921.
Wilson, R. E. 2008, *Astrophys. J.*, **672**, 575.
Wilson, R. E. 2012, *Astron. J.*, **144**, 73.
Wilson, R. E., and Devinney, E. J. 1971, *Astrophys. J.*, **166**, 605.
Wilson, R. E., and Van Hamme, W. 2014, *Astrophys. J.*, **780**, 151.
Wilson, R. E., Van Hamme, W., and Terrell D. 2010, *Astrophys. J.*, **723**, 1469.

Table 2. V385 Cam observations, ΔB, ΔV, ΔR$_c$, and ΔI$_c$, variable star minus comparison star.

ΔB	BHJD 2457370+	ΔB	BHJD 2457370+	ΔB	BHJD 2457370+	ΔB	BHJD 2457370+	ΔB	BHJD 2457370+
0.3550	2.4803	0.7350	2.6559	0.4400	2.8537	0.4950	3.5058	0.3390	4.9119
0.3440	2.4822	0.7020	2.6582	0.4340	2.8561	0.5070	3.5094	0.3530	4.9142
0.3390	2.4840	0.6430	2.6615	0.4190	2.8584	0.5270	3.5117	0.3500	4.9171
0.3590	2.4859	0.6170	2.6638	0.4390	2.8607	0.5230	3.5163	0.3510	4.9193
0.3490	2.4886	0.6200	2.6661	0.4470	2.8639	0.5070	3.5201	0.3180	4.9238
0.3720	2.4905	0.5900	2.6685	0.4410	2.8662	0.5270	3.5224	0.3300	4.9267
0.3720	2.4923	0.5980	2.6722	0.4560	2.8685	0.5340	3.5455	0.3200	4.9290
0.3670	2.4942	0.5800	2.6745	0.4490	2.8708	0.9680	4.4816	0.3180	4.9312
0.3620	2.4974	0.5450	2.6769	0.4600	2.8758	0.8400	4.4861	0.3280	4.9334
0.3850	2.4993	0.5480	2.6792	0.4780	2.8781	0.8580	4.4901	0.3290	4.9364
0.3950	2.5011	0.5130	2.6836	0.4720	2.8804	0.7820	4.4941	0.3320	4.9386
0.3660	2.5030	0.4890	2.6859	0.4870	2.8827	0.7620	4.4982	0.3050	4.9409
0.4030	2.5057	0.4960	2.6882	0.5050	2.8859	0.6960	4.5025	0.3010	4.9431
0.3760	2.5076	0.4930	2.6906	0.5010	2.8882	0.6840	4.5048	0.3190	4.9464
0.3930	2.5095	0.4870	2.6956	0.5110	2.8906	0.5900	4.5070	0.3440	4.9486
0.3980	2.5113	0.4680	2.6979	0.5160	2.8929	0.6380	4.5093	0.3260	4.9531
0.3930	2.5139	0.4850	2.7003	0.5260	2.8961	0.5970	4.5151	0.3830	4.9581
0.3860	2.5157	0.4800	2.7026	0.5390	2.8984	0.5730	4.5173	0.3370	4.9631
0.4010	2.5176	0.4660	2.7059	0.5370	2.9007	0.5400	4.5196	0.3630	4.9682
0.4030	2.5194	0.4640	2.7083	0.5500	2.9031	0.5080	4.5219	0.4080	4.9706
0.3990	2.5221	0.4640	2.7106	0.5450	2.9068	0.5060	4.5265	0.3700	4.9729
0.4260	2.5240	0.4460	2.7129	0.5560	2.9091	0.5130	4.5290	0.3680	5.8398
0.4220	2.5258	0.4390	2.7164	0.5520	2.9114	0.5010	4.5315	0.3520	5.8426
0.4200	2.5277	0.4330	2.7188	0.5430	2.9138	0.4760	4.5340	0.3530	5.8452
0.4240	2.5329	0.4360	2.7211	0.5510	2.9179	0.4950	4.5387	0.3750	5.8481
0.4510	2.5348	0.4290	2.7234	0.5330	2.9202	0.4650	4.5405	0.3550	5.8519
0.4450	2.5366	0.4080	2.7316	0.5420	2.9225	0.4680	4.5424	0.3600	5.8547
0.4310	2.5385	0.4050	2.7339	0.5570	2.9248	0.4490	4.5443	0.3650	5.8573
0.4380	2.5420	0.4080	2.7362	0.5560	2.9287	0.4780	4.5511	0.3640	5.8598
0.4460	2.5443	0.3960	2.7385	0.5590	2.9310	0.4600	4.5529	0.3750	5.8634
0.4560	2.5466	0.3990	2.7437	0.5520	2.9333	0.4470	4.5547	0.3590	5.8663
0.4640	2.5490	0.3960	2.7460	0.5420	2.9356	0.4210	4.5566	0.3670	5.8689
0.4830	2.5535	0.3860	2.7483	0.5650	2.9388	0.4690	4.5611	0.3860	5.8714
0.4860	2.5558	0.3790	2.7506	0.5620	2.9411	0.4400	4.5629	0.3660	5.8749
0.4940	2.5581	0.3780	2.7544	0.5600	2.9435	0.4420	4.5648	0.3650	5.8775
0.5020	2.5604	0.3820	2.7567	0.5570	2.9458	0.3950	4.5666	0.3620	5.8800
0.5200	2.5648	0.3860	2.7590	0.5470	2.9494	0.4260	4.5711	0.3700	5.8827
0.5440	2.5671	0.3720	2.7613	0.5530	2.9517	0.4270	4.5730	0.3790	5.8859
0.5640	2.5694	0.3640	2.7653	0.5430	2.9540	0.4010	4.5748	0.3870	5.8885
0.5780	2.5718	0.3790	2.7676	0.5250	2.9563	0.3950	4.5766	0.3680	5.8912
0.6180	2.5756	0.3730	2.7699	0.5100	2.9595	0.3930	4.5833	0.3790	5.8937
0.6460	2.5779	0.3710	2.7723	0.5160	2.9618	0.3890	4.5852	0.3750	5.8970
0.6600	2.5803	0.3700	2.7775	0.5110	2.9642	0.3980	4.5885	0.3980	5.8996
0.6930	2.5826	0.3760	2.7798	0.5140	2.9665	0.3650	4.5903	0.3790	5.9021
0.7410	2.5877	0.3870	2.7821	0.4870	2.9700	0.4160	4.5921	0.3870	5.9048
0.7770	2.5900	0.3740	2.7844	0.4890	2.9723	0.3940	4.5940	0.3980	5.9105
0.8210	2.5924	0.3840	2.7875	0.4830	2.9747	0.3580	4.5976	0.4100	5.9138
0.8570	2.5947	0.3730	2.7898	0.4670	2.9770	0.3660	4.6071	0.3870	5.9172
0.9140	2.5982	0.3740	2.7921	0.4860	2.9801	0.3670	4.6149	0.4060	5.9207
0.9090	2.6005	0.3730	2.7945	0.4490	2.9825	0.3470	4.6180	0.4060	5.9252
0.9120	2.6029	0.3810	2.8001	0.4700	2.9848	0.3880	4.6557	0.4140	5.9287
0.9800	2.6052	0.3820	2.8024	0.4610	2.9871	0.4220	4.8317	0.4140	5.9321
0.9480	2.6092	0.3860	2.8048	0.4460	2.9907	0.4130	4.8435	0.4160	5.9354
0.9790	2.6115	0.3840	2.8071	0.4450	2.9930	0.4110	4.8506	0.4320	5.9403
1.0040	2.6138	0.3820	2.8124	0.4780	2.9953	0.4180	4.8590	0.4090	5.9437
0.9970	2.6162	0.4010	2.8147	0.4280	3.4574	0.3880	4.8693	0.4430	5.9470
0.9860	2.6201	0.4000	2.8170	0.4420	3.4662	0.3480	4.8759	0.4550	5.9504
1.0100	2.6225	0.4040	2.8194	0.3960	3.4685	0.3550	4.8781	0.4500	5.9545
1.0140	2.6248	0.4010	2.8223	0.4210	3.4708	0.3690	4.8815	0.4660	5.9579
0.9760	2.6271	0.4070	2.8246	0.4490	3.4782	0.3740	4.8837	0.4660	5.9613
0.9880	2.6309	0.4040	2.8270	0.4410	3.4805	0.3550	4.8881	0.5060	5.9645
0.9860	2.6332	0.4180	2.8293	0.4240	3.4828	0.3630	4.8904	0.5070	5.9691
0.9700	2.6356	0.4340	2.8333	0.4150	3.4852	0.3330	4.8926	0.5130	5.9724
0.9530	2.6379	0.4080	2.8356	0.4500	3.4882	0.3670	4.8948	0.5020	5.9757
0.9200	2.6412	0.4170	2.8379	0.4620	3.4905	0.3410	4.8979	0.5260	5.9838
0.8900	2.6436	0.4300	2.8402	0.4440	3.4929	0.3440	4.9001	0.3420	5.5720
0.8780	2.6459	0.4150	2.8433	0.4500	3.4952	0.3550	4.9023	0.3510	5.5774
0.8330	2.6482	0.4220	2.8456	0.4650	3.4988	0.3410	4.9046		
0.7740	2.6512	0.4290	2.8479	0.4950	3.5011	0.3210	4.9074		
0.7480	2.6536	0.4260	2.8502	0.4800	3.5035	0.3490	4.9097		

Table continued on following pages

Table 2. V385 Cam observations, ΔB, ΔV, ΔR$_c$, and ΔI$_c$, variable star minus comparison star, cont.

ΔV	VHJD 2457370+	ΔV	VHJD 2457370+	ΔV	VHJD 2457370+	ΔV	VHJD 2457370+	ΔV	VHJD 2457370+
0.3140	2.4809	0.6600	2.6567	0.3940	2.8545	0.4530	3.5066	0.3280	4.8712
0.3210	2.4827	0.6370	2.6590	0.3900	2.8569	0.4890	3.5101	0.3160	4.8766
0.3060	2.4846	0.6110	2.6623	0.3970	2.8592	0.4940	3.5125	0.3240	4.8788
0.3130	2.4865	0.5730	2.6646	0.3850	2.8615	0.4790	3.5148	0.3080	4.8823
0.3290	2.4892	0.5630	2.6669	0.3910	2.8647	0.4970	3.5171	0.2960	4.8845
0.3230	2.4910	0.5550	2.6692	0.3950	2.8670	0.5080	3.5209	0.3130	4.8889
0.3130	2.4929	0.4690	2.6730	0.4150	2.8693	0.8960	4.4665	0.3050	4.8911
0.3200	2.4948	0.5080	2.6753	0.3900	2.8716	0.9080	4.4690	0.2870	4.8934
0.3420	2.4980	0.4820	2.6777	0.4130	2.8765	0.8810	4.4715	0.3150	4.8956
0.3310	2.4998	0.4750	2.6800	0.4340	2.8789	0.8820	4.4754	0.2970	4.8986
0.3430	2.5017	0.4510	2.6844	0.4430	2.8812	0.8760	4.4777	0.2960	4.9009
0.3380	2.5036	0.4520	2.6867	0.4350	2.8835	0.8230	4.4801	0.2900	4.9031
0.3460	2.5063	0.4340	2.6890	0.4630	2.8867	0.8570	4.4824	0.2740	4.9053
0.3330	2.5082	0.4490	2.6914	0.4660	2.8890	0.8000	4.4915	0.2930	4.9082
0.3450	2.5100	0.4290	2.6964	0.4740	2.8914	0.7210	4.4956	0.2830	4.9104
0.3460	2.5119	0.4260	2.6987	0.4860	2.8937	0.6650	4.4996	0.2850	4.9127
0.3550	2.5144	0.4450	2.7011	0.4900	2.8969	0.6110	4.5033	0.3030	4.9149
0.3420	2.5163	0.4430	2.7034	0.5090	2.8992	0.5630	4.5055	0.2860	4.9178
0.3580	2.5181	0.4000	2.7067	0.5140	2.9015	0.5920	4.5078	0.2880	4.9201
0.3590	2.5200	0.4160	2.7090	0.5150	2.9038	0.5740	4.5101	0.2860	4.9223
0.3680	2.5227	0.3990	2.7114	0.5240	2.9076	0.5210	4.5158	0.2770	4.9246
0.3610	2.5245	0.4020	2.7137	0.5260	2.9099	0.4980	4.5181	0.2850	4.9275
0.3660	2.5264	0.3870	2.7172	0.5230	2.9122	0.5190	4.5204	0.2740	4.9297
0.3780	2.5282	0.3870	2.7196	0.5360	2.9146	0.4670	4.5227	0.3020	4.9320
0.3880	2.5334	0.3840	2.7219	0.5260	2.9186	0.4410	4.5273	0.2910	4.9342
0.3860	2.5353	0.3770	2.7242	0.5270	2.9210	0.4380	4.5298	0.3190	4.9371
0.3930	2.5372	0.3580	2.7324	0.5200	2.9233	0.4240	4.5323	0.2760	4.9394
0.3960	2.5390	0.3500	2.7347	0.5180	2.9256	0.4280	4.5348	0.2810	4.9416
0.3980	2.5428	0.3580	2.7370	0.5270	2.9295	0.4230	4.5392	0.2980	4.9438
0.4050	2.5451	0.3410	2.7393	0.5350	2.9318	0.3970	4.5411	0.2970	4.9471
0.4040	2.5474	0.3510	2.7445	0.5250	2.9341	0.4210	4.5430	0.3250	4.9494
0.4120	2.5497	0.3440	2.7468	0.5250	2.9364	0.4070	4.5449	0.3050	4.9539
0.4220	2.5543	0.3470	2.7491	0.5450	2.9396	0.4050	4.5516	0.3110	4.9599
0.4140	2.5566	0.3450	2.7514	0.5250	2.9419	0.3960	4.5535	0.3040	4.9649
0.4380	2.5589	0.3410	2.7552	0.5210	2.9443	0.3900	4.5553	0.3230	4.9690
0.4590	2.5613	0.3340	2.7575	0.5270	2.9466	0.3850	4.5571	0.3300	4.9714
0.4840	2.5656	0.3330	2.7598	0.5210	2.9502	0.3700	4.5616	0.3120	4.9737
0.5030	2.5679	0.3410	2.7621	0.5220	2.9525	0.3890	4.5635	0.3370	4.9761
0.5100	2.5702	0.3320	2.7661	0.4980	2.9548	0.3750	4.5653	0.3250	4.9803
0.5250	2.5726	0.3350	2.7684	0.4970	2.9571	0.3700	4.5671	0.3470	4.9895
0.5680	2.5764	0.3440	2.7707	0.4950	2.9603	0.3710	4.5717	0.3050	5.5733
0.5890	2.5787	0.3300	2.7730	0.4540	2.9626	0.3700	4.5735	0.3130	5.5787
0.6230	2.5811	0.3320	2.7783	0.4880	2.9650	0.3460	4.5754	0.3090	5.5826
0.6710	2.5834	0.3230	2.7806	0.4440	2.9673	0.3490	4.5772	0.3150	5.8406
0.7000	2.5885	0.3330	2.7829	0.4280	2.9708	0.3920	4.5802	0.3100	5.8434
0.7440	2.5908	0.3350	2.7852	0.4330	2.9731	0.3700	4.5821	0.2960	5.8460
0.7640	2.5932	0.3320	2.7883	0.4290	2.9754	0.3710	4.5839	0.3020	5.8488
0.8020	2.5955	0.3210	2.7906	0.4110	2.9778	0.3340	4.5857	0.3110	5.8527
0.8580	2.5990	0.3400	2.7929	0.4110	2.9809	0.3560	4.5890	0.3000	5.8555
0.8480	2.6013	0.3300	2.7953	0.3850	2.9833	0.3290	4.5927	0.3100	5.8581
0.8830	2.6037	0.3330	2.8009	0.4040	2.9856	0.3190	4.5945	0.3250	5.8606
0.8860	2.6060	0.3390	2.8032	0.3630	2.9879	0.3350	4.5985	0.3080	5.8642
0.9220	2.6100	0.3470	2.8056	0.4080	2.9915	0.3330	4.6049	0.3060	5.8671
0.8860	2.6123	0.3440	2.8079	0.3780	2.9938	0.3110	4.6080	0.3040	5.8697
0.8950	2.6146	0.3450	2.8132	0.3860	3.4581	0.3150	4.6127	0.3170	5.8722
0.9020	2.6170	0.3520	2.8155	0.3770	3.4669	0.3110	4.6159	0.3240	5.8757
0.9090	2.6209	0.3440	2.8178	0.3640	3.4692	0.3050	4.6190	0.3270	5.8783
0.9180	2.6232	0.3620	2.8202	0.3620	3.4716	0.3300	4.6222	0.3330	5.8808
0.9040	2.6256	0.3590	2.8231	0.3560	3.4739	0.3080	4.6336	0.3290	5.8835
0.9070	2.6279	0.3650	2.8254	0.3870	3.4789	0.2990	4.6372	0.3290	5.8867
0.9140	2.6317	0.3550	2.8278	0.3780	3.4813	0.3770	4.6584	0.3300	5.8893
0.9160	2.6340	0.3500	2.8301	0.3790	3.4836	0.3580	4.6918	0.3270	5.8920
0.9140	2.6363	0.3640	2.8341	0.3850	3.4859	0.3800	4.7022	0.3420	5.8945
0.8720	2.6387	0.3650	2.8364	0.3830	3.4889	0.4890	4.7475	0.3240	5.8978
0.8390	2.6420	0.3610	2.8387	0.4150	3.4913	0.4880	4.7607	0.3260	5.9004
0.8160	2.6444	0.3770	2.8410	0.4100	3.4936	0.4860	4.7961	0.3320	5.9029
0.7810	2.6467	0.3910	2.8440	0.4330	3.4959	0.3710	4.8365	0.3380	5.9056
0.7710	2.6490	0.3900	2.8464	0.4250	3.4996	0.3540	4.8455	0.3450	5.9117
0.7150	2.6520	0.3750	2.8487	0.4380	3.5019	0.3630	4.8525	0.3430	5.9150
0.6820	2.6544	0.3800	2.8510	0.4520	3.5042	0.3450	4.8629	0.3580	5.9184

Table continued on following pages

Table 2. V385 Cam observations, ΔB, ΔV, ΔR$_c$, and ΔI$_c$, variable star minus comparison star, cont.

ΔV	VHJD 2457370+	ΔV	VHJD 2457370+	ΔV	VHJD 2457370+	ΔV	VHJD 2457370+	ΔV	VHJD 2457370+
0.3630	5.9219	0.3880	5.9365	0.4040	5.9516	0.4530	5.9656	0.5030	5.9802
0.3670	5.9263	0.3670	5.9415	0.3980	5.9557	0.4640	5.9702	0.5000	5.9849
0.3640	5.9298	0.3920	5.9448	0.4180	5.9590	0.4660	5.9735	0.5180	5.9882
0.3520	5.9332	0.4030	5.9482	0.4420	5.9624	0.4930	5.9768		

ΔR	RHJD 2457370+	ΔR	RHJD 2457370+	ΔR	RHJD 2457370+	ΔR	RHJD 2457370+	ΔR	RHJD 2457370+
0.2600	2.4769	0.8430	2.6321	0.2900	2.8060	0.4050	2.9712	0.3580	4.5434
0.2810	2.4813	0.8250	2.6344	0.2950	2.8113	0.3820	2.9735	0.3400	4.5502
0.2820	2.4832	0.8380	2.6368	0.3030	2.8136	0.3680	2.9759	0.3470	4.5520
0.2850	2.4850	0.7950	2.6401	0.3180	2.8159	0.3860	2.9790	0.3530	4.5539
0.2760	2.4877	0.7710	2.6424	0.3100	2.8182	0.3490	2.9813	0.3440	4.5557
0.2600	2.4896	0.7550	2.6448	0.3040	2.8212	0.3550	2.9837	0.3320	4.5602
0.2730	2.4915	0.7130	2.6471	0.3100	2.8235	0.3540	2.9860	0.3400	4.5620
0.2800	2.4933	0.6640	2.6501	0.2990	2.8258	0.3390	2.9896	0.3400	4.5639
0.2650	2.4965	0.6400	2.6524	0.3170	2.8282	0.3160	3.4523	0.3230	4.5657
0.2870	2.4984	0.6290	2.6548	0.3290	2.8322	0.3000	3.4544	0.3210	4.5703
0.2920	2.5003	0.6060	2.6571	0.3390	2.8368	0.3090	3.4565	0.3150	4.5721
0.2920	2.5021	0.5480	2.6604	0.3280	2.8391	0.3140	3.4585	0.3180	2.8345
0.2860	2.5048	0.5350	2.6627	0.3220	2.8421	0.3030	3.4651	0.3010	4.5739
0.2940	2.5067	0.5160	2.6650	0.3250	2.8445	0.3170	3.4674	0.3100	4.5758
0.2980	2.5086	0.4540	2.6711	0.3320	2.8468	0.3080	3.4697	0.3190	4.5788
0.3070	2.5104	0.4390	2.6734	0.3270	2.8491	0.3270	3.4721	0.3080	4.5806
0.3020	2.5130	0.4330	2.6757	0.3290	2.8526	0.3380	3.4771	0.2900	4.5825
0.3210	2.5148	0.4260	2.6781	0.3320	2.8550	0.3420	3.4794	0.2950	4.5843
0.3020	2.5167	0.4220	2.6825	0.3290	2.8573	0.3410	3.4818	0.3020	4.5876
0.3230	2.5185	0.4040	2.6848	0.3300	2.8596	0.3430	3.4841	0.2770	4.5913
0.3200	2.5212	0.3890	2.6871	0.3400	2.8627	0.3500	3.4871	0.2770	4.5931
0.3150	2.5231	0.3950	2.6894	0.3590	2.8651	0.3650	3.4894	0.2740	4.5961
0.3360	2.5249	0.4000	2.6945	0.3510	2.8674	0.3660	3.4918	0.2870	4.5992
0.3210	2.5268	0.3870	2.6968	0.3570	2.8697	0.3630	3.4941	0.3100	4.6056
0.3320	2.5320	0.3930	2.6991	0.3670	2.8746	0.3920	3.4977	0.2700	4.6134
0.3370	2.5339	0.3660	2.7015	0.3970	2.8770	0.3940	3.5001	0.2710	4.6166
0.3430	2.5357	0.3760	2.7048	0.3960	2.8793	0.3870	3.5024	0.2720	4.6197
0.3460	2.5376	0.3830	2.7071	0.4020	2.8816	0.4030	3.5047	0.2660	4.6244
0.3580	2.5409	0.3680	2.7095	0.4050	2.8848	0.4060	3.5083	0.4790	4.7522
0.3530	2.5432	0.3590	2.7118	0.4170	2.8871	0.4210	3.5106	0.4780	4.7886
0.3610	2.5455	0.3450	2.7153	0.4140	2.8894	0.4420	3.5130	0.3430	4.8256
0.3650	2.5478	0.3350	2.7177	0.4400	2.8918	0.4470	3.5153	0.3170	4.8407
0.3740	2.5523	0.3330	2.7200	0.4460	2.8950	0.4820	3.5190	0.3190	4.8477
0.3790	2.5547	0.3310	2.7223	0.4540	2.8973	0.4790	3.5214	0.2930	4.8665
0.3880	2.5570	0.3150	2.7305	0.4610	2.8996	0.4630	3.5253	0.2770	4.8748
0.3990	2.5593	0.3170	2.7328	0.4720	2.9019	0.4650	3.5295	0.2860	4.8770
0.4070	2.5637	0.3130	2.7351	0.4940	2.9057	0.4830	3.5335	0.2910	4.8804
0.4300	2.5660	0.3150	2.7374	0.4950	2.9080	0.8080	4.4594	0.2870	4.8827
0.4430	2.5683	0.2920	2.7426	0.4990	2.9103	0.8000	4.4670	0.2770	4.8871
0.4580	2.5706	0.3100	2.7449	0.4990	2.9127	0.8370	4.4695	0.2930	4.8893
0.4990	2.5745	0.3010	2.7472	0.5040	2.9167	0.8210	4.4735	0.2740	4.8915
0.5170	2.5768	0.2960	2.7495	0.5050	2.9191	0.8250	4.4759	0.2600	4.8938
0.5370	2.5791	0.3030	2.7532	0.5140	2.9214	0.8270	4.4782	0.2670	4.8968
0.5690	2.5815	0.2790	2.7556	0.4890	2.9237	0.8250	4.4806	0.2680	4.8990
0.6300	2.5866	0.2970	2.7579	0.5050	2.9276	0.7920	4.4843	0.2560	4.9013
0.6750	2.5889	0.2970	2.7602	0.4910	2.9299	0.7630	4.4883	0.2660	4.9035
0.6990	2.5913	0.2800	2.7642	0.5030	2.9322	0.7020	4.4923	0.2340	4.9064
0.7140	2.5936	0.2870	2.7665	0.4900	2.9345	0.6540	4.4964	0.2740	4.9086
0.7810	2.5971	0.2830	2.7688	0.5000	2.9377	0.5990	4.5015	0.2630	4.9109
0.7880	2.5994	0.2860	2.7711	0.4860	2.9400	0.5710	4.5037	0.2390	4.9131
0.7890	2.6017	0.2910	2.7763	0.5020	2.9424	0.5360	4.5083	0.2700	4.9160
0.7770	2.6041	0.2790	2.7787	0.4990	2.9447	0.4790	4.5163	0.2620	4.9182
0.8090	2.6081	0.2850	2.7810	0.4880	2.9483	0.4370	4.5186	0.2470	4.9205
0.8210	2.6104	0.2890	2.7833	0.4800	2.9506	0.4340	4.5208	0.2510	4.9227
0.8290	2.6127	0.2800	2.7864	0.4840	2.9529	0.3910	4.5253	0.2570	4.9256
0.8340	2.6150	0.2960	2.7887	0.4680	2.9552	0.3970	4.5278	0.2680	4.9279
0.8310	2.6190	0.2820	2.7910	0.4520	2.9584	0.3810	4.5303	0.2440	4.9301
0.8360	2.6213	0.2760	2.7933	0.4370	2.9607	0.3780	4.5328	0.2530	4.9324
0.8310	2.6237	0.2930	2.7990	0.4240	2.9630	0.3760	4.5377	0.2480	4.9353
0.8460	2.6260	0.3000	2.8013	0.4220	2.9654	0.3670	4.5396	0.2400	4.9375
0.8310	2.6298	0.3050	2.8036	0.4020	2.9689	0.3380	4.5415	0.2770	4.9398

Table continued on next page

Table 2. V385 Cam observations, ΔB, ΔV, ΔR$_c$, and ΔI$_c$, variable star minus comparison star, cont.

ΔR	RHJD 2457370+	ΔR	RHJD 2457370+	ΔR	RHJD 2457370+	ΔR	RHJD 2457370+	ΔR	RHJD 2457370+
0.2520	4.9420	0.2460	4.9498	0.2670	4.9608	0.2970	4.9719	0.3150	4.9851
0.2410	4.9453	0.2490	4.9520	0.2810	4.9671	0.3050	4.9742	0.3590	4.9913
0.2520	4.9476	0.2540	4.9558	0.2960	4.9695	0.2940	4.9776		

ΔI	IHJD 2457370+	ΔI	IHJD 2457370+	ΔI	IHJD 2457370+	ΔI	IHJD 2457370+	ΔI	IHJD 2457370+
0.2210	2.4773	0.7320	2.6348	0.2480	2.8116	0.3430	2.9762	0.2670	4.5706
0.2260	2.4817	0.7390	2.6371	0.2500	2.8139	0.3420	2.9794	0.2770	4.5725
0.2110	2.4835	0.7150	2.6405	0.2550	2.8163	0.3240	2.9817	0.2770	4.5743
0.2140	2.4854	0.6900	2.6428	0.2740	2.8186	0.2990	2.9840	0.2740	4.5761
0.2340	2.4881	0.6640	2.6451	0.2530	2.8216	0.2930	2.9864	0.2350	4.5791
0.2480	2.4900	0.6370	2.6474	0.2530	2.8239	0.2820	2.9899	0.2650	4.5828
0.2170	2.4918	0.5910	2.6505	0.2690	2.8262	0.2700	2.9923	0.2460	4.5879
0.2420	2.4937	0.5540	2.6528	0.2610	2.8285	0.2540	2.9946	0.2540	4.5916
0.2350	2.4969	0.5280	2.6551	0.2820	2.8325	0.2560	2.9969	0.2360	4.5934
0.2360	2.4988	0.5070	2.6575	0.2610	2.8348	0.2850	3.4547	0.2230	4.5998
0.2310	2.5006	0.4870	2.6607	0.2810	2.8372	0.2800	3.4589	0.2180	4.6140
0.2320	2.5025	0.4890	2.6630	0.2750	2.8395	0.2580	3.4678	0.2320	4.6171
0.2330	2.5052	0.4260	2.6654	0.2600	2.8425	0.2570	3.4701	0.2580	4.6203
0.2510	2.5071	0.4260	2.6677	0.2780	2.8448	0.2570	3.4725	0.2630	4.6254
0.2410	2.5089	0.4060	2.6714	0.2650	2.8471	0.2880	3.4798	0.4450	4.7759
0.2520	2.5108	0.4000	2.6738	0.2830	2.8495	0.2720	3.4822	0.4520	4.7844
0.2540	2.5133	0.3650	2.6761	0.2990	2.8530	0.2930	3.4845	0.4560	4.7916
0.2530	2.5152	0.3690	2.6784	0.2960	2.8553	0.2840	3.4875	0.2850	4.8281
0.2470	2.5170	0.3250	2.6828	0.2750	2.8576	0.2970	3.4899	0.3000	4.8418
0.2740	2.5189	0.3330	2.6852	0.2930	2.8600	0.3100	3.4922	0.2470	4.8675
0.2590	2.5216	0.3410	2.6875	0.2900	2.8631	0.3100	3.4945	0.2270	4.8751
0.2790	2.5234	0.3200	2.6898	0.3000	2.8654	0.3340	3.4982	0.2120	4.8774
0.2550	2.5253	0.3100	2.6948	0.3000	2.8678	0.3340	3.5005	0.2040	4.8808
0.2750	2.5272	0.3120	2.6972	0.3240	2.8701	0.3540	3.5028	0.1910	4.8830
0.2890	2.5324	0.3080	2.6995	0.3320	2.8750	0.3510	3.5051	0.1940	4.8874
0.2840	2.5342	0.2890	2.7018	0.3310	2.8773	0.3860	3.5087	0.2030	4.8896
0.2960	2.5361	0.2980	2.7052	0.3630	2.8796	0.3930	3.5110	0.2140	4.8919
0.2980	2.5379	0.3050	2.7075	0.3670	2.8820	0.4140	3.5134	0.2000	4.8941
0.3000	2.5412	0.2940	2.7098	0.3800	2.8852	0.3990	3.5157	0.1990	4.8971
0.2830	2.5435	0.3040	2.7121	0.3940	2.8875	0.4110	3.5195	0.2090	4.8994
0.2910	2.5459	0.2810	2.7157	0.4030	2.8898	0.4190	3.5218	0.2140	4.9016
0.3100	2.5482	0.2900	2.7180	0.4040	2.8921	0.7730	4.4597	0.2000	4.9038
0.3160	2.5527	0.2780	2.7203	0.4360	2.8977	0.7710	4.4673	0.1960	4.9067
0.3240	2.5550	0.2830	2.7227	0.4450	2.9000	0.7790	4.4698	0.1960	4.9090
0.3250	2.5574	0.2690	2.7308	0.4620	2.9023	0.7640	4.4738	0.2060	4.9112
0.3380	2.5597	0.2520	2.7331	0.4710	2.9060	0.7530	4.4762	0.2020	4.9134
0.3570	2.5640	0.2770	2.7355	0.4660	2.9084	0.7360	4.4785	0.2010	4.9164
0.3570	2.5664	0.2560	2.7378	0.4600	2.9107	0.7130	4.4809	0.2080	4.9186
0.3930	2.5687	0.2520	2.7429	0.4620	2.9130	0.6880	4.4848	0.1930	4.9209
0.4120	2.5710	0.2420	2.7452	0.4840	2.9171	0.6440	4.4889	0.2150	4.9231
0.4160	2.5749	0.2440	2.7476	0.4700	2.9194	0.5910	4.4929	0.2070	4.9260
0.4560	2.5772	0.2540	2.7499	0.4680	2.9217	0.5440	4.4969	0.2030	4.9282
0.4700	2.5795	0.2590	2.7536	0.4670	2.9241	0.4970	4.5018	0.2250	4.9305
0.4960	2.5818	0.2380	2.7559	0.4750	2.9279	0.4920	4.5041	0.2150	4.9327
0.5600	2.5870	0.2330	2.7583	0.4710	2.9302	0.4680	4.5086	0.2220	4.9356
0.5990	2.5893	0.2260	2.7606	0.4620	2.9326	0.4080	4.5144	0.2100	4.9379
0.6160	2.5916	0.2310	2.7645	0.4840	2.9349	0.4020	4.5166	0.2040	4.9401
0.6550	2.5939	0.2250	2.7669	0.4590	2.9381	0.3690	4.5189	0.2380	4.9424
0.6810	2.5975	0.2380	2.7692	0.4600	2.9404	0.3350	4.5257	0.2440	4.9479
0.7130	2.5998	0.2330	2.7715	0.4670	2.9427	0.3290	4.5282	0.2040	4.9565
0.7100	2.6021	0.2230	2.7767	0.4660	2.9450	0.3010	4.5307	0.2270	4.9615
0.7460	2.6044	0.2280	2.7790	0.4570	2.9486	0.3010	4.5332	0.2510	4.9675
0.7520	2.6084	0.2440	2.7814	0.4430	2.9509	0.3050	4.5381	0.2240	4.9699
0.7380	2.6108	0.2390	2.7837	0.4600	2.9533	0.3000	4.5400	0.2580	4.9722
0.7380	2.6131	0.2350	2.7867	0.4420	2.9556	0.2810	4.5438	0.2690	4.9746
0.7310	2.6154	0.2180	2.7891	0.4030	2.9588	0.2800	4.5505	0.2490	4.9780
0.7480	2.6194	0.2520	2.7914	0.4130	2.9611	0.2850	4.5542	0.2450	4.9812
0.7470	2.6217	0.2290	2.7937	0.3780	2.9634	0.3000	4.5561	0.2580	4.9856
0.7590	2.6240	0.2550	2.7994	0.3620	2.9657	0.2870	4.5606		
0.7300	2.6263	0.2440	2.8017	0.3480	2.9692	0.2860	4.5624		
0.7480	2.6301	0.2490	2.8040	0.3500	2.9716	0.2780	4.5642		
0.7530	2.6325	0.2460	2.8063	0.3450	2.9739	0.2870	4.5661		

Table 3. O–C residual calculations for V385 Cam.

	Epoch 2400000+	Cycle	Linear Residuals	Quadratic Residuals	Weight	Error	Reference
1	51396.7983	−9712.5	0.0035	0.0042	0.5	0.0010	Shaw (1994)
2	51515.8450	−9519.0	−0.0048	−0.0044	0.5	0.0010	Khruslov (2006)
3	56892.0900	−781.0	−0.0008	−0.0017	0.2	0.0100	Shappee et al. (2014), Kohanek et al. (2017)
4	56920.0830	−735.5	−0.0027	−0.0035	0.2	0.0200	Shappee et al. (2014), Kohanek et al. (2017)
5	56960.9990	−669.0	−0.0022	−0.0030	0.2	0.0200	Shappee et al. (2014), Kohanek et al. (2017)
6	57462.7560	146.5	0.0010	0.0013	0.2	0.0200	Shappee et al. (2014), Kohanek et al. (2017)
7	57789.7710	678.0	−0.0007	0.0004	0.2	0.0040	Shappee et al. (2014), Kohanek et al. (2017)
8	57817.7670	723.5	0.0004	0.0016	0.2	0.0200	Shappee et al. (2014), Kohanek et al. (2017)
9	57994.0390	1010.0	−0.0028	−0.0012	0.2	0.0200	Shappee et al. (2014), Kohanek et al. (2017)
10	58022.0320	1055.5	−0.0047	−0.0030	0.2	0.0200	Shappee et al. (2014), Kohanek et al. (2017)
11	58170.9310	1297.5	−0.0013	0.0008	0.2	0.0100	Shappee et al. (2014), Kohanek et al. (2017)
12	56309.7391	−1727.5	0.0026	0.0007	1.0	0.0007	Diethelm (2013)
13	57372.6207	0.0	0.0030	0.0030	1.0	0.0003	Present observations
14	57372.9266	0.5	0.0012	0.0013	1.0	0.0011	Present observations
15	57374.4678	3.0	0.0042	0.0043	0.5	0.0043	Present observations
		r.m.s	0.00278	0.00266			

Table 4. Light curve characteristics for V385 Cam.

Filter	Phase 0.25	Magnitude Max. I	±σ	Phase 0.75	Magnitude Max. II	±σ
B		0.365	0.010		0.323	0.009
V		0.318	0.015		0.289	0.015
R		0.279	0.008		0.249	0.008
I		0.236	0.012		0.218	0.007

Filter	Phase 0.5	Magnitude Min. II	±σ	Phase 0.0	Magnitude Min. I	±σ
B		0.556	0.013		1.002	0.013
V		0.532	0.012		0.902	0.011
R		0.507	0.006		0.837	0.006
I		0.467	0.012		0.691	0.020

Filter	Min. I – Max. I	±σ	Max. I – Max. II	±σ	Min. I – Min. II	±σ
B	0.637	0.023	0.043	0.019	0.446	0.025
V	0.583	0.026	0.030	0.031	0.370	0.023
R	0.558	0.014	0.030	0.016	0.330	0.011
I	0.456	0.032	0.018	0.019	0.224	0.032

Filter	Max. I – Max. II	±σ	Min. II – Max. I	±σ
B	0.356	0.010	0.191	0.023
V	0.303	0.015	0.213	0.027
R	0.271	0.008	0.228	0.014
I	0.229	0.012	0.231	0.024

Table 5. B, V, R_c, I_c Wilson-Devinney program solution parameters.

Parameters	Values
$\lambda_B, \lambda_V, \lambda_R, \lambda_I$ (nm)	440, 550, 640, 790
g_1, g_2	0.32
A_1, A_2	0.5
Inclination (°)	85.16 ± 0.09
T_1, T_2 (K)	5500, 4520 ± 3
Ω_1, Ω_2	2.7717 ± 0.0020, 2.913 ± 0.002
$q(m_2/m_1)$	0.3904 ± 0.0007
Fill-outs: F_1, F_2 (%)	96.0 ± 0.1, 91.3 ± 0.1
$L_1/(L_1+L_2+L_3)_I$	0.8756 ± 0.066
$L_1/(L_1+L_2+L_3)_R$	0.8909 ± 0.067
$L_1/(L_1+L_2+L_3)_V$	0.9100 ± 0.075
$L_1/(L_1+L_2+L_3)_B$	0.9327 ± 0.077
JD_o (days)	2457372.6204 ± 0.0001
Period (days)	0.61498 ± 0.00007
$r_1/a, r_2/a$ (pole)	0.4147 ± 0.0015, 0.2387 ± 0.0010
$r_1/a, r_2/a$ (point)	0.4994 ± 0.0047, 0.2673 ± 0.0048
$r_1/a, r_2/a$ (side)	0.4385 ± 0.0020, 0.2448 ± 0.0030
$r_1/a, r_2/a$ (back)	0.4590 ± 0.0026, 0.2589 ± 0.0039
Spot 1 (Star 1)	
Co-Latitude	161.37 ± 0.03
Longitude	192.5 ± 0.1
Radius	31.80 ± 0.09
T-factor	1.151 ± 0.001 (6134 ± 33 K)
Spot 2 (Star 1)	
Co-Latitude	119 ± 2
longitude	97.54 ± 0.03
Radius	9.952 ± 0.002
T-factor	1.200 ± 0.006 (6134 ± 33 K)

Multi-filter Photometric Analysis of W Ursae Majoris (W UMa) Type Eclipsing Binary Stars KID 11405559 and V342 Boo

Tatsuya Akiba
University of Colorado Boulder, Department of Astrophysical and Planetary Sciences, 2000 Colorado Ave, Duane Physics Building, Rm. E226, Boulder, CO 80309; address email correspondence to gokhale@truman.edu

Andrew Neugarten
Subaru Telescope, National Astronomical Observatory of Japan, 650 North A'ohōkū Place, Hilo, HI 96720

Charlyn Ortmann
Vayujeet Gokhale
Truman State University, 100 E. Normal Street, Kirksville, MO 63501; gokhale@truman.edu

Received July 30, 2019; revised October 2, November 4, 2019; accepted November 12, 2019

Abstract We present light curve analysis of two variable stars, KID 11405559 and V342 Boo. KID 11405559 is selected from the Kepler Eclipsing Binary Catalog published by Kirk *et al.* (2016) and V342 Boo from a list of eclipsing binaries published by Kreiner (2004). In this paper, we present the light curves for these two objects using data collected at the 31-inch NURO telescope at Lowell Observatory in Flagstaff, Arizona, in three filters: Bessell B, V, and R. We generate truncated twelve-term Fourier fits for the light curves and quantify the O'Connell effect exhibited by these systems by calculating the difference ΔI in the heights of the primary and secondary maxima: the "Light Curve Asymmetry" (LCA) and the "O'Connell Effect Ratio" (OER). Additionally, we use the Fourier coefficients from the Fourier fit to confirm that KID 11405559 and V342 Boo are W UMa type eclipsing binary stars.

1. Introduction

The O'Connell Effect is the inequality in the out-of-eclipse maxima in the light curve of eclipsing binaries (O'Connell (1951). This inequality is unexpected, since one expects to receive the same amount of light from each of the components of the binary when the components are side-by-side, irrespective of which component is on which side. Several explanations have been proposed to explain the origin of this effect, but none of these theories are widely applicable to the systems exhibiting the O'Connell effect (see Wilsey and Beaky (2009)). The two models that have received attention recently are the "starspot" model and the "hotspot" model. The former model is based on introducing one or more starspots (regions cooler than the rest of the photosphere) on one, or both, components. The "hotspot" model is applicable for mass-transferring systems where the mass transfer impacts the accretion disk around the accreting star, resulting in an asymmetry in the luminosity of the binary system as viewed from Earth. Again, neither of these models is satisfactory—the "starspot" model essentially introduces an arbitrary number of free parameters with minimal, if any, constraints. Consequently, by adding any number of starspots of varying sizes, shapes, and temperatures it should be possible to fit any light curve. Also, the O'Connell effect has been observed in detached, over-contact, and semi-detached systems and so at best, the 'hotspot' model can only work for a certain class of systems exhibiting the O'Connell effect.

Eclipsing binary systems are classified into three main categories, based on the shape and variations in the observed light curves. Of the three types, W UMa-type systems are of particular interest to us since their structure and evolution are still not completely understood. Their light curves are characterized by continuous variability and similar depths of the two minima, caused by the ellipsoidal nature and the proximity of the two components. These systems are to be distinguished from β-Lyrae-type systems which display continuous variability but different depths of the minima, and Algol-type systems which exhibit a nearly constant brightness outside of eclipses. Algols and β-Lyraes are usually considered to be detached and semi-detached systems, respectively, whereas W UMa systems are considered near or over-contact systems. The continuously varying, smooth light curves of W UMa systems are ideal for Fourier analysis.

In this paper, we discuss light curve analysis of two eclipsing binary systems selected from Kirk *et al.* (2016) and Kreiner (2004), namely KID 11405559 and V342 Boo. Firstly, we use the criteria set by Rucinski (1997) to classify these systems as either Algol, β-Lyrae, or W UMa-type systems. We focus on the asymmetries in the light curves in each of the filters by calculating the difference in the heights of the primary and secondary maxima (ΔI), the "Light Curve Asymmetry" (LCA), and the "O'Connell Effect Ratio" (OER). The OER and LCA provide insight into the asymmetries of the light curve regions between the two eclipses (McCartney 1999). The OER is the ratio of the area under the curves between phases $\Phi = 0.0$ to $\Phi = 0.5$ and phases $\Phi = 0.5$ to $\Phi = 1.0$ (see section 3 below). An OER > 1 implies that the first half of the light curve has more total flux than the second half. The LCA, on the other hand, measures the deviance from symmetry of the two halves of the light curve. If both halves are perfectly symmetric, then we would expect the LCA to be zero. Note that the LCA and OER quantify different aspects of the O'Connell effect. For example, an OER equal to 1 does not necessarily imply an LCA equal to 0. The reason for this is that one can imagine a light curve with a tall but narrow peak maximum and short but wide secondary maximum. Although the areas under the curve of each region may be equal, the curves would be very asymmetric (Gardner 2015).

This project is part of an ongoing effort at Truman State University to introduce undergraduate students to differential aperture photometry by following several eclipsing binary systems per semester with the aim of generating and analyzing their light curves. As mentioned, we are focused on quantifying the asymmetries in the light curves and modeling the systems. System modeling is challenging without having spectroscopic data, since photometric data alone are not enough to constrain the system and determine a unique solution (Prša 2018). Consequently, we do not attempt to model the systems under consideration. However, we demonstrate that by superposing the two halves of an appropriately phased light curve, one can identify the phase at which the light curves are asymmetric (see section 3). In the starspot model, this phase information can be used to constrain the location and characteristics of the starspots. This information can be invaluable in testing the starspot model for EBs using high quality data over hundreds or thousands of orbital cycles —the kind of data we have access to from the Kepler (Prša 2011) and Transiting Exoplanet Survey Satellite (TESS) missions (Ricker et al. 2015). Indeed, our group is currently studying how the asymmetry in the light curve evolves over hundreds of orbits for several Kepler EBs (Koogler et al. 2019).

The outline of this paper is as follows: In the following section, we outline our observational data acquisition and data reduction methods, followed by an analysis of the light curves in section 3. We conclude in section 4 with a discussion on our results and our future plans.

2. Observations

We present BVR photometry of eclipsing variable stars KID 11405559 ($P = 0.284941$ d) and V342 Boo ($P = 0.29936$ d). The data were collected using the $2k \times 2k$ Loral NASACam CCD attached to the 31-inch National Undergraduate Research Observatory (NURO) telescope at Lowell Observatory, Flagstaff, Arizona (Table 1). The filters used are Bessell BVR. The images are processed by bias subtraction and (sky) flat fielding using the software package MAXIM DL (Diffraction Limited 2012). No dark subtraction was performed since for the nitrogen-cooled camera at NURO, the dark current is negligible. Differential photometry is then performed on the target with a suitable comparison and check star, using MAXIM DL. The aperture size was adjusted to match between 3 to 4 times the full width at half maximum (FWHM) of the brightest object on which photometry was performed for a given target. Similarly, the inner sky annulus was adjusted to about 5 times the FWHM. We searched for any comparison stars from the Tycho (Høg et al. 2000) and the American Association of Variable Star Observers (AAVSO) Photometric All-Sky Survey (APASS; Henden et al. 2016) catalogues that are present in the image frame, and used these stars to determine the B and V magnitudes of each of the targets (Table 2). Instrumental magnitudes were used for the R-filter since the R magnitudes for comparison and check stars were not listed. Stars of brightness comparable to the target star were chosen as check stars. We inspected and confirmed that the check and comparison stars show no variability in each of the filters for both objects. The error on a single observation in each of the filters was approximately 2 mmag for both objects.

Table 1. Observation dates, instrument, and filters for the targets.

Target	Date of Observation	Telescope	Filters
KID 11405559	05/21/2017	NURO	Bessell BVR
	05/22/2017	NURO	Bessell BVR
	05/23/2017	NURO	Bessell BVR
	05/24/2017	NURO	Bessell BVR
V342 Boo	05/21/2017	NURO	Bessell BVR
	05/22/2017	NURO	Bessell BVR

3. Analysis and results

3.1. Light curves

To plot a light curve, the time axis is phase-folded using the equation

$$\Phi = \frac{T - T_0}{P} - \mathrm{Int}\left(\frac{T - T_0}{P}\right) \quad (1)$$

where T is the time, P is the period of the object, and T_0 is an arbitrary epoch. The resulting light curves for each object in the B and V filters are shown in Figure 1. Only the B and V plots are shown since we could not identify a star with a reliable R magnitude in the image field. We calculate and set the epoch of the primary minimum, defined as the deeper of the two eclipses, as phase "0". This is achieved by setting the epoch, T_0 in Equation 1 to the epoch of the primary minimum.

In the B and V light curves, we see that the two systems have similar but slightly different depths of minima. The light curves for both objects are smoothly varying with comparable primary and secondary minima characteristic of W UMa systems.

Table 2. Target, comparison, and check star coordinates, and comparison star B and V magnitudes.

Star	Name	R.A. (J2000) h m s	Dec. (J2000) ° ′ ″	V	B
Target	KID 11405559	19 32 54.16	+49 14 33.2		
Comparison	KID 11352756	19 33 20.82	+49 17 54.4	11.22	12.41
Check	TYC 3564-1900-1	19 32 19.00	+49 10 47.7	11.67	14.07
Target	V342 Boo	13 59 53.52	+17 53 57.4		
Comparison	—	13 59 35.81	+17 50 42.8	13.116	13.724
Check	—	13 59 57.98	+17 56 48.0	13.777	14.92

A modest O'Connell effect is seen as well, with the secondary maxima (the maxima following the secondary eclipse) slightly higher than the primary maxima. This effect is even more prominent when the normalized flux is plotted, instead of the magnitude (see Figure 2). For each data point, the normalized flux (Warner and Harris 2006) is calculated from the magnitude by applying the equation

$$I(\Phi)_{obs} = 10^{-0.4 \times (m(\Phi) - m(max))} \quad (2)$$

where $m(\Phi)$ is the magnitude at a certain phase Φ and $m(max)$ is the maximum magnitude observed for the object. We perform Fourier fit analyses on the light curves of each object in each filter similar to Wilsey and Beaky (2009). A truncated twelve-term Fourier fit is given by

$$I(\Phi)_{fit} = a_0 + \sum_{n=1}^{12}(a_n \cos(2\pi n\Phi) + b_n \sin(2\pi n\Phi)) \quad (3)$$

where a_0, a_n, and b_n are the Fourier coefficients of the fit, and Φ is the phase (Hoffman *et al.* 2009). Note that the Fourier fits are generated after the data are adjusted to align the primary eclipse with phase "0". The light curves of the two objects in each filter along with their Fourier fits are presented in Figure 2.

We use the phased normalized flux plots (Figure 2) and the corresponding Fourier fits to classify the two systems, and then proceed to discuss the asymmetries in the light curve.

3.2. Classification of systems

In this paper, we apply the following criteria to classify the systems:

1. Distinguish the systems as either W UMa or β-Lyrae, or detached Algol type systems utilizing Rucinski's (1997) criterion:

   (a) if $a_4 > a_2(0.125 - a_2)$ then the system can be considered a W UMa or a β-Lyrae system.

   (b) if $a_4 < a_2(0.125 - a_2)$ then the system may be considered a detached eclipsing binary or an Algol.

   Note that both coefficients a_2 and a_4 are negative.

2. If condition (a) above is met, use the Fourier coefficient a_1 to distinguish between a W UMa or β-Lyrae system as follows Wilsey and Beaky (2009):

   (a) if $|a_1| < 0.05$ the system is classified as a W UMa type

   (b) if $|a_1| > 0.05$ the system is of the β-Lyrae type.

Rucinski (1973, 1993, 1997) provides an excellent overview of the use of the Fourier coefficients in determining the orbital elements of eclipsing binary stars. As discussed in Gardner *et al.* (2015), a_2 is a measure of the global distortion of the contact structure, whilst a_4 represents the more localized eclipse effects. On the other hand, the first cosine term from the Fourier fit (a_1) is the dominant term contributing to the difference in the primary and secondary eclipse magnitudes (Wilsey and Beaky 2009). Hence, the first cosine coefficient a_1 provides a measure

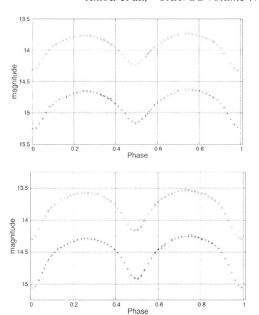

Figure 1. Light curves for KID 11405559 (upper plot) and V342 Boo (lower plot) in the B and V filters (blue and green, respectively). Note the depth of the primary and secondary minima is comparable—a distinctive characteristic of W UMa type systems. The average error on each measurement for both objects in each filter is 0.002 magnitude. Error bars are not shown for the sake of clarity.

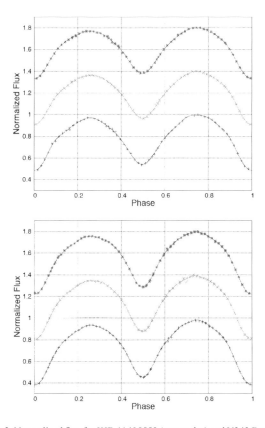

Figure 2. Normalized flux for KID 11405559 (upper plot) and V342 Boo (lower plot). The Fourier fits (continuous curves) are plotted along with the blue, green, and red curves corresponding to B, V, and R filters, respectively. The average error in the flux for each measurement for both objects in each filter is 0.001. Error bars are not shown for the sake of clarity.

of the magnitude difference of the two eclipses and is given by $-2a_1$. We experimented with our Fourier fit coefficients to determine that if the depth of the primary and secondary minima are exactly equal, then a_1 is identically zero. With increasing a_1, the difference between the magnitudes of the two eclipses becomes more prominent, which can be used to distinguish a W UMa system from a β-Lyrae system. When criterion (1a) above is clearly satisfied, and when $|a_1| \approx 0.05$, the depths of the minima differ by approximately 10%, which we take as the dividing line between a W UMa and a β-Lyrae type system.

We evaluated these relationships for each of our objects and the results are recorded in Table 3. The uncertainties of the Fourier coefficients are determined using the NonlinearModelFit function (Wolfram Res. Co. 2019) on MATHEMATICA, where each data point was weighted by the square of its signal-to-noise ratio (SNR).

3.3. Asymmetries in the light curve: quantifying the O'Connell Effect

The simplest way to quantify the O'Connell effect is to calculate the difference in the normalized flux near the primary (I_p) and secondary (I_s) maxima:

$$\Delta I \equiv I_p - I_s$$

A positive ΔI implies that the peak after the primary eclipse is brighter than the peak after the secondary eclipse. A negative ΔI implies the opposite.

We calculate ΔI in two ways:

1. Firstly, we found the values of the two peak magnitudes of the Fourier fit functions. The difference between the two maxima gives us a measure of the the O'Connell effect. These values are presented in Table 4 as "ΔI (Fourier)."

$$\Delta I_{fit} = I_{pfit} - I_{sfit} \quad (4)$$

2. Secondly, all of the data points within ± 0.05 phase of each maximum are averaged. The difference between the two average values for the maxima is calculated, yielding another measure of ΔI. These values are recorded in Table 4 as "ΔI (Average)."

$$\Delta I_{Ave} = \langle I_p \rangle - \langle I_s \rangle \quad (5)$$

In both systems, and in all filters, both methods result in a negative ΔI (Table 4), which is consistent with Figure 2.

Table 4. Quantifying the O'Connell Effect in terms of difference in maxima. Please see the text (section 3.3) for details.

Target	Filter	$|2b_1|$	ΔI (Fourier)	ΔI (Average)
KID 11405559	B	0.032 ± 0.003	-0.024 ± 0.008	-0.029 ± 0.002
	V	0.031 ± 0.001	-0.036 ± 0.004	-0.0321 ± 0.0008
	R	0.029 ± 0.002	-0.031 ± 0.005	-0.035 ± 0.002
V342 Boo	B	0.041 ± 0.002	-0.044 ± 0.005	-0.045 ± 0.002
	V	0.040 ± 0.002	-0.044 ± 0.006	-0.040 ± 0.005
	R	0.036 ± 0.002	-0.038 ± 0.005	-0.0373 ± 0.0007

The first sine term in the Fourier series has extrema at phase $\Phi = 0.25$ and $\Phi = 0.75$, the locations of the out-of-eclipse maxima in the light curves. This makes it the dominant component accounting for the asymmetry in the peak magnitudes (Wilsey and Beakey 2009, Gardner et al. 2015). The coefficient, b_1, associated with the first sine term is the half-amplitude of the sine wave and consequently, $|2b_1|$ is a good approximation to ΔI. The calculated values of $|2b_1|$, ΔI (Fourier) and ΔI (Average) are shown in Table 4. The uncertainties for $|2b_1|$ and ΔI_{fit} are calculated by propagating the uncertainties on the Fourier coefficients (See Appendices B and C). Note that the uncertainty on ΔI_{Ave} is calculated from the addition of uncertainties in quadrature of the data points used in the average (Appendix C), and consequently is a function of the signal-to-noise ratio (i.e. how good our data are) close to the two maxima. This leads to significant variation in the errors on ΔI_{Ave}.

We also quantified the O'Connell effect by calculating the O'Connell Effect Ratio (OER) and the "Light Curve Asymmetry" (LCA) as described by McCartney (1999). As mentioned, The OER is simply the ratio of the area under the curves between phases $\Phi = 0.0$ to $\Phi = 0.5$ and phases $\Phi = 0.5$ to $\Phi = 1.0$, whilst the LCA measures the deviance from symmetry of the two halves of the light curve. The OER and LCA are given by:

$$OER = \frac{\int_{0.0}^{0.5} (I(\Phi)_{fit} - I(\Phi)_{fit}) d\Phi}{\int_{0.5}^{1.0} (I(\Phi)_{fit} - I(\Phi)_{fit}) d\Phi} \quad (6)$$

And

$$LCA = \sqrt{\int_{0.0}^{0.5} \frac{(I(\Phi)_{fit} - I(1.0 - \Phi)_{fit})^2}{I(\Phi)_{fit}^2} d\Phi} \quad (7)$$

where, $I(\Phi)_{fit}$ is given by Equation 3. The values for these parameters are tabulated in Table 5. The uncertainties of the

Table 3. Classification of systems based on Fourier coefficients.

Target	Filter	a_1	a_2	a_4	$a_2(0.125 - a_2)$	Classification
KID 11405559	B	0.015 ± 0.001	-0.218 ± 0.001	-0.0426 ± 0.0011	-0.0748 ± 0.0005	W UMa
	V	0.0172 ± 0.0005	-0.2075 ± 0.0005	-0.0445 ± 0.0006	-0.0690 ± 0.0002	W UMa
	R	0.0152 ± 0.0007	-0.2009 ± 0.0008	-0.0426 ± 0.0008	-0.0655 ± 0.0003	W UMa
V342 Boo	B	0.0383 ± 0.0007	-0.2490 ± 0.0008	-0.0589 ± 0.0007	-0.0931 ± 0.0003	W UMa
	V	0.0375 ± 0.0007	-0.2436 ± 0.0008	-0.0608 ± 0.0008	-0.0898 ± 0.0004	W UMa
	R	0.0341 ± 0.0007	-0.2399 ± 0.0007	-0.0604 ± 0.0007	-0.0875 ± 0.0003	W UMa

OER and LCA are calculated by first defining an "uncertainty function" for each of the integrands (see Appendix B) following the calculus approach of error propagation presented in Hughes and Hase (2010). From this, we obtain the uncertainty of each integrand as a function of phase, which we then integrate to compute the uncertainties of the OER and LCA (Appendices D and E).

Finally, we superpose the two halves of the phase-folded light curves to provide a visual demonstration of the asymmetries in the light curves. We calculate

$$\Delta I(\Phi)_{fit} = I(\Phi)_{fit} - I(1-\Phi)_{fit} \qquad (8)$$

and plot this function against Φ ranging from 0 to 0.5. As an example, in Figure 3, we plot $\Delta I(\Phi)_{fit}$ for the "B" filter for both objects under consideration, and show the difference in the normalized flux at geometrically equivalent points in the orbit of the binary in the bottom panels of each plot. Figure 4 shows just the difference in the normalized flux in all three filters for KID 11405559 and V342 Boo. We note that for both objects, the asymmetry is greatest between the phases 0.15 and 0.25. Also, for the most part, the flux under the secondary half of the phase is greater than that in the primary half. This is consistent with a negative value for ΔI and for an OER < 1, as discussed previously. Note also that the magnitude of the asymmetry is marginally greater in V342 Boo than in KID 11405559.

4. Discussion

We have quantified the asymmetries in the light curves of the two objects under consideration: KID 11405559 and V342 Boo. Both systems exhibit a small O'Connell effect, with the asymmetry being the greatest around a phase of about 0.2 (or 0.8). In both systems, ΔI is negative and the OER < 1, which implies that the secondary half of the light curve is brighter than the primary half. It is reasonable to assume that the magnitude and the location of this asymmetry is related, within the starspot model, to the location and characteristics of the starspots on one or the other component of the binary. For a single light curve, phase-folded over one or two orbital cycles of a binary, this information may not be of much consequence. But for EBs in the Kepler and TESS field, where data for several hundred if not several thousand orbital periods are available, this information can prove invaluable in setting constraints or in overruling the starspot model, even without having access to spectroscopic data. For example, we are in the process (Koogler *et al.* in preparation) of analyzing the light curves of several

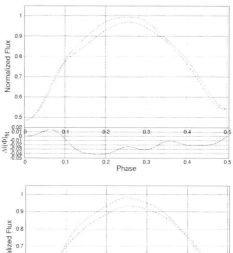
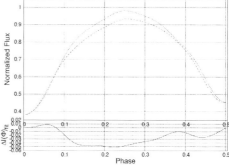

Figure 3. Superposed phased plots of the primary half (dashed line) and the secondary half (dotted line) of the light curves in the B filter for KID 11405559 (upper plot) and V342 Boo (lower plot). The bottom panel (solid blue) shows the difference between the two halves of the light curve. In the absence of any asymmetry, the two curves should coincide, and the solid blue curve in the bottom panel would be a flat line at '0'.

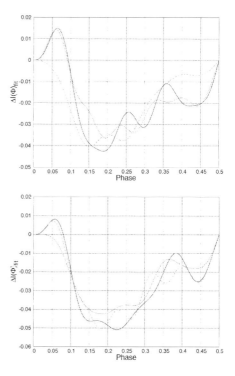

Figure 4. Difference in normalized flux in the B (blue solid curve), V (green dashed), and R (red dotted) filters for KID 11405559 (upper plot) and V342 Boo (lower plot).

Table 5. Quantifying the O'Connell Effect in terms of OER and LCA. Please see text (section 3.3) for the definitions of the OER and LCA.

Target	Filter	OER	LCA
KID 11405559	B	0.94 ± 0.02	0.020 ± 0.004
	V	0.94 ± 0.01	0.019 ± 0.002
	R	0.94 ± 0.02	0.018 ± 0.003
V342 Boo	B	0.93 ± 0.01	0.027 ± 0.003
	V	0.93 ± 0.02	0.025 ± 0.004
	R	0.94 ± 0.01	0.022 ± 0.003

hundred EB systems, for which we can analyze hundreds, if not thousands, of consecutive orbits using Kepler data. This gives us information on how the various asymmetry parameters like the OER, LCA, and ΔI evolve in time for hundreds of objects over thousands of orbital cycles. We plan on generating starspot models for a given orbital cycle and propagating this model to fit subsequent orbits in order to test the starspot model. This should enable us to derive constraints on the size, temperature, and lifetimes of the starspots.

5. Acknowledgements

We have made extensive use of the tools available on the AAVSO website (https://www.aavso.org), in particular, the VSP tool to generate star charts, and the VSTAR and VPHOT tools to perform photometry. In addition, we have used the SIMBAD database (http://simbad.u-strasbg.fr/simbad/), operated at CDS, Strasbourg, France, and NASA's Astrophysics Data System. We are thankful for the support provided by the Office of Student Research at Truman State University, and to the Missouri Space Grant Consortium. This research was made possible through the use of the AAVSO Photometric All-Sky Survey (APASS), funded by the Robert Martin Ayers Sciences Fund. The authors would also like to thank the anonymous referee for useful comments and suggestions, which greatly improved the manuscript.

References

Diffraction Limited. 2012, MAXIMDL image processing software (http://www.cyanogen.com).

Gardner, T., Hahs, G., and Gokhale, V. 2015, *J. Amer. Assoc. Var. Star Obs.*, **43**, 186.

Henden, A. A., Templeton, M., Terrell, D., Smith, T. C., Levine, S., and Welch, D. 2016, VizieR Online Data Catalog: AAVSO Photometric All Sky Survey (APASS) DR9, II/336.

Hoffman, D. I., Harrison, T. E., and McNamara, B. J. 2009, *Astron. J.*, **138**, 466.

Høg, E., et al. 2000, *Astron. Astrophys.*, **355**, L27.

Hughes, I., and Hase, T. 2010, *Measurements and their Uncertainties: A Practical Guide to Modern Error Analysis*, Oxford University Press, Oxford, 43–44.

Kirk, B. et al. 2016, *Astron. J.*, **151**, 68

Koogler, B. R., Shroyer, J. E., and Gokhale, V. M., in preparation.

Kreiner, J. M. 2004, *Acta Astron.*, **54**, 207.

McCartney, S. A. 1999, Ph.D. dissertation, University of Oklahoma.

O'Connell, D. J. K. 1951, *Publ. Riverview Coll. Obs.*, **2**, 85.

Prša, A. 2018, *Modeling and Analysis of Eclipsing Binary Stars, The Theory and Design Principles of PHOEBE*, IOP Publishing, Bristol, U. K.

Prša, A., et al. 2011, *Astron. J.*, **141**, 83.

Ricker, G. R., et al. 2015, *J. Astron. Telesc. Instrum. Syst.*, **1**, 014003.

Rucinski, S. M. 1973, *Acta Astron.*, **23**, 79.

Rucinski, S. M. 1993, *Publ. Astron. Soc. Pacific*, **105**, 1433.

Rucinski, S. M. 1997, *Astron. J.*, **113**, 407.

Warner, B. D., and Harris, A. W. 2006, *A Practical Guide to Lightcurve Photometry and Analysis*, Springer, New York.

Wilsey N. J., and Beaky M. M. 2009, in *The Society for Astronomical Sciences 28th Annual Symposium on Telescope Science*, The Society for Astronomical Sciences, Rancho Cucamonga, CA, 107.

Wolfram Research Co. 2019, How to Fit Models with Measurement Errors (https://reference.wolfram.com/language/howto/FitModelsWithMeasurementErrors.html).

Appendix A: Obtaining uncertainties on the Fourier coefficients

The uncertainty on each data point is computed as the reciprocal of the signal-to-noise ratio (SNR). Those uncertainties in magnitude are converted into uncertainties in the normalized flux using the formula:

$$I(\Phi)_{obs} = 10^{-0.4 \times (m(\Phi) - m(\max))}, \quad (A1)$$

where appropriate.

Using a twelve-term truncated Fourier fit, we use MATHEMATICA's NonlinearModelFit function to run a weighted least squares calculation on our data, weighing each data point by the reciprocal of its uncertainty squared. MATHEMATICA outputs the best estimate, standard error, t-statistic, and p-value for each of the 25 Fourier coefficients. We use the standard error of each Fourier coefficient as its uncertainty in the rest of our uncertainty calculations.

Appendix B: Uncertainty propagation on the Fourier series

The uncertainty δf for a general function $f = f(x, y, \ldots)$ is given by:

$$\delta f = \sqrt{\left(\frac{\partial f}{\partial x}\delta x\right)^2 + \left(\frac{\partial f}{\partial y}\delta y\right)^2 + \ldots}, \quad (B1)$$

where δx is the uncertainty on x, δy is the uncertainty on y, and so on (Hughes and Hase 2010). Using this, we generate an "uncertainty function" which gives us the uncertainty of the normalized flux as a function of phase.

Now, the Fourier series is given by:

$$I(\Phi) = a_0 + \sum_i a_i \cos(2\pi i \Phi) + \sum_i b_i \sin(2\pi i \Phi)$$

And so,

$$\frac{\partial L}{\partial a_0} = 1, \quad \frac{\partial L}{\partial a_i} = a_i \cos(2\pi i \Phi) \text{ and, } \frac{\partial L}{\partial b_i} = \sin(2\pi i \Phi)$$

Therefore, using Equation B1, the uncertainty on the normalized flux at each phase is given by:

$$\delta I(\Phi) = (\delta a_0)^2 + \sqrt{\sum_i [\delta a_i \cos(2\pi i \Phi)]^2 + \sum_i [\delta b_i \sin(2\pi n \Phi)]^2} \quad (B2)$$

We refer to Equation B2 as the "uncertainty function" in section 3.3.

Appendix C: Uncertainty propagation for ΔI

We calculate ΔI (Fourier) and ΔI (Average) using Equations 4 and 5 (Table 4). ΔI (Fourier) is calculated from the difference in normalized flux between the two maxima from our Fourier fit. We denote the two phases of maxima (one after the primary eclipse and the other after the secondary eclipse) as Φ_p and Φ_s, respectively. We calculate the uncertainty of ΔI (Fourier) by adding $\delta I(\Phi_p)$ and $\delta I(\Phi_s)$ in quadrature, as follows:

$$\delta(\Delta I_{fit}) = [\delta I(\Phi_p)]^2 + [\delta I(\Phi_s)]^2 \quad (C1)$$

where $\delta I(\Phi_p)$ and $\delta I(\Phi_s)$ are calculated by using Equation B2.

In order to calculate ΔI (Average), we average data points within ± 0.05 phase of each maxima and then take the difference between the two maxima. The uncertainty on each average is calculated by adding the uncertainties of data points used in quadrature and dividing that by the number of data points:

$$\delta(\langle I_p \rangle) = \frac{\sqrt{(\delta x_1)^2 + (\delta x_2)^2 + \cdots + (\delta x_n)^2}}{n}$$

with a similar expression for $\delta I(\Phi_s)$.

The uncertainty on ΔI (Average) then, is given by adding these uncertainties of the averages in quadrature:

$$\delta(\Delta I_{Ave}) = [\delta(\langle I_p \rangle)]^2 + [\delta(\langle I_s \rangle)]^2 \quad (C2)$$

Appendix D: Error propagation for the O'Connell Effect Ratio (OER)

The OER is given by Equation 6:

$$OER = \frac{\int_{0.0}^{0.5} (I(\Phi)_{fit} - I(\Phi)_{fit}) d\Phi}{\int_{0.5}^{1.0} (I(\Phi)_{fit} - I(\Phi)_{fit}) d\Phi}$$

From Equation B2,

$$\delta I(\Phi) = (\delta a_0)^2 + \sqrt{\sum_i [\delta a_i \cos(2\pi i \Phi)]^2 + \sum_i [\delta b_i \sin(2\pi n \Phi)]^2}$$

And adding $\delta I(\Phi)$ and $\delta I(0.0)$ in quadrature,

$$\delta(I(\Phi) - I(0.0)) = \sqrt{[\delta I(\Phi)]^2 + [\delta I(0.0)]^2} =$$

$$\sqrt{[2(\delta a_0)^2 + \sum_i [\delta a_i]^2 + \sum_i [\delta a_i \cos(2\pi i \Phi)]^2 + \sum_i [\delta b_i \sin(2\pi i \Phi)]^2}$$

In order to evaluate the errors on the integrals involved in calculating the OER, we make the following approximations:

$$\delta \int_{0.0}^{0.5} [I(\Phi) - I(0.0)] d\Phi \approx \delta \int_{0.0}^{0.5} [\delta(I(\Phi) - I(0.0))] d\Phi, \text{ and}$$

$$\delta \int_{0.5}^{1.0} [I(\Phi) - I(0.0)] d\Phi \approx \delta \int_{0.5}^{1.0} [\delta(I(\Phi) - I(0.0))] d\Phi,$$

and treating $\delta(I(\Phi) - I(0.0))$ as a function of Φ, we integrate over the appropriate limits to obtain the uncertainty for both the numerator and denominator in our expression for the OER.

We then evaluate the uncertainty of the OER, quoted in Table 5, using:

$$\frac{\delta(OER)}{OER} = \sqrt{\left(\frac{\delta\int_{0.0}^{0.5}[I(\Phi)-I(0.0)]d\Phi}{\int_{0.0}^{0.5}[I(\Phi)-I(0.0)]d\Phi}\right)^2 + \left(\frac{\delta\int_{0.5}^{1.0}[I(\Phi)-I(0.0)]d\Phi}{\int_{0.5}^{1.0}[I(\Phi)-I(0.0)]d\Phi}\right)^2}$$

(D1)

Appendix E: Error propagation for the Light Curve Asymmetry (LCA)

The LCA is given by Equation 7:

$$\text{LCA} = \sqrt{\int_{0.0}^{0.5} \frac{(I(\Phi)_{\text{fit}} - I(1.0-\Phi)_{\text{fit}})^2}{I(\Phi)_{\text{fit}}^2} d\Phi}$$

The error analysis of the LCA is a bit more involved. For the sake of clarity, let

$$J(\Phi) = I(\Phi) - I(1.0 - \Phi), \quad K(\Phi) = \frac{J(\Phi)}{I(\Phi)} \text{ and, } L(\Phi) = [K(\Phi)]^2$$

so that $L(\Phi)$ is the integrand, and we are interested in calculating an expression for $\delta L(\Phi)$.

Note that

$$I(1.0 - \Phi) = a_0 + \sum_i a_i \cos[2\pi i(1.0 - \Phi)] + \sum_i b_i \sin[2\pi i(1.0 - \Phi)].$$

and since,

$$\cos[2\pi i(1.0 - \Phi)] = \cos(2\pi i - 2i\Phi) = \cos(2\pi i\Phi) \text{ and,}$$
$$\sin[2\pi i(1.0 - \Phi)] = \sin(2\pi i - 2i\Phi) = -\sin(2\pi i\Phi),$$

we have

$$I(1.0 - \Phi) = a_0 + \sum_i a_i \cos(2\pi i\Phi) - \sum_i b_i \sin(2\pi i\Phi).$$

Thus,

$$J(\Phi) = I(\Phi) - I(1.0 - \Phi) =$$
$$= \left[a_0 + \sum_i a_i \cos(2\pi i\Phi) + \sum_i b_i \sin(2\pi i\Phi) \right]$$
$$- \left[a_0 + \sum_i a_i \cos(2\pi i\Phi) + \sum_i b_i \sin(2\pi i\Phi) \right]$$
$$= 2 \sum_i b_i \sin(2\pi i\Phi).$$

Now, taking the partial derivatives of $L(\Phi)$ with respect to the Fourier coefficients, we get

$$\frac{\partial L}{\partial a_0} = 2K(\Phi) J(\Phi) [-I(\Phi)^2]$$
$$= -2K(\Phi) J(\Phi) I(\Phi)^2. \quad (E1)$$

$$\frac{\partial L}{\partial a_i} = 2K(\Phi) J(\Phi) [-I(\Phi)^2] \cos(2\pi i\Phi)$$
$$= -2K(\Phi) J(\Phi) I(\Phi)^2 \cos(2\pi i\Phi), \text{ and} \quad (E2)$$

$$\frac{\partial L}{\partial b_i} = 2K(\Phi) \left[\frac{2\sin(2\pi i\Phi) - J(\Phi)\sin(2\pi i\Phi)}{I(\Phi)^2} \right]$$
$$= \frac{2K(\Phi)[2I(\Phi) - J(\Phi)]\sin(2\pi i\Phi)}{I(\Phi)^2}. \quad (E3)$$

And so as before (Hughes and Hase 2010),

$$\delta L(\Phi) = \sqrt{\left(\frac{\partial L}{\partial a_0} \delta a_0 \right)^2 + \sum_i \left(\frac{\partial L}{\partial a_i} \delta a_i \right)^2 + \left(\frac{\partial L}{\partial b_i} \delta b_i \right)^2} \quad (E4)$$

where each of the partial derivatives are given by Equations E1, E2, and E3.

Using the same approximation as before, we have:

$$\delta \int_{0.0}^{0.5} L(\Phi) d\Phi \approx \int_{0.0}^{0.5} \delta L(\Phi) d\Phi,$$

We integrate $\delta L(\Phi)$ over the appropriate limits. We then calculate the uncertainty of the LCA, quoted in Table 5, using:

$$\frac{\delta(LCA)}{LCA} = \frac{1}{2} \left[\frac{\delta \int_{0.0}^{0.5} L(\Phi) d\Phi}{\int_{0.0}^{0.5} L(\Phi) d\Phi} \right]$$
$$\approx \frac{1}{2} \left[\frac{\delta \int_{0.0}^{0.5} L(\Phi) d\Phi}{\int_{0.0}^{0.5} L(\Phi) d\Phi} \right] \quad (E5)$$

Observations and Preliminary Modeling of the Light Curves of Eclipsing Binary Systems NSVS 7322420 and NSVS 5726288

Matthew F. Knote
Department of Aerospace, Physics, and Space Sciences, Florida Institute of Technology, 150 West University Boulevard, Melbourne, FL, 32901, and Department of Physics and Astronomy, Ball State University, 2000 West University Avenue, Muncie, IN 47306; Visiting Astronomer at SARA Observatory; mknote2015@my.fit.edu

Ronald H. Kaitchuck
Department of Physics and Astronomy and SARA, Ball State University, 2000 West University Avenue, Muncie, IN 47306; rkaitchu@bsu.edu

Robert C. Berrington
Department of Physics and Astronomy and SARA, Ball State University, 2000 West University Avenue, Muncie, IN 47306; rberring@bsu.edu

Received June 24, 2019; revised August 9, 2019; accepted September 20, 2019

Abstract We present new photometric observations of the β Lyrae-type eclipsing binary systems NSVS 7322420 and NSVS 5726288. These observations represent the first multi-band photometry performed on these systems. The light curves were analyzed with PHOEBE, a front-end GUI based on the Wilson-Devinney program, to produce models to describe our observations. Our preliminary solutions indicate that NSVS 7322420 is a primary filling semi-detached system with unusual features warranting further study. These features include a pronounced O'Connell effect, a temporal variance in the light curve, and an unusual "kink" in the light curve around the secondary eclipse. The cause of these features is unknown, but one possibility is the transfer of mass between the component stars. Meanwhile, NSVS 5726288 is probably a typical detached system.

1. Introduction

Eclipsing binary systems are an important class of variable star as they allow the determination of several characteristics that are otherwise difficult or impossible to determine without resolving the stellar components. Photometric measurements allow the determination of the orbital inclination of the system as well as the shapes and radii of the component stars for totally eclipsing systems, while radial velocity measurements allow the determination of the absolute masses of the components. With both types of measurements, a nearly complete description of the system and its components can be obtained.

We chose two eclipsing binary candidates identified by Hoffman *et al.* (2008), who identified 409 candidate Algol- and β Lyrae-type eclipsing binary systems in the Northern Sky Variability Survey (NSVS). The NSVS was a survey of the northern sky conducted to record stellar variations of faint objects (Woźniak *et al.* 2004). The survey, which was conducted between April 1, 1999, and March 30, 2000, at Los Alamos National Laboratory, imaged 14 million objects north of declination −38° with magnitudes between 8 and 15.5. There were typically a few hundred measurements taken for each object, creating a strong base set for searching for variability. Many systems in the NSVS have not had follow up studies performed due to the enormous volume of data, but works such as Hoffman *et al.* (2008) have made progress in cataloguing systems based on the NSVS data set itself. The two targets we chose were NSVS 7322420 and NSVS 5726288; details of these two systems and the stars used to analyze them are given in Table 1, while finder charts for the systems are shown in Figure 1. The magnitudes given in Table 1 are taken from the AAVSO Photometric All Sky Survey (APASS) Data Release 9 (Henden *et al.* 2016); the Sloan r' and i' magnitudes given by APASS were converted into Cousins R and I magnitudes using transformation equations found in Table 1 of Jester *et al.* (2005).

Section 2 of this paper describes our acquisition and reduction of the observational data. Section 3 outlines the methods used to analyze our data as well as the results of our analysis. Finally, section 4 contains the discussion of our results as well as details of potential future work.

Table 1. List of the designation, coordinates, and B, V, R, and I magnitudes of each variable star, comparison star, and check star.

Star	Name	R.A. (J2000) h m s	Dec. (J2000) ° ′ ″	B	V	R	I
V	NSVS 7322420	08 16 12.90	+26 41 13.67	11.675	10.983	10.538	10.122
	NSVS 5726288	20 19 11.65	+44 15 47.05	11.607	11.297	11.058	10.831
C	TYC 1936-1001-1	08 17 03.36	+26 45 14.52	11.215	10.695	10.344	10.014
	TYC 3163-221-1	20 18 51.80	+44 12 44.81	11.112	10.781	10.549	10.328
K (check)	TYC 1936-113-1	08 17 05.13	+26 43 06.50	12.228	11.207	10.654	10.139
	BD+43 3570	20 19 56.71	+44 11 19.09	11.021	10.894	10.745	10.600

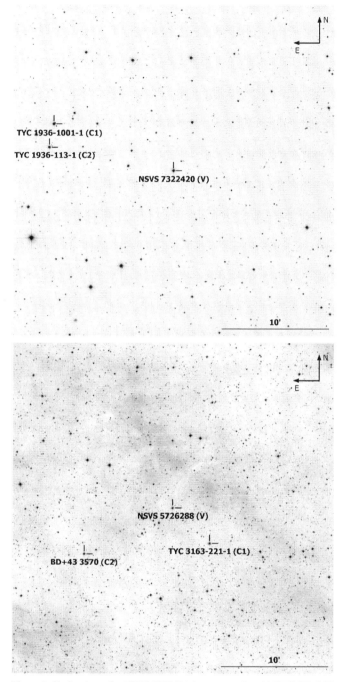

Figure 1. Finder charts for NSVS 7322420 (upper panel) and NSVS 5726288 (lower panel) as well as their comparison stars (labeled C1 and C2).

2. Observations

Our survey of the two systems was conducted between March 2013 and November 2016. We conducted our observations at two separate sites: the Ball State University Observatory (BSUO) in Muncie, Indiana, and Kitt Peak National Observatory (KPNO). The BSUO observations were conducted with a STXL-6303E CCD camera cooled to –10° C mounted on a 16-inch Meade LX200 while the KPNO observations were conducted with a CCD custom built by Astronomical Research Cameras, Inc. (ARC) cooled to –110° C mounted on the 0.9-meter SARA-KP telescope (Keel *et al.*

Table 2. A list of dates of observation along with the location and filters used.

Target	Observation Date	Telescope	Filters
NSVS 7322420	03/22/2013	BSUO	B, V, R, I
	03/30/2013	BSUO	B, V, R, I
	04/02/2013	BSUO	B, V, R, I
	04/03/2013	BSUO	B, V, R, I
	04/04/2013	BSUO	B, V, R, I
	04/05/2013	BSUO	B, V, R, I
	04/14/2013	BSUO	V, R, I
	04/16/2013	KPNO	V, R, I
	04/21/2013	BSUO	V, R, I
	04/22/2013	BSUO	V, R, I
	04/30/2013	BSUO	B, V, R, I
	05/01/2013	BSUO	B, V, R, I
	05/19/2013	BSUO	B, V, R, I
	05/27/2013	KPNO	B, V, R, I
	11/29/2013	KPNO	B, V, R, I
	01/17/2014	KPNO	B, V, R, I
	01/28/2014	KPNO	B, V, R, I
	01/29/2014	KPNO	B, V, R, I
	10/15/2014	KPNO	B, V, R, I
	10/27/2014	KPNO	B, V, R, I
NSVS 5726288	05/15/2013	BSUO	B, V, R, I
	05/19/2013	BSUO	B, V, R, I
	05/25/2013	BSUO	B, V, R, I
	06/03/2013	BSUO	B, V, R, I
	06/04/2013	BSUO	B, V, R, I
	06/11/2013	BSUO	B, V, R, I
	06/14/2013	BSUO	B, V, R, I
	06/15/2013	BSUO	B, V, R, I
	06/17/2013	BSUO	B, V, R, I
	06/19/2013	BSUO	B, V, R, I
	06/20/2013	BSUO	B, V, R, I
	07/17/2013	BSUO	B, V, R, I
	09/22/2014	BSUO	B, V, R, I
	09/23/2014	BSUO	B, V, R, I
	09/25/2014	BSUO	B, V, R, I
	10/15/2014	KPNO	B, V, R, I
	05/29/2015	BSUO	B, V, R
	11/14/2016	KPNO	B, V, R, I
	11/15/2016	KPNO	B, V, R, I

Note: BSUO, Ball State University Observatory; KPNO, Kitt Peak National Observatory.

2017). Observations were conducted in the Bessell B, V, R, and I filters on all nights excepting brief periods in late April 2013 and May 2015; the Bessell filters (Bessell 1990) closely approximate the Johnson-Cousins photometric system. A list of the nights of observation is presented in Table 2.

Reduction of the obtained images was performed using the CCDRED package contained in the Image Reduction and Analysis Facility (IRAF, Tody 1993). The software ASTROIMAGEJ (Collins *et al.* 2017) was then used to perform photometry on the calibrated images. We used the method of differential photometry to obtain the differential magnitude of our system. The comparison star has a similar color index to the variable so that the two will be similarly affected by atmospheric extinction, although the choice of comparison is limited due to the field of view of our instruments (approximately 15 arcminutes). The comparison star is compared to a check star to ensure that the comparison is non-variable. Once the photometry has been performed, the data is converted to a function of orbital phase (Φ) rather than absolute time using Equations 3 and 4 for

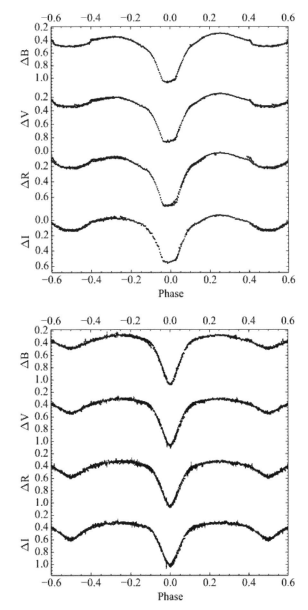

Figure 2. Phased light curves for B, V, R, and I-band data on NSVS 7322420 (upper panel) and NSVS 5726288 (lower panel). Error bars are plotted for all data points. The vertical axis shows the differential magnitude in the given band, while the horizontal axis shows the orbital phase of the system. The data for NSVS 7322420 are only the subset of data taken from January 17, 2014, to January 29, 2014.

NSVS 7322420 and NSVS 5726288, respectively. The phased light curves that were produced and used in our analysis are shown in Figure 2. The unphased standardized magnitude data were uploaded to the AAVSO International Database (Kafka 2015) where they are available for download.

3. Analysis

3.1. Methodology
3.1.1. Ephemeris, period determination, and O–C calculation

The time of minimum for each primary eclipse was calculated using the method described by Kwee and van Woerden (1956). This process was applied to each eclipse in each filter, and the obtained values for each individual eclipse were averaged together to give the reported time of minimum. Period determination was done using the program PERANSO (Paunzen and Vanmunster 2016) rather than by examination of the times of minimum light. PERANSO allows the user to input a set of observations and use one of a variety of methods to calculate the period. We chose to use the analysis of variance (ANOVA) method described by Schwarzenberg-Czerny (1996). This method, which uses periodic orthogonal polynomials to fit the observed data, excels at detecting and discarding aliases of the true period.

The observed times of minimum light were then compared to a calculated time of minimum. This calculated value for the nth epoch is found using the equation:

$$T_{cal,n} = T_{obs,0} + E_n P \qquad (1)$$

where $T_{obs,0}$ is an observed reference time of minimum, E_n is the number of periods elapsed since $T_{obs,0}$, and P is the orbital period. The calculated value was then obtained by subtracting it from the corresponding observed value to obtain O–C. The error in O–C is given by the equation:

$$\sigma_{O-C,n} = \sqrt{\sigma_{T_{obs,n}}^2 + \sigma_{T_{obs,0}}^2} \qquad (2)$$

where σ denotes the error of the subscripted parameter.

3.1.2. Magnitude calibration

The output file generated by ASTROIMAGEJ gives the brightness of the system in relative flux, from which the differential magnitude can be calculated. Estimating the temperature of the stellar components, however, requires the apparent magnitude of the system. The apparent magnitude was obtained by adding the calibrated apparent magnitude of the comparison star given by Henden et al. (2016) to the differential magnitude obtained from ASTROIMAGEJ. We performed an analysis to determine the effect the atmosphere had on our determination of the magnitude; we found that the correction factor was negligible and thus our determined magnitude very closely approximates the true apparent magnitude. The error in magnitude was then calculated by adding in quadrature the error in the apparent magnitude of the comparison star (also provided by Henden et al. 2016) and the error determined from ASTROIMAGEJ. We performed the magnitude calibration and error determination on all data points.

3.1.3. Temperature estimation

A rough estimation for the temperature of the primary component was made by calculating the (B–V) color index of the system at phase 0.5. Since the observations in the B and V filters did not occur simultaneously, the observational values from the B filter were linearly interpolated to coincide with the values for the V filter. The difference between the interpolated B data and observed V data were then calculated and displayed as a function of phase. These raw values were corrected for interstellar reddening based on the work by Schlafly and Finkbeiner (2011), which provides E(B–V) values for given coordinates based on the assumption that we are observing

Table 3. O–C and observed and calculated times of primary minimum for NSVS 7322420 and NSVS 5726288.

Target	Epoch	Time of Minimum (HJD) Observed	Time of Minimum (HJD) Calculated	O–C (days)
NSVS 7322420	0	2456413.632618 ± 0.000201	—	0 ± 0.000201
	454	2456625.861477 ± 0.000027	2456625.862636 ± 0.000201	–0.001159 ± 0.000203
	559	2456674.944067 ± 0.000100	2456674.946671 ± 0.000201	–0.002604 ± 0.000225
NSVS 5726288	0	2456457.738889 ± 0.000178	—	0 ± 0.000178
	543	2456922.677753 ± 0.000231	2456922.685354 ± 0.000178	–0.007601 ± 0.000292
	570	2456945.796265 ± 0.000127	2456945.804239 ± 0.000178	–0.007974 ± 0.000219

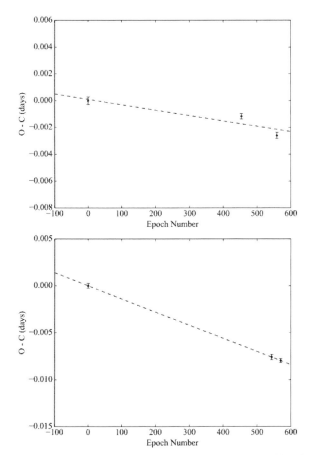

Figure 3. Observed minus calculated (O–C) times of minimum light for NSVS 7322420 (upper panel) and NSVS 5726288 (lower panel) plotted against epoch number. The dashed lines indicate the best linear fit to the data.

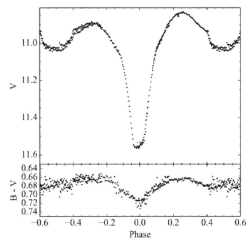

Figure 4. (B–V) color curve for NSVS 7322420. The apparent V-band magnitude is plotted in the top panel, while the (B–V) color curve is plotted in the lower panel. Error bars are not plotted for clarity.

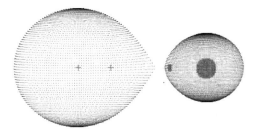

Figure 5. Three-dimensional visualization of the model for NSVS 7322420. The phase of the system in this figure is 0.25, with the primary component on the left and the secondary component on the right; the primary is moving toward the observer. The red markers on the left and right represent the center of each star while the red marker in the center represents the center of mass of the binary system. Two of the three hot spots are clearly visible on the secondary component.

objects beyond our galaxy. While this assumption does not hold for our stars, this provides a first approximation correction.

Once the corrected (B–V) color curve was obtained, the temperature of the primary component was estimated based upon the (B–V) color index at phase 0.5. At this phase, the primary component mostly or completely obscures the secondary and is therefore the dominant contributor to the observed flux. We used the work presented in Flower (1996) to convert the (B–V) color index at phase 0.5 to temperature under the assumption that the stars are on the main sequence. The secondary temperature was determined through modeling.

3.1.4. Light curve analysis

We used the program PHOEBE (PHysics Of Eclipsing BinariEs) v0.31a (Prša and Zwitter 2005) to analyze the light curves of the systems and produce a best-fit model. PHOEBE is a graphical user interface built on the Wilson-Devinney (WD) program introduced by Wilson and Devinney (1971). This model includes many parameters that affect the synthetic light curve produced by the program. These parameters are altered using an iterative least-squares analysis known as differential correction to produce the best fit to the observed data.

We input the phased apparent magnitude curves from all four filters into PHOEBE. Data from all filters were fit simultaneously to better constrain the system parameters. The primary temperature was held fixed at the previously estimated value throughout the modeling process while the secondary temperature was allowed to vary. The fact that all of our target stars had temperatures suggesting convective envelopes allowed us to set the gravity brightening and surface albedos to their theoretical values of

Table 4. Results of the modeling process and analysis for NSVS 7322420 and NSVS 5726288. The errors are the formal errors given by PHOEBE.

Parameter	NSVS 7322420	NSVS 5726288
Orbital Period (d)	0.467467 ± 0.000015	0.856255 ± 0.000007
T_{eff} of Primary Component (K, fixed)	5,700	7,300
T_{eff} of Secondary Component (K)	3,361 ± 29	5,632 ± 37
Orbital Inclination (°)	91.66 ± 0.04	79.60 ± 0.02
Surface Potential of Primary Component	—	3.1608 ± 0.0035
Surface Potential of Secondary Component	2.6085 ± 0.0088	3.1791 ± 0.0039
Mass Ratio	0.3388 ± 0.0025	0.5789 ± 0.0014
Magnitude Difference ($M_2 - M_1$)	3.4974	1.7431
Spot 1 Temperature (K)	6,227 ± 89	—
Spot 1 Latitude (°, fixed)	0	—
Spot 1 Longitude (°, fixed)	90	—
Spot 1 Radius (°, fixed)	16	—
Spot 2 Temperature (K, fixed)	6,722 ± 58	—
Spot 2 Latitude (°, fixed)	0	—
Spot 2 Longitude (°, fixed)	30	—
Spot 2 Radius (°, fixed)	5	—
Spot 3 Temperature (K, fixed)	6,722 ± 58	—
Spot 3 Latitude (°, fixed)	0	—
Spot 3 Longitude (°, fixed)	340	—
Spot 3 Radius (°, fixed)	5	—

0.32 (Lucy 1967) and 0.5 (Ruciński 1969), respectively. We performed a mass ratio search (as described in Qian *et al.* 2007) to determine the best fit mass ratio.

Following this, we allowed other parameters of the system to vary. These parameters include the Kopal surface potential (Ω), the mass ratio of the system ($q = m_2/m_1$), the orbital inclination (i) of the system, and a phase shift (which was negligible for both systems). Throughout the process, we continued to normalize the luminosities of the component stars and interpolate limb darkening coefficients based on the work by van Hamme (1993). The limb darkening coefficients were interpolated using a logarithmic law as the temperature for all stars is less than 9,000 K.

3.2. NSVS 7322420

Our period analysis for NSVS 7322420 yielded a period of 0.467467 ± 0.000015 day, which is quite close to the value of 0.46740 day published by Hoffman *et al.* (2008). We determined the average time of minimum on three primary eclipses, and these and the O–C values are listed in Table 3 while a diagram showing the best linear fit to these values is displayed in Figure 3. We used the first time of minimum as the reference minimum, resulting in a linear ephemeris of:

$$T_{min}(HJD) = 2456413.6326(2) + 0.467467(15) d E \quad (3)$$

The (B–V) color curve we obtained for this system is shown in Figure 4, from which we estimate a (B–V) color index of 0.68 ± 0.07 during the secondary eclipse (the large error is due to the uncertainty in the apparent magnitudes of the comparison star). Schlafly and Finkbeiner (2011) estimate an E(B–V) of 0.03 at the coordinates of this system, giving a (B–V) color index of 0.65 ± 0.07 during the secondary eclipse. This corresponds to a temperature for the primary component of 5,700 ± 230 K and indicates a spectral type of G3 ± 3 (Fitzgerald 1970).

The system proved difficult to model, due at least in part to the presence of maxima of unequal height: the maximum following primary eclipse is 0.062 magnitude brighter than the maximum following secondary eclipse in the B band. This is known as the O'Connell effect (O'Connell 1951, Milone 1968) and may be explained by a hot or cool spot on one of the stars or from gas streams creating one or more hot spots (Wilsey and Beaky 2009). The model we present matches the observed light curve marginally well, but it is highly artificial as it requires three hot spots in order to adequately explain the observed features in the light curve. While the hot spot located at 90° longitude can be interpreted as the impact of a matter stream onto the stellar surface (Wilsey and Beaky 2009), the two spots located close to 0° longitude have no clear physical interpretation. Spots introduce a considerable amount of degeneracy into the solution, and only the temperature factor of the first spot could be allowed to vary without the program diverging. As a result of these factors, this model should not be interpreted as a completely accurate physical description of the system, and while the general system characteristics are most likely accurate, the specific cause of the light curve irregularities is likely not accurately explained by this model.

The model indicates that the system is a primary filling semi-detached system with a secondary component of spectral type M. Details of this model are given in the middle column of Table 4 while a three-dimensional visualization (produced by BINARY MAKER 3, Bradstreet and Steelman 2002) of the model is given in Figure 5. A comparison of the observed data and synthetic light curves is presented in Figure 8.

3.3. NSVS 5726288

Our period analysis for NSVS 5726288 yielded a period of 0.856255 ± 0.000007 day, which differs quite significantly from the value of 0.59935 day published by Hoffman *et al.* (2008). This 0.59935-day period is almost exactly seven-tenths of our calculated period, indicating that it is likely an alias caused by poor temporal coverage. We determined the average time of minimum on three primary eclipses, and these and the O–C values are listed in Table 3 while a diagram showing the best linear fit to these values is displayed in Figure 3. We used the first time of minimum as the reference minimum, resulting in a linear ephemeris of:

$$T_{min}(HJD) = 2456457.738889(178) + 0.856255(7) d E \quad (4)$$

The (B–V) color curve we obtained for this system is shown in Figure 6, from which we estimate a (B–V) color index of 0.28 ± 0.04 during the secondary eclipse. Schlafly and Finkbeiner (2011) estimate a E(B–V) of 1.55 at the coordinates of this system, giving a (B–V) color index of -1.27 ± 0.04 during the secondary eclipse. This gives a non-physical value for the effective temperature, preventing us from using this method to determine an interstellar extinction correction for this system. The reason for such a large correction is that the system lies in the plane of the Milky Way and is therefore subject to significant extinction. With no better way to determine the reddening of the system, we chose to use the raw color index (0.28 ± 0.04) as a basis for temperature estimation. This corresponds to a temperature of $7,300 \pm 210$ K and indicates a spectral type of $A8 \pm 1$ (Fitzgerald 1970).

A single, well-fitting model was produced for this system, the details of which can be found in the rightmost column of Table 4. The model indicates a detached system with a secondary component of spectral type G. It should be noted that, due to the lack of correction for interstellar extinction, the values for the temperature are a lower bound. A three-dimensional visualization of the model is given in Figure 7 while a comparison of the observed data and synthetic light curves is presented in Figure 9.

4. Discussion

NSVS 7322420 is a semi-detached system with the primary component filling its critical lobe. The system contains a G3 primary and an M secondary, and the model indicates that the stars are of significantly dissimilar mass. The light curve of the system displays a pronounced O'Connell effect and an unusual "kink" as the system enters and exits secondary eclipse. The system is also rapidly evolving, and data taken in 2013 and late 2014 could not be combined with data taken in early 2014 without producing obvious discontinuities that rendered modeling the system impossible. This light curve variability forced us to use only a subset of the data taken on the system that was relatively close temporally. The model produced is artificial and likely does not accurately explain these unusual features; a more sophisticated modeling program may be necessary.

NSVS 5726288 is a rather standard detached system containing two components of dissimilar mass in a moderately inclined orbit. The system contains an A8 primary and an early G secondary. The determined values for the temperature are lower bounds due to the lack of correcting for interstellar extinction, so it is possible that these stars are significantly hotter than our model indicates. The period of 0.856255 day differs significantly from the value published by Hoffman *et al.* (2008), which is most likely a symptom of poor temporal coverage of the system by the NSVS.

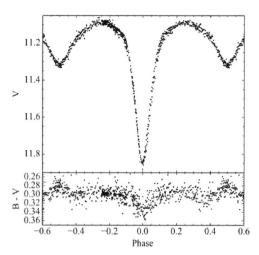

Figure 6. (B–V) color curve for NSVS 5726288. The apparent V-band magnitude is plotted in the top panel, while the (B–V) color curve is plotted in the lower panel. Error bars are not plotted for clarity.

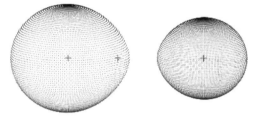

Figure 7. Three-dimensional visualization of the model for NSVS 5726288. The phase of the system in this figure is 0.25, with the primary component on the left and the secondary component on the right; the primary is moving toward the observer. The red markers on the left and right represent the center of each star while the red marker in the center represents the center of mass of the binary system.

The O'Connell effect observed in NSVS 7322420 is a poorly understood phenomenon. Wilsey and Beaky (2009) describe several ideas regarding the cause of the effect, but they note that none of these ideas can adequately explain all known examples. The most prevalent explanation of the O'Connell effect is that hot or cool spots caused by chromospheric activity on one of the components creates a difference in the observed flux for different hemispheres of the star. However, the spots may need to be unrealistically large in size in order to fully describe the observed O'Connell effect, and spots also suffer from the ability to explain almost any deviation from a synthetic light curve if placed correctly, which reduces confidence in the reliability of such models. Furthermore, Drake *et al.* (2014) find "no evidence for changes in the maxima that are expected as star spot numbers or sizes vary," further diminishing the theory that star spots are the cause of the O'Connell effect. An alternate hypothesis proposed by Liu and Yang (2003) suggests that the effect is caused by material surrounding the stars impacting the components as they orbit, resulting in heating of the leading hemispheres. At this time, however, we lack the observational evidence to suggest a plausible source for such material in NSVS 7322420, and we consider it unlikely to explain the effect in the system.

Another hypothesis proposes that the asymmetry in observed flux is caused by a hot spot created by the impact of a matter

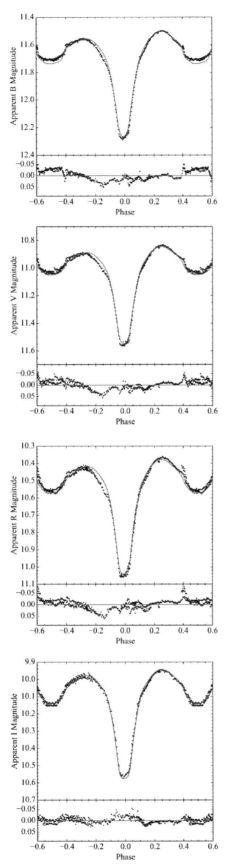

Figure 8. Observed B (top panel), V (second panel), R (third panel), and I (fourth panel) data for NSVS 7322420 plotted against synthetic light curves. The upper part of each panel plots the apparent magnitude of the observed and synthetic light curves against the phase of the system. The residuals for the model are shown in the bottom part of each panel.

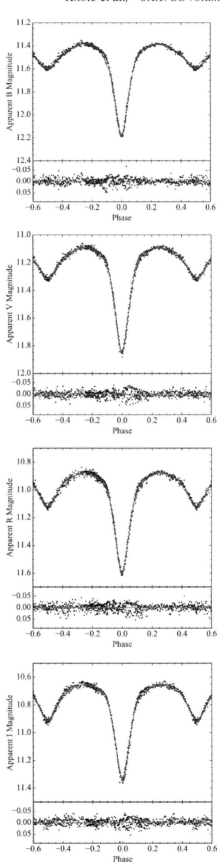

Figure 9. Observed B (top panel), V (second panel), R (third panel), and I (fourth panel) data for NSVS 5726288 plotted against synthetic light curves. The upper part of each panel plots the apparent magnitude of the observed and synthetic light curves against the phase of the system. The residuals for the model are shown in the bottom part of each panel.

stream on either the stellar surface or on a circumstellar accretion disk. This hypothesis has been used to explain the O'Connell effect observed in GR Tauri (Gu et al. 2004), a system that exhibits light curve variability similar to what we observed in NSVS 7322420. Due to this and the previously mentioned irregularities in the light curve of the system (including an asymmetric primary minimum similar to RY Scuti, a system that was modeled with an accretion disk by Djurašević et al. 2008), we believe that the mass transfer theory represents the most likely explanation of the O'Connell effect in NSVS 7322420.

Unfortunately, photometry alone cannot provide a full description of these systems. Radial velocity data obtained from spectra of these systems would allow us to determine the absolute masses of the components, and therefore the absolute sizes and luminosities of the stars. Spectroscopic data would also provide a direct way to measure the temperatures of the components, which would further refine the parameters of the models. Finally, spectroscopic analysis of NSVS 7322420 could provide clues as to the cause of the unusual features seen in its light curve. Further photometric and spectroscopic observations are being conducted on NSVS 732240 and will be described in a future paper. The system is also serving as an archetype for a study of systems suspected of undergoing mass transfer.

5. Acknowledgements

We would like to thank Drs. Eric Perlman and Saida Caballero-Nieves of FIT for their comments and suggestions regarding this paper. We would also like to thank the Indiana Space Grant Consortium, which partially funded this research project. This paper is based on observations obtained with the SARA Observatory 0.9-meter telescope at Kitt Peak, which is owned and operated by the Southeastern Association for Research in Astronomy (saraobservatory.org). The authors are honored to be permitted to conduct astronomical research on Iolkam Du'ag (Kitt Peak), a mountain with particular significance to the Tohono O'odham Nation.

References

Bessell, M. S. 1990, *Publ. Astron. Soc. Pacific*, **102**, 1181.
Bradstreet, D. H., and Steelman, D. P. 2002, *Bull. Amer. Astron. Soc.*, **34**, 1224.
Collins, K. A., Kielkopf, J. F., Stassun, K. G., and Hessman, F. V. 2017, *Astron. J.*, **153**, 77.
Djurašević, G., Vince, I., and Atanacković, O. 2008, *Astron. J.*, **136**, 767.
Drake, A. J., et al. 2014, *Astrophys. J., Suppl. Ser.*, **213**, 9.
Fitzgerald, M. P. 1970, *Astron. Astrophys.*, **4**, 234 .
Flower, P. J. 1996, *Astrophys. J.*, **469**, 355.
Gu, S.-H., Chen, P.-S, Choy, Y.-K., Leung, K.-C., Chung, W.-K., and Poon, T.-S. 2004, *Astron. Astrophys.*, **423**, 607.
Henden, A. A., Templeton, M., Terrell, D., Smith, T. C., Levine, S., and Welch, D. 2016, VizieR Online Data Catalog: AAVSO Photometric All Sky Survey (APASS) DR9, II/336.
Hoffman, D. I., Harrison, T. E., Coughlin, J. L., McNamara, B. J., Holtzman, J. A., Taylor, G. E., and Vestrand, W. T. 2008, *Astron. J.*, **136**, 1067.
Jester, S., et al. 2005, *Astron. J.*, **130**, 873.
Kafka, S. 2015, variable star observations from the AAVSO International Database (https://www.aavso.org/aavso-international-database).
Keel, W. C., et al. 2017, *Publ. Astron. Soc. Pacific*, **129**, 5002.
Kwee, K. K., and van Woerden, H. 1956, *Bull. Astron. Inst. Netherlands*, **12**, 327.
Liu, Q.-Y., and Yang, Y.-L. 2003, *Chin. J. Astron. Astrophys.*, **3**, 142.
Lucy, L. B. 1967, *Z. Astrophys.*, **65**, 89.
Milone, E. E. 1968, *Astron. J.*, **73**, 708.
O'Connell, D. J. K. 1951, *Publ. Riverview Coll. Obs.*, **2**, 85.
Paunzen, E., and Vanmunster, T. 2016, *Astron. Nachr.*, **337**, 239.
Prša, A., and Zwitter, T. 2005, *Astrophys. J.*, **628**, 426.
Qian, S.-B., Yuan, J.-Z., Soonthornthum, B., Zhu, L.-Y., He, J.-J., and Yang, Y.-G. 2007, *Astrophys. J.*, **671**, 811.
Ruciński, S. M. 1969, *Acta Astron.*, **19**, 245.
Schlafly, E. F., and Finkbeiner, D. P. 2011, *Astrophys. J.*, **737**, 103.
Schwarzenberg-Czerny, A. 1996, *Astrophys. J., Lett.*, **460**, L107.
Tody, D. 1993, in *Astronomical Data Analysis Software and Systems II*, eds. R. J. Hanisch, R. J. V. Brissenden, J. Barnes, ASP Conf. Ser. 52, Astronomical Society of the Pacific, San Francisco, 173.
van Hamme, W. 1993, *Astron. J.*, **106**, 2096.
Wilsey, N. J., and Beaky, M. M. 2009, in *The Society for Astronomical Sciences 28th Annual Symposium on Telescope Science*, Society for Astronomical Sciences, Rancho Cucamonga, CA, 107.
Wilson, R. E., and Devinney, E. J. 1971, *Astrophys. J.*, **166**, 605.
Woźniak, P. R., et al. 2004, *Astron. J.*, **127**, 2436.

Period Analysis of All-Sky Automated Survey for Supernovae (ASAS-SN) Data on Pulsating Red Giants

John R. Percy
Lucas Fenaux
Department of Astronomy and Astrophysics, and Dunlap Institute for Astronomy and Astrophysics, University of Toronto, 50 St. George Street, Toronto, ON M5S 3H4, Canada; john.percy@utoronto.ca

Received June 24, 2019; revised July 30, 2019; accepted August 5, 2019

Abstract The All-Sky Automated Survey for Supernovae (ASAS-SN) has recently used over 2,000 days of sustained photometric data to identify more than 50,000 variable stars, automatically classify these, determine periods and amplitudes for those that are periodic—part of a remarkable project to classify 412,000 known variable stars and determine their basic properties. This information about the newly-discovered variables, along with the photometric data, is freely available on-line, providing an outstanding resource for both science and education. In this paper, we analyze ASAS-SN V data on two small random samples of pulsating red giants (PRGs) in detail, and compare our results with those found by ASAS-SN. For the majority of a sample of 29 mostly semi-regular (SR) PRGs, the ASAS-SN results are incorrect or incomplete: either the ASAS-SN periods are exactly 2, 3, or 4 times the actual period, or the ASAS-SN period is a "long secondary period" with a shorter pulsation period present, or the star is multi-periodic or otherwise complex, or the star's data and analysis are contaminated by instrumental effects. For almost all of a sample of 20 of the *longest-period* Mira stars (period 640 days or more), the ASAS-SN period is exactly 2 or more times the actual period. The results are not surprising, given the very complex behavior of PRGs.

1. Introduction

Red giant stars are unstable to pulsation. In the *General Catalogue of Variable Stars* (GCVS; Samus *et al.* 2017), pulsating red giants (PRGs) are classified according to their light curves. Mira (M) stars have reasonably regular light curves, with visual ranges greater than 2.5 magnitudes. Semi-regular (SR) stars are classified as SRa if there is appreciable periodicity, and SRb if there is little periodicity. Irregular (L) stars have very little or no periodicity.

Mira stars have periods which "wander" by a few percent; this wandering can be described and modelled as random, cycle-to-cycle fluctuations (Eddington and Plakidis 1929). Their maximum magnitudes vary from cycle to cycle, as observers of Mira itself know. The variability of SR stars is even more complex. Some stars are multiperiodic; two or more pulsation modes are excited (e.g. Kiss *et al.* 1999). About a third show long secondary periods (LSPs), 5 to 10 times the pulsation period (Wood 2000); their cause is unknown. The amplitudes of PRGs vary by up to a factor of 10 on time scales of 20 to 30 pulsation periods (Percy and Abachi 2013). There may also be very slow variations in mean magnitude (Percy and Qiu 2019). In a very few stars, thermal pulses cause large, secular changes in period, amplitude, and mean magnitude (Templeton *et al.* 2005 and references therein).

Our previous studies of PRGs have used long-term visual and sometimes photoelectric observations from the American Association of Variable Star Observers International Database (AID; Kafka 2019). Now, an important and very useful new source of data is available: the All-Sky Automated Survey for Supernovae (ASAS-SN).

ASAS-SN uses a network of up to 24 telescopes around the world to survey the entire visible sky every night down to about 18th magnitude (Shappee *et al.* 2014; Jayasinghe *et al.* 2018, 2019a). It has been doing so for over 2000 days (since about JD 2456500). ASAS-SN has identified over 50,000 variable stars, determined periods and amplitudes for those that are periodic, classified these using machine learning, and made this information and the data available on-line (asas-sn.osu.edu/variables). It has also used machine learning to uniformly classify 412,000 known variables (Jayasinghe *et al.* 2019a).

ASAS-SN used three period-search techniques: Generalized Lomb-Scargle, Multi-Harmonic Analysis of Variance, and the Box Least Squares. For classification purposes, they began with an open-source random forest classifier *Upsilon*, trained using OGLE and EROS-2 data. They note that the performance for the *Upsilon* classifier is low for SR variables—only 36 percent. They then built a new classifier based on ASAS-SN data, using a set of 16 features of the light or phase curve as classification criteria. The precision of the new classifier, for SR variables, is given as 63 percent. They note also that the classifier has difficulty distinguishing between SR and L variables.

The purpose of the present project was to analyze a small, random sample of the PRG data in more detail, and investigate the reliability of the ASAS-SN classifications and periods and amplitudes of PRGs, given our previous knowledge of and experience with such stars. Jayasinghe *et al.* (2018) comment only briefly on the ASAS-SN classification of these very complex variables.

Vogt *et al.* (2016) have recently carried out a related study: analysis of 2,875 Mira stars observed in the original ASAS project, which extended from 2000 to 2009. They used a semi-automatic method based on the observed times of maximum light. They found that, whereas their periods agreed with those in the VSX Catalogue (Watson *et al.* 2014) in more than 95 percent of the stars, their periods agreed with those obtained by Richards *et al.* (2012), who used an automatic machine-learning method, in only 76 percent of the stars. Most often, the latter periods differed from the Vogt *et al.* (2016) periods by a ratio of small whole numbers.

2. Data and analysis

For our initial project, we analyzed a sample of 22 stars classified as SR, five as M, and three as L. They were randomly chosen around a random position on the sky. The SR and L stars were chosen to have ASAS-SN amplitudes of at least 0.5 magnitude (with one accidental exception), so that the results would not be unduly affected by noise, and would therefore be meaningful. The datasets were approximately 2,000 days in length. For each of the SR and M stars, ASAS-SN provides a period and amplitude, a light curve and phase curve, and a quantity T which is a statistical measure of the confidence of the period; lower values indicate higher confidence. The data were downloaded, and analyzed using the AAVSO VSTAR time-series package (Benn 2013), which includes a Fourier analysis routine.

The results were interesting, so we carried out a subsidiary project, to analyze a sample of 20 Mira stars with the *longest* ASAS-SN periods—longer than 639 days. Miras with such long periods would be especially interesting and important, astrophysically.

3. Results

Of the 29 stars in our initial project, 7 were acceptably analyzed (e.g. Figure 1). For 9 stars, the ASAS-SN period was exactly 2, 3, or 4 times the actual period; the phase curve had not one, but 2, 3, or 4 cycles in it. While this is a mathematical possibility, it is unphysical for a pulsator such as a PRG to have such a phase curve. These periods might be considered as "sub-harmonics" of the correct periods. Figure 2 shows an example with three cycles per unit phase. In the figures, we have chosen to show the ASAS-SN light and phase curves as they were when this project was carried out, rather than new plots, since the former are more relevant to the present project.

For five stars, the light curve also showed periodic variability on a time scale 5 to 10 times shorter than the ASAS-SN period. The latter was clearly a "long secondary period," whereas the shorter period was the pulsation period. Figure 3 shows an example in which the LSP is actually half the ASAS-SN period of 1,022 days. The shorter pulsation period of 55 ± 2 days is clearly visible.

For a very few stars, the light curve included some faint, highly-discordant data, and it appeared that ASAS-SN had included these data in the analysis. Figure 4 shows an example. These discordant photometric points are probably due to astrometry problems and their effect on the image-subtraction process (Kochanek and Jayasinghe 2019). The non-discordant data show periods of 423.9 and 66.9 days, with V amplitudes of 0.09 and 0.07, respectively, rather than the (spurious) ASAS-SN amplitude of 2.5. The longer period is probably a long secondary period.

For a few other stars, the variability appears to be either bimodal or more complex. Figure 5 shows an example in which there may be periods of 469 days (the ASAS-SN period) and about half that value—typical of PRGs which are pulsating in the fundamental and first overtone modes (e.g. Kiss *et al.*

1999). Bimodal pulsators can be useful for determining the physical properties of the stars. Figure 6 shows the light curve of a star which ASAS-SN classifies as irregular (L type) but which clearly shows some periodicity, and is SR; we obtain a best period of 121 days. Figure 7 shows a star with an unusual light curve. There are two maxima which take the form of slow "eruptions." They may, however, be maxima of a faint Mira star, with the 15th-magnitude points being background "noise" limits.

Table 1 lists the results of the initial project. It gives: the ASAS-SN name of the star, minus ASAS-SN-V J; the ASAS-SN classification; the period PA in days given by ASAS-SN; the pulsation period PP in days obtained by us; the V amplitude ΔV; the mean V magnitude <V>; the (J–K) color; and the following notes: x2, x3, x4: the ASAS-SN period is exactly 2, 3, or 4 times the correct pulsation period; lsp: the ASAS-SN period is a long secondary period, and a shorter pulsation period can be seen in the light curve; tpp: the star shows evidence of two pulsation periods, differing by a factor of approximately two; dd: the analysis is affected by contamination by discordant data (see above); spp: the pulsational phase curve is more sawtooth than sinusoidal; OK: the ASAS-SN analysis is correct; *: see the figures, or "Notes on individual stars," below. This Table and Figures 1–7 show the remarkable diversity of results and behavior which occur in a sample of less than 30 stars.

In the 20 long-period Mira stars in the subsidiary project, the ASAS-SN period was in almost every case exactly 2, 3, 4, or 5 times the actual period. Table 2 lists the 20 Mira stars with mean V magnitudes between 12 and 14, and with the longest periods. The magnitude range was chosen because it is optimal for ASAS-SN data. They are listed in order of decreasing ASAS-SN period. The columns list: the name of the star, minus ASAS-SN-V J; the period PA in days given by ASAS-SN; the mean V magnitude <V>; the ASAS-SN amplitude ΔVA; and the following notes: x2, x3, x4, x5: the ASAS-SN period is exactly 2, 3, 4, or 5 times the correct pulsation period; dd: the analysis seems to have been complicated by discordant data (see above); OK: the ASAS-SN analysis is correct; *: see the figures, or "Notes on individual stars," below.

3.1. Notes on individual stars in Table 1

Figures 1–7 and their captions provide both light/phase curves and notes about seven illustrative stars in the sample. Sections 3.1 and 3.2 include stars which are not specifically mentioned in the previous section.

ASAS-SN-V J054606.99-694202.8 The light curve is unusual; it is non-sinusoidal, and there are two maxima in the 544-day cycle. It is not clear whether the behavior is periodic.

ASAS-SN-V J053035.52-685923.2 The light curve shows a slow decline, with some cyclic variations superimposed; their time scale is about 200 days. The slow decline could be part of a long secondary period.

ASAS-SN-V J054110.62-693804.1 The star has a double-humped maximum.

ASAS-SN-V J045337.64-691811.2 Unlike the other stars in the sample, this star had a very small amplitude, but it was possible to show that the actual period is exactly 1/5 of the ASAS-SN period.

Table 1. Analysis of ASAS-SN observations of 29 pulsating red giants.

Name—ASAS-SN-V	Type	PA(d)	PP(d)	ΔV	<V>	J–K	Notes (see text)
J053227.48-691652.8	SR	469	227	1.2	12.7	1.112	x2?, Figure 5
J055444.75-694714.7	SR	544	264	1.5	12.86	0.951	x2, lsp
J061214.08-694558.6	SR	643	318	1.5	16.82	1.725	x2
J054102.00-704309.9	SR	702	702	1.2	15.78	1.357	OK, spp
J191920.70-195042.1	SR	82	81	0.8	13.43	1.201	OK, tpp?
J054747.21-602210.3	SR	418	139	2.5	13.27	0.944	x3, Figure 2
J191639.10-215848.8	SR	25	38	0.3	11.61	1.245	
J205350.26-593921.1	SR	168	168	1.2	11.81	1.209	tpp
J054606.99-694202.8	SR	544	538	1.5	15.92	1.381	OK, spp, *
J191715.66-200034.1	SR	59	59	0.8	12.75	1.191	OK
J051623.43-690014.3	SR	466	233	0.7	14.95	1.287	x2
J053035.52-685923.2	SR	643	295	0.9	13.13	1.194	lsp?, tpp, *
J052011.96-694029.4	SR	662	662	1.1	15.1	1.494	OK, spp, Figure 7
J052337.99-694445.8	SR	636	400	1.2	16.38	1.228	
J054036.77-692620.6	SR	505	458	1.0	13.39	1.172	tpp
J054110.62-693804.1	SR	702	694	1.4	12.65	1.158	OK, *
J045337.64-691811.2	SR	430	80	0.2	13.3	1.115	x5, *
J045412.77-701708.6	SR	437	204	0.5	13.58	1.181	tpp
J050354.98-721652.3	SR	138	138?	1.5	16.21	0.898	OK, Figure 1
J171247.59+265024.8	M	889	—	0.0	13.4	0.785	
J195424.95-114932.2	M/SR	848	424	0.2	13.5	1.269	lsp, Figure 4
J175514.90+184006.9	M	815	204	2.7	13.68	1.186	x4
J182825.60+171943.2	M	728	243	2.35	13.23	1.538	x3
J185653.55-392537.4	M	645	215	2.30	13.76	1.272	x3, *
J020359.53+141132.4	L/SR	irr	389	1.25	12.01	1.148	SR, lsp?
J181616.35-281634.1	L/SR	irr	121	1.38	13.69	1.149	SR, Figure 6
J194755.85-611127.5	L/SR	irr	400	1.14	13.07	1.151	SR
J042630.05+255344.6	SR	1022	30	0.95	13.66	1.759	x2, lsp, Figure 3
J082819.18-143319.3	SR	1020	78/128	0.63	11.66	1.251	tpp

ASAS-SN-V J185653.55-392537.4 The pulsation amplitude is slowly decreasing during the time of observation.

ASAS-SN-V J042630.05+255344.6 One-half the ASAS-SN period is a long secondary period. A shorter pulsation period is also present.

3.2, Notes on individual stars in Table 2

ASAS-SN-V J171247.59+265024.8 There are a few points between JD 2457850 and JD 2457896 which are four magnitudes fainter than the rest, which are almost constant; these are presumably due to instrumental effects, as discussed above, rather than due to an eclipse. For the rest of the points, the highest peak has an amplitude of only 0.017 mag.

ASAS-SN-V J195424.95-114932.2 There are discordant points.

ASAS-SN-V J181958.07-395457.8 There are a few discordant points.

ASAS-SN-V J182346.68-363942.1 There are discordant points, which have caused ASAS-SN to classify this as a large-amplitude Mira star. For the rest, periods of 158 ± 8 days and 83 ± 4 days are present, with small amplitudes (Figure 8). They may possibly be the fundamental and first overtone pulsation periods.

ASAS-SN-V J144304.69-753418.9 The light curve is unusual; there are variations on a time scale of about 100 days, superimposed on irregular long-term variations (Figure 9). The ASAS-SN period of 641.8 days is unlikely.

Table 2. Analysis of ASAS-SN observations of 20 long period Mira stars.

Name—ASAS-SN-V	PA(d)	<V>	ΔVA	Notes (see text)
J171247.59+265024.8	888.8	13.5	3.05	*
J195424.95-114932.2	848.4	13.6	2.5	Figure 4, lsp?
J175514.90+184006.9	814.7	13.7	2.7	x4
J182825.60+171943.2	814.7	13.2	2.35	x3
J065708.96+473521.9	725.8	13.48	2.02	x2, QX Aur
J202918.27+125429.1	721.0	13.31	2.22	x5, XZ Del
J190214.90+471259.7	716.1	13.49	2.71	x2, WZ Lyr
J175727.78+243018.0	695.1	13.31	2.56	x2
J184802.27-293034.0	675.7	13.27	2.36	dd
J184706.22-314645.6	675.3	13.62	4.61	x3, V962 Sgr
J181958.07-395457.8	665.9	12.75	2.78	*
J082915.17+182307.3	655.8	13.66	2.1	x2
J124209.54-435503.3	645.6	13.93	2.79	OK, V1132 Cen
J182037.28-385833.5	645.5	13.8	2.02	x4
J182346.68-363942.1	645.5	13.73	2.39	Figure 8
J185653.55-392537.4	645.0	13.76	2.29	x3, AB CrA, *
J144304.69-753418.9	641.8	12.25	2.5	Figure 9
J141547.57-480350.7	641.0	13.65	3.02	x3
J175730.94-744810.7	640.5	13.45	2.05	x4
J184614.49-301856.4	639.6	13.64	2.02	x3, V1935 Sgr

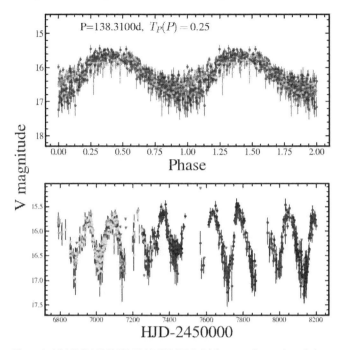

Figure 1. ASAS-SN-V J050354.98-721652.3: Light curve (bottom), and phase curve (top) using the ASAS-SN period of 138.3 days. This period satisfactorily represents the data. In this and the following figures, T is a statistical measure of confidence in the star's period. Source: ASAS-SN website.

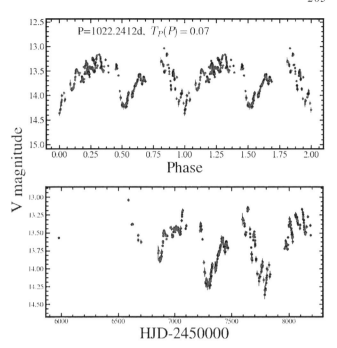

Figure 3. ASAS-SN-V J042630.05+255344.6: Light curve (bottom), and phase curve (top) using the ASAS-SN period of 1022.2 days. There are two (long) cycles in the phase curve, rather than one, and there are also more rapid variations with a period of 55 ± 2 days. This is presumably the pulsation period, and the long secondary period is 511.1 days—exactly half the ASAS-SN period. Source: ASAS-SN website.

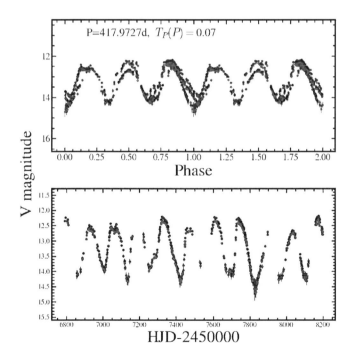

Figure 2. ASAS-SN-V J054747.21-602210.3: Light curve (bottom), and phase curve (top) using the ASAS-SN period of 418.0 days. The actual period is exactly one-third of this; there are three cycles in the phase curve, rather than one. Source: ASAS-SN website.

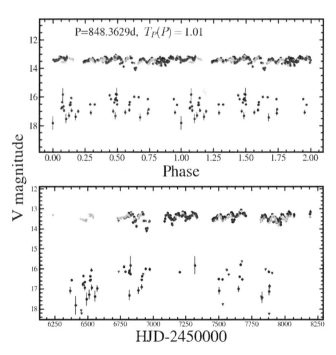

Figure 4. ASAS-SN-V J195424.95-114932.2: Light curve (bottom), and phase curve (top) using the ASAS-SN period of 848.4 days. The ASAS-SN analysis has been complicated by the fainter discordant points, which are spurious. Analysis of the brighter V data gives periods of 423.9 days (V amplitude 0.09) and 66.9 days (V amplitude 0.07). The former period (half the ASAS-SN period) may be a long secondary period, and the latter may be a pulsation period. Source: ASAS-SN website.

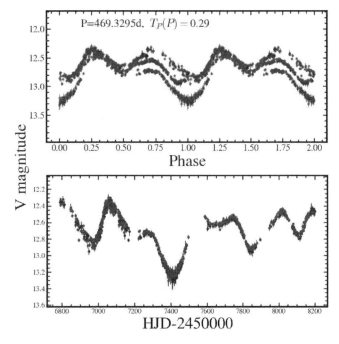

Figure 5. ASAS-SN-V J053227.48-691652.8: Light curve (bottom), and phase curve (top) using the ASAS-SN period of 469.3 days. The star may pulsate in two modes, with the second period being about half of the first period. Source: ASAS-SN website.

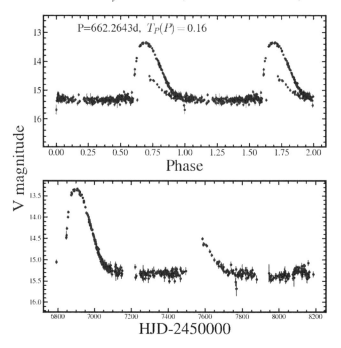

Figure 7. ASAS-SN-V J052011.96-694029.4: Light curve (bottom), and phase curve (top) using the ASAS-SN period of 662.3 days. The light curve shows two "eruptions," 662 days apart. On the other hand, these could be maxima of a faint Mira star, with the 15th-magnitude points being a background noise limit. Source: ASAS-SN website.

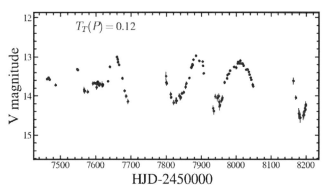

Figure 6. ASAS-SN-V J181616.35-281634.1: This star is considered irregular (type L) in the ASAS-SN catalogue, but the above light curve suggests that it has a period of 121 days, and is therefore an SR star. Source: ASAS-SN website.

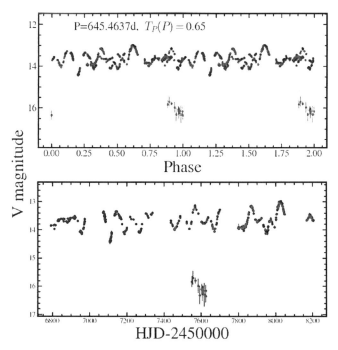

Figure 8. ASAS-SN-V J182346.68-363942.1: Light curve (bottom), and phase curve (top) using the ASAS-SN period of 645.5 days. The ASAS-SN amplitude of 2.39 occurs because of the presence of the fainter discordant data. Our analysis of the rest of the data gives periods of 153 ± 8 and 83 ± 4 days, both with amplitudes of 0.23. This may be a bimodal pulsator. Source: ASAS-SN website.

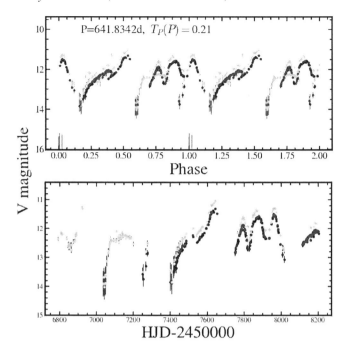

Figure 9. ASAS-SN-V J144304.69-753418.9: Light curve (bottom), and phase curve (top) using the ASAS-SN period of 641.8 days. The light curve is highly unusual and interesting. There are variations on time scales from 100 to 300 days. Source: ASAS-SN website.

4. Discussion

The ASAS-SN data begin about JD 2456500 so, as of the time of carrying out this project (October 2018 to February 2019), there were only about 2000 days of data. This is adequate for studying many aspects of PRG variability, but not the very long-term variations in period, amplitude, and mean magnitude which are known to occur in these stars. Only the visual data can presently do that.

It is interesting to note that, when Vogt *et al.* (2016) compared their results with those of Richards *et al.*'s (2012) results which were obtained using a machine-learning approach, the discrepancy was most often by a ratio of small whole numbers, such as 2 or 1/2. We find a similar result.

Pulsating red giants are understandably a challenge for automated analysis and classification. ASAS-SN carries out a comprehensive search for the best period for each star, and then uses this and other parameters of the light/phase curve to arrive at a final analysis and classification. As noted above, however, SR variables are not strictly periodic; they have "wandering" pulsation periods, variable pulsation amplitudes, additional periods (including LSPs), and residual irregularity. It is difficult to define a single period for these stars, and the phase curve will be constantly variable with time.

Jayasinghe and his colleagues (2019b) have made unspecified refinements to the analysis and classification procedure, and have provided a list of updated periods for the stars in Table 1. About two-thirds now agree with our values. However, 12 of the 20 long-period Mira stars in Table 2 still have incorrect or incomplete analyses and/or classifications.

Most of this project was carried out by undergraduate math major LF. It illustrates the great educational potential of the ASAS-SN data, with its immense quantity and variety. We can envision a large number and variety of projects which could be carried out by students at the college level, and perhaps even at the high school level, using ASAS-SN data. The AAVSO VSTAR time-series analysis package is well-suited for use with these and other data.

5. Conclusions

We have analyzed ASAS-SN observations of pulsating red giants (mostly semi-regular and Mira stars) and compared our results with the periods, amplitudes, and classifications given by ASAS-SN. For many stars, the actual periods are a small integral fraction of the ASAS-SN period ("sub-harmonics"), because the ASAS-SN phase curve incorrectly contains two or more cycles of variability, rather than one. In other cases, the ASAS-SN period is a long secondary period; the shorter pulsation period is visible in the light curve. For a few stars, the ASAS-SN analysis is complicated by the presence of faint data which are spurious and due to instrumental problems. In a few others, the star is bimodal or otherwise complex. The few irregular (type L) stars that we analyzed were probably semi-regular (type SR).

Given the complexity of pulsating red giants as noted above, it is not surprising that the ASAS-SN automatic analysis procedure produced incorrect or incomplete results. Perhaps the procedure can be trained to "solve" these very complex stars! Indeed, the ASAS-SN variable star data and website have been significantly updated and improved in the weeks since we completed this project in February 2019, and some (but not all) of the problems with the PRG analysis and classification have been alleviated.

The ASAS-SN data on PRGs can be exceptionally useful for analyzing these stars, and are invaluable for both scientific and educational purposes. The data for individual PRGs in the ASAS-SN catalogue should, however, be confirmed by careful inspection of the light and phase curves, and by more detailed analysis if necessary, to avoid the types of problems shown in Figures 2–9.

6. Acknowledgements

This paper made use of ASAS-SN photometric data. We thank the ASAS-SN project team for their remarkable contribution to stellar astronomy, and for making the data freely available on-line. Thanks also to Chris Kochanek and especially Tharindu Jayasinghe for helpful comments. We acknowledge and thank the University of Toronto Work-Study Program for financial support. The Dunlap Institute is funded through an endowment established by the David Dunlap Family and the University of Toronto.

References

Benn, D. 2013, VSTAR data analysis software (http://www.aavso.org/vstar-overview).

Eddington, A. S., and Plakidis, S. 1929, *Mon. Not. Roy. Astron. Soc.*, **90**, 65.

Jayasinghe, T., *et al.* 2018, *Mon. Not. Roy. Astron. Soc.*, **477**, 3145.

Jayasinghe, T., *et al.* 2019a, *Mon. Not. Roy. Astron. Soc.*, **486**, 1907.

Jayasinghe, T., *et al.* 2019b, private communication (April 10)

Kafka, S. 2019, variable star observations from the AAVSO International Database (https://www.aavso.org/aavso-international-database).

Kiss, L. L., Szatmary, K., Cadmus, R. R., Jr., and Mattei, J. A. 1999, *Astron. Astrophys.*, **346**, 542.

Kochanek, C., and Jayasinghe, T. 2019, private communication (April 2).

Percy, J. R., and Abachi, R. 2013, *J. Amer. Assoc. Var. Star Obs.*, **41**, 93.

Percy, J. R., and Qiu, A. L. 2019, *J. Amer. Assoc. Var. Star Obs.*, **47**, 76.

Richards, J. W., Starr, D. L., Miller, A. A., Bloom, J. S., Butler, N. R., Brink, H., and Crellin-Quick, A. 2012, *Astrophys. J., Suppl. Ser.*, **203**, 32.

Samus, N. N., *et al.* 2017, *General Catalogue of Variable Stars*, Sternberg Astronomical Institute, Moscow (GCVS; http://www.sai.msu.ru/gcvs/gcvs/index.htm).

Shappee, B. J., *et al.* 2014, *Astrophys. J.*, **788**, 48.

Templeton, M. R., Mattei, J. A., and Willson, L. A. 2005, *Astron. J.*, **130**, 776.

Vogt, N., *et al.* 2016, *Astrophys. J., Suppl. Ser.*, **227**, 6.

Watson, C., Henden, A. A., and Price, C. A. 2014, AAVSO International Variable Star Index VSX (Watson+, 2006–2014; http://www.aavso.org/vsx).

Wood, P. R. 2000, *Publ. Astron. Soc. Australia*, **17**, 18.

A Photometric Study of Five Low Mass Contact Binaries

Edward J. Michaels
Stephen F. Austin State University, Department of Physics, Engineering and Astronomy, P.O. Box 13044, Nacogdoches, TX 75962; emichaels@sfasu.edu

Received August 16, 2019; revised September 1, 2019; accepted September 1, 2019

Abstract Presented are precision multicolor photometry and the results of light curve modeling for five recently discovered contact binary systems: V338 Dra, NSVS 6133550, V1377 Tau, NSVS 3917713, and V2802 Ori. These systems all have orbital periods of less than 0.3 d and stellar effective temperatures less than solar. The photometric solutions derived from the Wilson-Devinney program resulted in contact configurations for each system. The solution mass ratios should be well determined since the light curves of each system exhibited total eclipses. A reliable mass ratio allowed absolute parameters of the component stars to be estimated. Light curve asymmetries were apparent in all five stars. These were attributed to magnetic activity and were modeled as cool and hot spots on the stars.

1. Introduction

1.1. Background

W UMa contact binaries are surprisingly common in the solar neighborhood. Recent surveys have discovered many new candidates in this W-subtype group (ASAS, Pojmański 1997; NSVS, Woźniak et al. 2004; CRTS, Drake et al. 2014). Most of these candidates have not received follow-up photometric studies. A number of these eclipsing systems were placed on an observing program at the Waffelow Creek Observatory to obtain precision multi-band CCD photometry. Several studies have already been completed: BN Ari, PY Boo, V958 Mon, V2790 Ori, V737 Cep, and V384 Ser (Michaels 2015, 2016a, 2016b, 2016c, 2018; Michaels et al. 2019). These stars are all low mass contact binaries (LMCB) with orbital periods of less than 0.3 d.

Presented in this paper are the results from recent photometric observations for five additional LMCB. A brief history of each system is given in the next subsection, with the photometric observations presented in section 2. New minima times, ephemerides, observed properties of each system, and analysis of the light curves using the Wilson-Devinney model are presented in section 3. Discussion of the results and conclusions are given in section 4.

1.2. Notes on individual stars

V338 Dra Khruslov (2006) identified V338 Dra (GSC 04182-01259, 2MASS J15491114+6038029) as a W-subtype eclipsing binary from The Northern Sky Variability (NSVS) data base. It was also classified as a W-subtype by Hoffman et al. (2009) using NSVS and by Drake et al. (2014) using data from the Catalina Real-time Transient Survey (CRTS). Two times of minima and light elements were published by Hoňková et al. (2013) who gives an orbital period of P = 0.235149 d. A third minima time is given by Hübscher and Lehmann (2015). Using all-sky spectrally matched Tycho-2 stars, Pickles and Depagne (2010) found a spectral type of K4V.

NSVS 6133550 Located in the constellation Lacerta, NSVS 6133550 (GSC 03223-01180, TYC 3223-1180-1) was first identified as a variable star in the NSVS survey (Woźniak et al. 2004). It was classified as a W-type eclipsing binary in Gettel et al.'s (2006) catalogue of 1,022 bright contact binaries which gives an orbital period of 0.27511 d. This star is also listed as a contact binary in a number of additional catalogues (Hoffman et al. 2009; Terrell et al. 2012; Drake et al. 2014). McDonald et al. (2017) give an effective temperature of T_{eff} = 5467 K and Ammons et al. (2006) give a similar value of T_{eff} = 5474 K. Using all-sky spectrally matched Tycho-2 stars, Pickles and Depagne (2010) determined a spectral type of K0V. Only two light minima were found in the literature for this star (Hübscher and Lehmann 2013).

V1377 Tau The variability of V1377 Tau (GSC 00067-00348, ASAS J033959+0314.5) was identified in the All-Sky Automated Survey (ASAS) and was classified as a contact eclipsing system (EC) with an orbital period of P = 0.282709 d. Automated variable star classification techniques using NSVS and ASAS observations also classified it as a W UMa close contact binary (Hoffman et al. 2009; Richards et al. 2012). Data from the ROSAT X-ray satellite indicates this star is an X-ray source (Appenzeller et al. 1998; Szczygiel et al. 2008). In a study of ROSAT late-type stars, the spectral type and distance to V1377 Tau were determined. Using medium and high-resolution spectroscopy gives a spectral type of K2, a distance of 194 pc, and a rotational velocity of 63 km/s for the primary star (Zickgraf et al. 2005). A total of eight light minima were found in the literature (Diethelm 2010, 2012; Nagai 2012).

NSVS 3917713 NSVS 3917713 (GSC 03280-00990, 2MASS J01541484+4618008) is located in the constellation of Andromeda. It was discovered in the NSVS survey data and classified as an EW-type binary star with an orbital period of P = 0.289222 d and has a maximum V magnitude of 13.087 (Gettel et al. 2006). It was also identified as a contact binary by Hoffman et al. (2009). Observations of this star were acquired by the All-Sky Automated Survey for Supernovae (ASAS-SN; Kochanek et al. 2017) and by SuperWASP (Butters et al. 2010). Light curves from these data are also indicative of a contact binary. A literature search did not locate any minima times for this star.

V2802 Ori V2802 Ori (GSC 00103-00894, TYC 103-894-1) was first identified as a variable star in the ASAS survey (Pojmański 1997). It is listed in the fifth part of their *Catalog of Variable Stars* which gives a period of 0.295706 d, a visual magnitude of V = 11.25, and a ΔV = 0.70 (Pojmański *et al.* 2002). An NSVS automated classification technique identified this star as a W UMa contact binary (Hoffman *et al.* 2009). Pilecki and Stępień (2012) performed light curve modeling on this star using ASAS data.

2. Photometric observations

Multi-band photometric observations for all stars were acquired using a 0.36-m Ritchey-Chrétien robotic telescope at the Waffelow Creek Observatory (http://obs.ejmj.net/index.php). All images were acquired using a SBIG-STXL camera equipped with a cooled KAF-6303E CCD (–30° C). The telescope and camera have an image scale of 0.65 arcsec / pixel and a field of view 33.7' × 22.5'. Each star was imaged in three passbands, Sloan g', r', and i'. In addition, observations in the Johnson B and V passbands were obtained for NSVS 6133550 and Johnson V for V2802 Ori. The observation season, number of images acquired, and the number of nights needed to complete the multi-color light curves are given in Table 1. Bias, dark, and flat frames were obtained on each night for image calibration. MIRA software (Mirametrics 2015) was used for image calibration and to perform the ensemble differential aperture photometry of the light images. The finder charts, Figures 1–5, show the locations of the comparison and check stars, which should be useful for future observers. Table 2 lists the comparison and check stars and their standard magnitudes taken from the AAVSO Photometric All-Sky Survey data base (APASS; Henden *et al.* 2015). The instrumental magnitudes for each star were converted to standard magnitudes using these calibrated comparison star magnitudes. The Heliocentric Julian Date of each observation was converted to orbital phase (φ) using the new epochs and orbital periods given in Table 5. The folded light curves for each star are shown in Figures 6–10. All light curves in this paper were plotted from orbital phase –0.6 to 0.6 with negative phase defined as φ – 1. Error bars were omitted from the plotted points for clarity. The Sloan r' check-star magnitudes are plotted on the bottom panel of Figures 6–10. The check-star magnitudes were plotted and inspected each night, but no significant variability was found. The standard deviations for all check star observations are in Table 2. The standard error of a single observation ranged from 5 to 10 mmag. The light curve properties for each star are given in Table 3 (Min I, Min II, Max I, Max II, Δm, and total eclipse duration). The observations for each star can be accessed from the AAVSO International Database (Kafka 2017).

3. Analysis

3.1. Ephemerides

New times of minima for each star were determined using the Kwee and van Woerden (1956) method. These minima and other minima found in the literature are compiled in Table 4. For each star, the initial epoch was taken from the first primary minimum listed in Table 4 and the orbital period from The International Variable Star Index (VSX, Watson *et al.* 2006–2014). These light elements are shown in Table 5 and were used to calculate the (O–C) residuals in Table 4. Table 5 also gives new linear elements computed by least-squares solution using these residuals.

3.2. Color, temperature, spectral type, absolute magnitude, luminosity

For all systems in this paper the primary star is the hotter and lower mass component of each binary. The spectral type of V1377 Tau was determined from medium and high-resolution spectroscopy (Zickgraf *et al.* 2005). Its K2 spectral type gives an effective temperature of $T_{eff} = 5080 \pm 200$ K (Pecaut and Mamajek 2013). The effective temperatures of the other stars were determined from the average of their observed colors. To find the observed color, the phase and magnitude of the g' and r' observations were binned with a phase width of 0.01. The phases and magnitudes in each bin interval were averaged. The binned r' magnitudes were subtracted from the linearly interpolated g' magnitudes, resulting in an observed (g'–r') color at each phase point. Figure 11 shows the binned Sloan r' light curve of V338 Dra with the (g'–r') color plot in the bottom panel. All five systems show only small color changes over their entire orbital phase range due to the small temperature differences between the component stars. The observed colors were corrected for color excess using three-dimensional maps of interstellar dust reddening based on Pan-STARRS 1 and 2MASS photometry and Gaia parallaxes (Green *et al.* 2018; Gaia 2016, 2018). The color excesses for these stars were very small, which is likely due to their proximity to Earth and their locations above or below the galactic plane. Using the corrected colors, the effective temperature of each star was interpolated from tables of Covey *et al.* (2007) and Pecaut and Mamajek (2013). The resulting average observed colors, color excesses, effective temperatures and spectral types are shown in Table 6. The absolute visual magnitude (M_V) of each star was computed using the following equation:

$$M_V = V - A_V - 5 \log(d/10), \quad (1)$$

where V is the apparent magnitude at quadrature (φ = 0.75), A_V is the extinction (E(B–V) × 3.086), and d the distance in parsecs. For the stars without B and V observations, the (B–V) color was determined from the transformation equation of Bilir *et al.* (2005):

$$B - V = \frac{(g'-r') + 0.25187}{1.12431}, \quad (2)$$

and the V magnitude from the transformation equation of Jester *et al.* (2005):

$$V = g' - 0.59(g'-r') - 0.01. \quad (3)$$

The distances in parsecs were determined from Gaia DR2 parallaxes (Bailer-Jones *et al.* 2018; Gaia 2016, 2018). The luminosity of each star in solar units can now be determined using the following equation:

$$M_V + BC_V = 4.74 - 2.5 \log(L/L_\odot), \quad (4)$$

where BC_V is the star's bolometric correction as interpolated

Table 1. Observation log.

System	Dates	No. Nights	Images Acquired				
			B	V	Sloan g'	Sloan r'	Sloan i'
V338 Dra	2017 May/Jun	8	—	—	690	626	669
NSVS 6133550	2017 Sep	9	764	1355	700	691	684
V1377 Tau	2017 Nov	4	—	—	388	444	449
NSVS 3917713	2017 Oct	8	—	—	734	1256	1336
V2802 Ori	2017 Dec	4	—	625	553	716	570

Table 2. APASS comparison and check star magnitudes.

System	R.A. (2000) h	Dec (2000) °	B	V	g'	r'	i'
V338 Dra	15.81976	+60.63414					
GSC 04182-01005 (C1)	15.84121	+60.78514	—	—	12.633	12.279	12.191
GSC 04182-00409 (C2)	15.82763	+60.79534	—	—	13.127	12.657	12.499
GSC 04182-00968 (C3)	15.79748	+60.79923	—	—	13.461	12.716	12.430
GSC 04182-01121 (C4)	15.80262	+60.57336	—	—	14.011	13.501	13.293
GSC 04182-00793 (C5)	15.82888	+60.72875	—	—	14.197	13.760	13.597
GSC 04182-01317 (C6)	15.85001	+60.70778	—	—	14.230	13.638	13.430
GSC 04182-00069 (K)	15.83814	+60.53532	—	—	13.675	13.400	13.315
Standard deviation of observed K-star magnitudes					± 0.011	± 0.009	± 0.011
NSVS 6133550	22.92147	+42.27613					
GSC 03223-01990 (C1)	22.93187	+42.31625	11.954	11.427	11.667	11.314	11.229
GSC 03223-01816 (C2)	22.93552	+42.15188	12.533	12.014	12.224	11.895	11.753
GSC 03223-01366 (C3)	22.92861	+42.24221	13.032	12.411	12.701	12.257	12.115
GSC 03223-01720 (C4)	22.92168	+42.41160	13.232	12.669	12.979	12.513	12.372
GSC 03223-02743 (C5)	22.93296	+42.24837	12.438	11.307	11.858	10.968	10.621
GSC 03223-01801 (C6)	22.90797	+42.30702	13.069	12.526	12.742	12.436	12.338
GSC 03223-03029 (C7)	22.93831	+42.21739	13.045	12.285	12.644	12.086	11.913
GSC 03223-03377 (K)	22.93798	+42.14120	13.251	11.859	12.517	11.430	10.945
Standard deviation of observed K-star magnitudes			± 0.016	± 0.008	± 0.010	± 0.007	± 0.008
V1377 Tau	3.66640	+3.24181					
GSC 00068-00333 (C1)	3.67456	+3.21767	—	—	13.189	12.789	12.666
GSC 00068-00903 (C2)	3.67394	+3.38703	—	—	12.842	12.413	12.284
GSC 00068-00029 (C3)	3.67622	+3.13630	—	—	13.248	12.757	12.570
TYC 67-331-1 (K)	3.65294	+3.10299	—	—	12.008	11.517	11.507
Standard deviation of observed K-star magnitudes					± 0.005	± 0.005	± 0.005
NSVS 3917713	1.90413	+46.30014					
GSC 03280-00679 (C1)	1.89989	+46.29019	—	—	13.633	12.943	12.621
GSC 03280-00905 (C2)	1.91095	+46.25715	—	—	13.113	12.574	12.370
GSC 03280-00996 (C3)	1.90056	+46.22029	—	—	13.413	12.694	12.421
GSC 03280-01188 (C4)	1.89365	+46.33531	—	—	13.161	12.560	12.340
GSC 03280-00594 (C5)	1.90456	+46.38871	—	—	13.912	12.971	12.548
GSC 03280-00799 (C6)	1.89889	+46.20799	—	—	13.114	12.294	11.974
GSC 03280-01340 (C7)	1.91112	+46.40029	—	—	13.799	13.241	13.031
GSC 03280-00987 (K)	1.89692	+46.21784	—	—	13.164	12.755	12.611
Standard deviation of observed K-star magnitudes					± 0.006	± 0.007	± 0.007
V2802 Ori	5.14771	+2.82089					
GSC 00103-00935 (C1)	5.14937	+2.84725	—	12.039	12.252	11.911	11.840
GSC 00103-01035 (C2)	5.15145	+2.85907	—	11.976	12.473	11.590	11.278
GSC 00103-00635 (C3)	5.15774	+2.76971	—	12.044	12.230	11.908	11.871
GSC 00103-01151 (C4)	5.14075	+2.70028	—	12.078	12.310	11.926	11.812
GSC 00103-00761 (K)	5.15299	+2.89693	—	11.231	11.756	10.827	10.493
Standard deviation of observed K-star magnitudes			± 0.005	± 0.006	± 0.006	± 0.006	

Table 3. Average light curve properties.

System		Min I Mag.	Min II Mag.	Max I Mag.	Max II Mag.	Δ Mag. MagMax II – Min I	Total Eclipse Duration (minutes)
V338 Dra	g'	14.71 ± 0.03	14.71 ± 0.02	14.00 ± 0.02	13.92 ± 0.02	0.79 ± 0.03	≈ 10
	r'	13.75 ± 0.01	13.74 ± 0.01	13.09 ± 0.03	13.04 ± 0.03	0.71 ± 0.04	—
	i'	13.33 ± 0.02	13.33 ± 0.02	12.70 ± 0.05	12.67 ± 0.05	0.66 ± 0.06	—
NSVS 6133550	B	12.51 ± 0.01	12.48 ± 0.01	12.05 ± 0.01	12.03 ± 0.01	0.49 ± 0.02	≈20
	V	11.74 ± 0.01	11.72 ± 0.01	11.32 ± 0.01	11.28 ± 0.01	0.46 ± 0.01	—
	g'	12.11 ± 0.01	12.08 ± 0.01	11.67 ± 0.02	11.64 ± 0.02	0.48 ± 0.02	—
	r'	11.53 ± 0.01	11.50 ± 0.01	11.11 ± 0.01	11.08 ± 0.01	0.45 ± 0.01	—
	i'	11.31 ± 0.01	11.28 ± 0.01	10.91 ± 0.02	10.88 ± 0.02	0.43 ± 0.02	—
V1377 Tau	g'	12.55 ± 0.01	12.45 ± 0.01	11.70 ± 0.02	11.66 ± 0.02	0.89 ± 0.02	≈12
	r'	11.79 ± 0.01	11.71 ± 0.01	10.97 ± 0.01	10.97 ± 0.01	0.83 ± 0.02	—
	i'	11.53 ± 0.01	11.46 ± 0.01	10.75 ± 0.01	10.72 ± 0.01	0.81 ± 0.01	—
NSVS 3917713	g'	14.02 ± 0.02	14.06 ± 0.01	13.41 ± 0.01	13.41 ±0.01	0.61 ± 0.02	≈21
	r'	13.35 ± 0.02	13.38 ± 0.01	12.77 ± 0.02	12.77 ± 0.02	0.58 ± 0.02	—
	i'	13.07 ± 0.02	13.08 ± 0.01	12.51 ± 0.02	12.50 ± 0.02	0.57 ± 0.03	—
V2802 Ori	V	11.80 ± 0.01	11.75 ± 0.01	11.19 ± 0.02	11.17 ± 0.02	0.63 ± 0.02	≈25
	g'	12.15 ± 0.01	12.10 ± 0.01	11.52 ± 0.02	11.50 ± 0.02	0.65 ± 0.02	—
	r'	11.53 ± 0.01	11.47 ± 0.01	10.93 ± 0.02	10.91 ± 0.02	0.61 ± 0.02	—
	i'	11.32 ± 0.01	11.27 ± 0.01	10.74 ± 0.02	10.73 ± 0.02	0.59 ± 0.02	—

from the tables of Pecaut and Mamajek (2013). Each systems absolute and apparent visual magnitudes at quadrature, stellar distance, and observed luminosity are shown in Table 6.

3.3. Light curve modeling

For light curve modeling, the standard magnitudes were binned in both phase and magnitude as detailed in section 3.2. The resulting binned magnitudes were converted to relative flux units. The first step employed in modeling each star was to attain preliminary fits to the observed light curves using the program BINARYMAKER3.0 (BM3; Bradstreet and Steelman 2002). Standard convective parameters were employed in the model, with the limb darkening coefficients taken from van Hamme's (1993) tabular values. The asymmetries in the light curves were not modeled initially with BM3. Once a good fit was obtained between the synthetic and observed light curves, the resulting parameters for each passband were averaged. These values were used as the input parameters for the Wilson-Devinney (WD) program where computations were done simultaneously in all passbands (Wilson and Devinney 1971; van Hamme and Wilson 1998). The light curve morphology of each star indicates a contact configuration, with the stars having a common convective envelope. The WD program was configured for overcontact binaries (Mode 3) and the Kurucz stellar atmosphere model was applied (Kurucz 2002). Each binned input data point was assigned a weight equal to the number of observations forming that point. The fixed inputs included standard convective parameters: gravity darkening, $g_1 = g_2 = 0.32$ (Lucy 1968) and albedo value $A_1 = A_2 = 0.5$ (Ruciński 1969). The temperature of the primary stars, T_1, were fixed at values given in Table 6 (see section 3.2). Logarithmic limb darkening coefficients were calculated by the program from tabulated values using the method of van Hamme (1993). The solution's adjustable parameters include the inclination (i), mass ratio ($q = M_2 / M_1$), potential ($\Omega_1 = \Omega_2$), temperature of the secondary star (T_2), the normalized flux for each wavelength (L), third light (l), and phase shift.

Low mass, rapidly revolving contact binaries are often magnetically active and thus spotted. Except for NSVS 3917713, the light curves for each star show an O'Connell effect with Max II ($\varphi = 0.75$) brighter than Max I ($\varphi = 0.25$), which is indicative of spotting (O'Connell 1951). In addition, the primary total eclipses of NSVS 6133550 and V2802 Ori are not flat, but having a small but obvious slope, which is also likely caused by star spots. Once the initial WD solutions were obtained, the resulting parameter values were transferred into BM3 to model the asymmetries in the light curves caused by star spots. Four of the stars required one or two cool spots on the larger secondary star to minimize the asymmetries. A single hot spot on the primary star of NSVS 3917713 provided the best fit. Two systems, NSVS 6133550 and NSVS 3917713, required a small third light (l) to fit the minima. The resulting best-fit BM3 parameters were then incorporated into new WD solutions that included adjustable spot parameters (co-latitude, longitude, radius, and temperature factor). The final spotted WD solution parameters for each star are shown in Table 7. The filling factor in Table 7 was computed using the method of Lucy and Wilson (1979) given by:

$$f = \frac{\Omega_{inner} - \Omega}{\Omega_{inner} - \Omega_{outer}}, \quad (5)$$

where Ω_{inner} and Ω_{outer} are the inner and outer critical equipotential surfaces and Ω is the equipotential that describes the stellar surface. Figures 12–16 show the normalized light curves overlaid by the synthetic solution curves (solid line), with the residuals shown in the bottom panel. A BM3 graphical representation of each system solution is shown in Figure 17 (Bradstreet and Steelman 2002).

Table 4. Times of minima and O-C residuals.

Epoch HJD 2400000+	Error	Cycle	O-C	References	Epoch HJD 2400000+	Error	Cycle	O-C	References
V338 Dra					55573.93480	—	1349.0	0.00136	Nagai 2012
56421.36484	0.00007	0.0	0.00000	Hoňková 2013	55574.92300	—	1352.5	0.00008	Nagai 2012
56421.48221	0.00005	0.5	–0.00020	Hoňková 2013	55862.86560	0.00040	2371.0	0.00356	Diethelm 2012
56771.50050	0.00630	1489.0	–0.00120	Hübscher and Lehman 2015	56282.68990	0.00030	3856.0	0.00500	Diethelm 2013
57888.69135	0.00007	6240.0	–0.00325	present paper	58078.74551	0.00005	10209.0	0.01033	present paper
57888.80810	0.00006	6240.5	–0.00408	present paper	58078.88644	0.00002	10209.5	0.00990	present paper
57898.68440	0.00007	6282.5	–0.00404	present paper	58080.72452	0.00009	10216.0	0.01038	present paper
57898.80270	0.00005	6283.0	–0.00331	present paper	58080.86551	0.00003	10216.5	0.01001	present paper
57899.74390	0.00007	6287.0	–0.00271	present paper	58081.71352	0.00003	10219.5	0.00989	present paper
57911.73590	0.00005	6338.0	–0.00331	present paper	58081.85522	0.00003	10220.0	0.01024	present paper
57911.85280	0.00005	6338.5	–0.00398	present paper	58084.68258	0.00003	10230.0	0.01051	present paper
57912.67650	0.00006	6342.0	–0.00330	present paper	58084.82341	0.00003	10230.5	0.00999	present paper
57912.79340	0.00006	6342.5	–0.00398	present paper					
57913.73390	0.00007	6346.5	–0.00407	present paper	NSVS 3917713				
57913.85241	0.00008	6347.0	–0.00314	present paper	56897.98995	—	–3941.0	–0.00577	Jayasinghe 2018
57915.73370	0.00006	6355.0	–0.00304	present paper	58037.76839	0.00008	0.0	0.00000	present paper
57915.85030	0.00007	6355.5	–0.00401	present paper	58037.91343	0.00008	0.5	0.00044	present paper
					58038.63572	0.00008	3.0	–0.00030	present paper
NSVS 6133550					58038.78066	0.00008	3.5	0.00004	present paper
56159.42270	0.00100	–0.5	0.00336	Hübscher and Lehman 2013	58038.92498	0.00007	4.0	–0.00025	present paper
56159.55690	0.00080	0.0	0.00000	Hübscher and Lehman 2013	58039.64826	0.00008	6.5	0.00001	present paper
58010.66296	0.00027	6728.5	0.00824	present paper	58039.79283	0.00007	7.0	–0.00002	present paper
58011.62795	0.00031	6732.0	0.01034	present paper	58039.93739	0.00007	7.5	–0.00007	present paper
58011.76341	0.00029	6732.5	0.00824	present paper	58040.66044	0.00007	10.0	–0.00004	present paper
58012.72852	0.00029	6736.0	0.01045	present paper	58040.80540	0.00008	10.5	0.00032	present paper
58012.86400	0.00027	6736.5	0.00838	present paper	58043.69731	0.00009	20.5	0.00014	present paper
58013.68921	0.00059	6739.5	0.00825	present paper	58049.62573	0.00009	41.0	–0.00023	present paper
58013.82914	0.00028	6740.0	0.01062	present paper	58049.77078	0.00009	41.5	0.00022	present paper
58015.61510	0.00010	6746.5	0.00835	present paper	58049.91507	0.00009	42.0	–0.00010	present paper
58015.75541	0.00011	6747.0	0.01110	present paper	58050.63822	0.00007	44.5	0.00003	present paper
58017.68097	0.00010	6754.0	0.01087	present paper	58050.78261	0.00007	45.0	–0.00018	present paper
58019.60753	0.00012	6761.0	0.01164	present paper	58050.92732	0.00007	45.5	–0.00008	present paper
58019.74275	0.00012	6761.5	0.00930	present paper					
58019.88242	0.00007	6762.0	0.01141	present paper	V2802 Ori				
58020.70756	0.00010	6765.0	0.01122	present paper	55526.87930	0.00020	–8690.5	0.01688	Diethelm 2011
58020.84325	0.00007	6765.5	0.00935	present paper	55882.90720	0.00040	–7486.5	0.01476	Diethelm 2012
58021.66867	0.00013	6768.5	0.00943	present paper	56246.91670	0.00040	–6255.5	0.01017	Diethelm 2013
58021.80829	0.00008	6769.0	0.01149	present paper	58096.69541	0.00006	0.0	0.00000	present paper
					58096.84306	0.00006	0.5	–0.00020	present paper
V1377 Tau					58097.87840	0.00006	4.0	0.00017	present paper
55192.55900	0.00030	0.0	0.00000	Diethelm 2010	58097.73031	0.00007	3.5	–0.00007	present paper
55544.67420	0.00050	1245.5	0.00114	Diethelm 2010	58098.76518	0.00007	7.0	–0.00017	present paper
55568.98700	—	1331.5	0.00097	Nagai 2012	58100.83534	0.00006	14.0	0.00005	present paper
55571.95560	—	1342.0	0.00112	Nagai 2012	58100.68708	0.00007	13.5	–0.00036	present paper

4. Discussion and conclusions

Provisional absolute stellar parameters were calculated for each system using the mass ratio and an estimate of the secondary star mass (M_2). The light curves of each system displayed a total primary eclipse, which provided the necessary constrains for an accurate determination of the mass ratios (q) (Wilson 1978; Terrell and Wilson 2005; Hambálek and Pribulla 2013). The secondary star masses were calculated using Gazeas and Stępień (2008) period-mass relation for contact binaries:

$$\log M_2 = (0.755 \pm 0.059) \log P + (0.416 \pm 0.024), \quad (6)$$

where P is the orbital period. The primary star's mass (M_1) was calculated using the solution mass ratio. The separation between the mass centers was then calculated using Kepler's Third Law. Using this semi-major axis value, the WD light curve program (LC) computed the radius and bolometric magnitude of each star. The luminosity of each stellar component was then calculated from the bolometric magnitudes. The values for the absolute stellar parameters in solar units are shown in Table 8. The computed solution luminosities are in good agreement with the observed luminosities to within the margin of errors (see Table 6). A comparison was made between the provisional absolute parameters of these systems and the mass and radius distribution of 112 contact binaries in a study by Gazeas and Stępień (2008, see their Figures 1–3). The geometrical and physical properties for those 112 stars were well determined from both photometric and radial velocity measurements. Two of their plots were reproduced in Figures 18 and 19. Figure 18 shows that the stellar radii derived from the photometric solutions are in good agreement with the period-radius relation

Table 5. Linear elements with errors.

System	Initial Elements		New Elements from Least-squares	
	Epoch (HJD)	P_{orb} (days)	Epoch (HJD)	P_{orb} (days)
V338 Dra	2456421.3648	0.235149	2457915.7332 (3)	0.23514847 (5)
NSVS 6133550	2456159.5569	0.275113	2458021.807 (1)	0.2751142 (2)
V1377 Tau	2455192.5590	0.282709	2458084.6823 (3)	0.28271000 (3)
NSVS 3917713	2458037.7684	0.289209	2458050.7828 (1)	0.28921045 (5)
V2802 Ori	2458096.6954	0.295706	2458100.8351 (3)	0.29570410 (6)

Table 6. Observed color, color excess, effective temperature, spectral type, apparent and absolute visual magnitudes at quadrature, distance and luminosity.

System	(g'–r')	E(g'–r')	T_{eff} (K)	SP	m_V	M_V	Dist. (pc)	$L_\odot$
V338 Dra	0.92 ± 0.04	0.01 ± 0.02	4749 ± 63	K3	13.39 ± 0.04	6.17 ± 0.09	274 ± 1	0.41 ± 0.04
NSVS 6133550	0.57 ± 0.02	0.00+0.01	5493 ± 54	G8	11.28 ± 0.01	4.98 ± 0.04	182 ± 2	0.94 ± 0.03
V1377 Tau	0.72 ± 0.02	0.00+0.02	5040 ± 200*	K2	11.24 ± 0.03	5.34 ± 0.06	152 ± 1	0.74 ± 0.09
NSVS 3917713	0.66 ± 0.02	0.04 ± 0.02	5301 ± 122	K0	13.02 ± 0.03	5.14 ± 0.09	359 ± 10	0.85 ± 0.07
V2802 Ori	0.60 ± 0.02	0.00+0.01	5359 ± 109	G9	11.17 ± 0.02	4.97 ± 0.02	174 ± 1	0.98 ± 0.03

*Determined spectroscopically.

(dashed lines). The dashed line in Figure 19 is a power-law fit to the more massive secondary stars (Gazeas and Stępień 2008). This is very similar to the mass-radius relation for single main-sequence (MS) stars with masses less than 1.8 $M_\odot$ (Gimenez and Zamorano 1985). Figure 19 shows that the massive components of each binary pair in this study are MS stars. The smaller, hotter primary stars of contact binaries do not follow a mass-radius relation but are considerably oversized and over-luminous when compared to MS stars of similar masses.

The final WD solutions gave fill-outs for the stars between 9% and 41%. This is consistent with a contact binary where both primary and secondary stars exceed their critical lobes. These systems are W-subtype eclipsing binaries where the more massive cooler secondary star has a lower surface brightness than its companion. The stars are all cooler than the sun, with spectral types from G8 to K3. There were asymmetries in the light curves of each system due to spotting, which indicates magnetically active stars. This is not unexpected, given the low temperature and rapid rotation of the stars.

V338 Dra has a very short orbital period for a LMCB at 0.235 d (5.64 hours). This period is close to the cut-off of ~ 0.22 d in the period distribution for contact binaries (Ruciński 1992). The components of this system are in shallow contact with a fill-out of 9%. V338 Dra has the coolest and smallest stars of the five systems in this study and the lowest observed luminosity. Two cool spots were modeled on the larger secondary star to account for the asymmetries in the light curves. The WD solution result showed negligible third light. There are very few contact binaries with orbital periods less than 0.24 d that have well determined physical parameters. RW Com is one of these stars and happens to be remarkably similar to V338 Dra (Deb and Singh 2011). A comparison of the physical and orbital properties of RW Com and V338 Dra is shown in Table 9. Both secondary stars in these systems have masses that are higher than expected, given the temperature and luminosities determined for these main-sequence stars.

The stars of NSVS 6133550 are in moderately deep contact with a fill-out of 41%. The small primary star is about half the radius of its cooler companion. This star has the most extreme mass ratio of the five systems at $q(M_1/M_2) = 0.20$. Light curve asymmetries were modeled by adding two cool spots on the larger cooler component of this magnetically active system. A third light contribution of about 3–4% of the total system flux was necessary to obtain a good fit between the observed and synthetic light curves.

V1377 Tau has a secondary star radius that is only 13% larger than the primary star. The component stars are in shallow contact with a fill-out = 16%. This binary has the deepest primary eclipse of the five systems, with a Δm = 0.89 in the Sloan g' passband. A single cool spot was modeled on the smaller primary star in the final WD solution iterations. This is the only star in the sample with enough eclipse timings for a preliminary (O–C) period analysis. A least-squares solution using the (O–C) residuals in Table 4 yields the following quadratic ephemeris:

$$\text{HJD Min I} = 2458084.6822(4) + 0.2827095(5)\text{E} - 5.6(2) \times 10^{-11} \text{E}^2. \quad (7)$$

This ephemeris suggests the orbital period is decreasing at a rate of $dP/dt = -1.4(2) \times 10^{-10}$ d yr^{-1} or 1.2 seconds per century. The quadratic fit to the O–C residuals is shown in Figure 20. Additional eclipse timings will be required to confirm this result.

The stars of the NSVS 3917713 system differ in temperature by about 100 K, giving nearly equal minima depths. These stars may be in thermal equilibrium due to a long evolutionary contact phase. The light curves show little evidence for the O'Connell effect, but a deficit of light was noted after Max II and before Max I. In the final WD iterations, a single hot spot was modeled on the hotter primary star, which greatly improved the fit between the observed and model light curves (residuals four times smaller). This system also required a small third light contribution, about 3% of the total flux, to fit the minima.

V2802 Ori is very similar to NSVS 3917713 in terms of mass ratio, fill-out, and the masses and radii of the individual stars.

Table 7. Results derived from light curve modeling.

Parameter	V338 Dra	NSVS 6133550	V1377 Tau	NSVS 3917713	V2802 Ori
phase shift	–0.0009 ± 0.0002	–0.0045 ± 0.0003	–0.0019 ± 0.0001	0.0002 ± 0.0001	–0.0024 ± 0.0002
filling factor	9%	41%	16%	21%	18%
i (°)	83.7 ± 0.3	77.6 ± 0.1	87.7 ± 0.8	88.4 ± 1.2	85.1 ± 0.3
T_1 (K)	[1] 4749	[1] 5493	[1] 5040	[1] 5301	[1] 5359
T_2 (K)	4633 ± 28	5267 ± 27	4858 ± 6	5203 ± 4	5068 ± 13
$\Omega_1 = \Omega_2$	5.46 ± 0.03	8.85 ± 0.04	4.85 ± 0.03	6.14 ± 0.02	6.62 ± 0.02
$q (M_2/M_1)$	2.19 ± 0.03	4.96 ± 0.04	1.78 ± 0.02	2.74 ± 0.02	3.09 ± 0.02
$L_1/(L_1+L_2)$ (B)	—	0.2495 ± 0.0037	—	—	—
$L_1/(L_1+L_2)$ (V)	—	0.2369 ± 0.0026	—	—	0.3317 ± 0.0027
$L_1/(L_1+L_2)$ (g')	0.3737 ± 0.0059	0.2438 ± 0.0032	0.4347 ± 0.0028	0.3141 ± 0.0017	0.3427 ± 0.0030
$L_1/(L_1+L_2)$ (r')	0.3597 ± 0.0038	0.2310 ± 0.0021	0.4171 ± 0.0024	0.3073 ± 0.0013	0.3216 ± 0.0024
$L_1/(L_1+L_2)$ (i')	0.3546 ± 0.0031	0.2259 ± 0.0018	0.4099 ± 0.0019	0.3045 ± 0.0011	0.3130 ± 0.0021
l_3 (B)	—	[2] 0.04 ± 0.02	—	—	—
l_3 (V)	—	[2] 0.03 ± 0.02	—	—	—
l_3 (g')	—	[2] 0.04 ± 0.02	—	[2] 0.035 ± 0.005	—
l_3 (r')	—	[2] 0.03 ± 0.01	—	[2] 0.018 ± 0.005	—
l_3 (i')	—	[2] 0.03 ± 0.01	—	[2] 0.017 ± 0.005	—
r_1 side	0.3109 ± 0.0011	0.2551 ± 0.0008	0.3278 ± 0.0009	0.2968 ± 0.0010	0.2854 ± 0.0006
r_2 side	0.4721 ± 0.0051	0.5639 ± 0.0039	0.4525 ± 0.0044	0.4918 ± 0.0030	0.4989 ± 0.0029
Spot Parameters					
Spot 1	$Star_2$	$Star_2$	$Star_2$	$Star_1$	$Star_2$
colatitude (°)	44 ± 15	52 ± 14	123 ± 32	131 ± 8	43 ± 10
longitude (°)	101 ± 4	121 ± 4	122 ± 3	5 ± 1	117 ± 3
spot radius (°)	24 ± 7	26 ± 7	17 ± 9	32 ± 5	20 ± 4
temp. factor	0.86 ± 0.04	0.95 ± 0.03	0.82 ± 0.07	1.11 ± 0.04	0.82 ± 0.03
Spot 2	$Star_2$	$Star_2$			$Star_2$
colatitude (°)	58 ± 11	88 ± 3	—	—	49 ± 9
longitude (°)	356 ± 4	4 ± 1	—	—	325 ± 9
spot radius (°)	20 ± 6	21 ± 3	—	—	15 ± 3
temp. factor	0.90 ± 0.04	0.91 ± 0.03	—	—	0.79 ± 0.05

[1] Assumed. [2] Third lights are the percent of light contributed at orbital phase 0.25. The subscripts 1 and 2 refer to the star being eclipsed at primary and secondary minimum, respectively. Note: The errors in the stellar parameters result from the least–squares fit to the model. The actual uncertainties are considerably larger.

Table 8. Provisional absolute parameters (solar units).

System	A	M_1	M_2	R_1	R_2	L_1	L_2	$L_1 + L_2$
V338 Dra	1.74 ± 0.04	0.40 ± 0.04	0.87 ± 0.09	0.56 ± 0.01	0.79 ± 0.02	0.14 ± 0.01	0.26 ± 0.03	0.40 ± 0.04
NSVS 6133550	1.88 ± 0.05	0.20 ± 0.02	0.98 ± 0.09	0.51 ± 0.01	1.02 ± 0.03	0.22 ± 0.01	0.72 ± 0.06	0.93 ± 0.08
V1377 Tau	2.10 ± 0.05	0.56 ± 0.05	1.00 ± 0.09	0.72 ± 0.02	0.93 ± 0.02	0.30 ± 0.06	0.43 ± 0.13	0.73 ± 0.18
NSVS 3917713	2.06 ± 0.05	0.37 ± 0.03	1.02 ± 0.09	0.64 ± 0.01	0.99 ± 0.03	0.29 ± 0.03	0.65 ± 0.10	0.94 ± 0.14
V2802 Ori	2.08 ± 0.05	0.34 ± 0.03	1.04 ± 0.09	0.62 ± 0.01	1.02 ± 0.03	0.28 ± 0.03	0.62 ± 0.08	0.90 ± 0.11

Note: The calculated values in this table are provisional. Radial velocity observations are not available for direct determination of M_1, M_2, and a.

Table 9. Comparison of physical and orbital parameters (solar units).

Star	P (day)	A	M_1	M_2	R_1	R_2	T_1 (K)	T_2 (K)	L_1	L_2
RW Com	0.2373	1.72	0.39	0.83	0.56	0.79	4830	4517	0.15	0.23
V338 Dra	0.2351	1.74	0.40	0.87	0.56	0.79	4749	4633	0.14	0.26

The photometric solution values for V2802 Ori, inclination, temperature ratio (T_2/T_1), and mass ratio (i=85.1°, T_2/T_1=0.98, q=3.09) were in good agreement with light curve modeling using ASAS survey data (i=83.5°, T_2/T_1=0.94, q=2.86) (Pilecki and Stępień 2012).

Evolutionary models of LMCB indicate these systems should be undergoing mass transfer from the currently less massive star to their companions (Stępień 2006, 2011). This would result in the lengthening of their orbital periods. Concurrently, angular momentum and mass loss due to magnetized winds would have the opposite effect, decreasing the orbital period. The angular momentum loss (AML) should dominate in LMCB, causing the orbits to contract, and eventually leading to the binary overflowing their outer critical surface. The binary would then merge, forming a single, rapidly rotating star. The extreme mass ratio of NSVS 6133550 was likely the result of this evolutionary driven combination of mass transfer and AML. Except for V1377 Tau, the eclipse timings currently available for these stars are few. Determining whether these stars are undergoing secular period changes will require additional observations. Cyclic variations in the period are also possible due to an orbiting third body. The third light modeled in the WD solutions for NSVS 6133550 and NSVS 3017713 could be the result of a low mass third star in these two systems. New eclipse timings over many years may provide evidence to support this supposition. In addition, a spectroscopic study of these systems would also be invaluable to confirming the provisional absolute stellar parameters presented here.

5. Acknowledgements

This research was made possible through the use of the AAVSO Photometric All-Sky Survey (APASS), funded by the Robert Martin Ayers Sciences Fund. This research has made use of the SIMBAD database and the VizieR catalogue access tool, operated at CDS, Strasbourg, France. This work has made use of data from the European Space Agency (ESA) mission Gaia (https://www.cosmos.esa.int/gaia), processed by the Gaia Data Processing and Analysis Consortium (DPAC, https://www.cosmos.esa.int/web/gaia/dpac/consortium). Funding for the DPAC has been provided by national institutions, in particular the institutions participating in the Gaia Multilateral Agreement. This paper makes use of data from the first public release of the WASP data (Butters *et al.* 2010) as provided by the WASP consortium and services at the NASA Exoplanet Archive, which is operated by the California Institute of Technology, under contract with the National Aeronautics and Space Administration under the Exoplanet Exploration Program.

References

Ammons, S. M., Robinson, S. E., Strader, J., Laughlin, G., Fischer, D., and Wolf, A. 2006, *Astrophys. J.*, **638**, 1004.
Appenzeller, I., *et al.* 1998, *Astrophys. J., Suppl. Ser.*, **117**, 319.
Bailer-Jones, C. A. L., Rybizki, J., Fouesneau, M., Mantelet, G., and Andrae, R. 2018, *Astron. J.*, **156**, 58.
Bilir, S. Karaali, S., and Tunçel, S. 2005, *Astron. Nachr.*, **326**, 321.
Bradstreet, D. H., and Steelman, D. P. 2002, *Bull. Amer. Astron. Soc.*, **34**, 1224.
Butters, O. W., *et al.* 2010, *Astron. Astrophys.*, **520**, L10.
Covey, K. R., *et al.* 2007, *Astron. J.*, **134**, 2398.
Deb, S., and Singh, H. P. 2011, *Mon. Not. Roy. Astron. Soc.*, **412**, 1787.
Diethelm, R. 2010, *Inf. Bull. Var. Stars*, No. 5920, 1.
Diethelm, R. 2011, *Inf. Bull. Var. Stars*, No. 5960, 1.
Diethelm, R. 2012, *Inf. Bull. Var. Stars*, No. 6011, 1.
Diethelm, R. 2013, *Inf. Bull. Var. Stars*, No. 6042, 1.
Drake, A. J., *et al.* 2014, *Astrophys. J., Suppl. Ser.*, **213**, 9.
Gaia Collaboration, *et al.* 2016, *Astron. Astrophys.*, **595A**, 1.
Gaia Collaboration, *et al.* 2018, *Astron. Astrophys.*, **616A**, 1.
Gazeas, K., and Stępień, K. 2008, *Mon. Not. Roy. Astron. Soc.*, **390**, 1577.
Gettel, S. J., Geske, M. T., and McKay, T. A. 2006, *Astron. J.*, **131**, 621.
Gimenez A., and Zamorano J. 1985, *Astrophys. Space Sci.*, **114**, 259.
Green, G. M., *et al.* 2018, *Mon. Not. Roy. Astron. Soc.*, **478**, 651.
Hambálek L., and Pribulla, T. 2013, *Contrib. Astron. Obs. Skalnaté Pleso*, **43**, 27.
Henden, A. A., *et al.* 2015, AAVSO Photometric All-Sky Survey, data release 9, (https://www.aavso.org/apass).
Hoffman, D., Harrison, T. E., and McNamara, B. J. 2009, *Astron. J.*, **138**, 466.
Hoňková, K., *et al.* 2013, *Open Eur. J. Var. Stars*, **160**, 1.
Hübscher, J., and Lehmann, P. B. 2013, *Inf. Bull. Var. Stars*, No. 6070, 1.
Hübscher, J., and Lehmann, P. B. 2015, *Inf. Bull. Var. Stars*, No. 6149, 1.
Jayasinghe, K., *et al.* 2018, *Mon. Not. Roy. Astron. Soc.*, **477**, 3145.
Jester, S., *et al.* 2005, *Astron. J.*, **130**, 873.
Kafka, S. 2017, variable star observations from the AAVSO International Database (https://www.aavso.org/aavso-international-database).
Khruslov, A. V. 2006, *Perem. Zvezdy Prilozh.*, **6**, 16 (http://www.astronet.ru/db/varstars/msg/eid/PZP-06-0016).
Kochanek, C. S., *et al.* 2017, *Publ. Astron. Soc. Pacific*, **129**, 104502.
Kurucz, R. L. 2002, *Baltic Astron.*, **11**, 101.
Kwee, K. K., and van Woerden, H. 1956, *Bull. Astron. Inst. Netherlands*, **12**, 327.
Lucy, L. B. 1968, *Astrophys. J.*, **151**, 1123.
Lucy, L. B., and Wilson, R. E. 1979, *Astrophys. J.*, **231**, 502.
Michaels, E. J. 2015, *J. Amer. Assoc. Var. Star Obs.*, **43**, 231.
Michaels, E. J. 2016a, *J. Amer. Assoc. Var. Star Obs.*, **44**, 30.
Michaels, E. J. 2016b, *J. Amer. Assoc. Var. Star Obs.*, **44**, 53.
Michaels, E. J. 2016c, *J. Amer. Assoc. Var. Star Obs.*, **44**, 137.
Michaels, E. J. 2018, *J. Amer. Assoc. Var. Star Obs.*, **46**, 27.
Michaels, E. J., Lanning, C. M., and Self, S. N. 2019, *J. Amer. Assoc. Var. Star Obs.*, **47**, 43.
McDonald I., Zijlstra, A. A., and Watson, R. A. 2017, *Mon. Not. Roy. Astron. Soc.*, **471**, 770.
Mirametrics. 2015, Image Processing, Visualization, Data Analysis, (https://www.mirametrics.com).

Nagai, K. 2012, *Bull. Var. Star Obs. League Japan*, **53**, 1.
O'Connell, D. J. K. 1951, *Publ. Riverview Coll. Obs.*, **2**, 85.
Pecaut, M. J., and Mamajek, E. E. 2013, *Astrophys. J., Suppl. Ser.*, **208**, 9 (http://www.pas.rochester.edu/~emamajek/EEM_dwarf_UBVIJHK_colors_Teff.txt).
Pickles, A., and Depagne, E. 2010, *Publ. Astron. Soc. Pacific*, **122**, 1437.
Pilecki, B., and Stępień, K. 2012, *Inf. Bull. Var. Stars*, No. 6012, 1.
Pojmański, G. 1997, *Acta Astron.*, **47**, 467.
Pojmański, G. *et al.* 2002, *Acta Astron.*, **52**, 397.
Richards, J. W., Starr, D. L., Miller, A. A., Bloom, J. S., Butler, N. R., Brink, H., and Crellin-Quick, A. 2012, *Astrophys. J., Suppl. Ser.*, **203**, 32.
Ruciński, S. M. 1969, *Acta Astron.*, **19**, 245.
Ruciński, S. M. 1992, *Astron. J.*, **103**, 960.
Stępień, K., 2006, *Acta Astron.*, **56**, 199.
Stępień, K., 2011, *Acta Astron.*, **61**, 139.
Szczygiel, D. M., Socrates, A., Paczyński, B., Pojmański, G., and Pilecki, B. 2008, *Acta Astron.*, **58**, 405.
Terrell, D., Gross, J., and Cooney, W. R. 2012, *Astron. J.*, **143**, 99.
Terrell, D., and Wilson, R. E. 2005, *Astrophys. Space Sci.*, **296**, 221.
van Hamme, W. 1993, *Astron. J.*, **106**, 2096.
van Hamme, W., and Wilson, R. E. 1998, *Bull. Amer. Astron. Soc.*, **30**, 1402.
Watson, C., Henden, A. A., and Price, C. A. 2014, AAVSO International Variable Star Index VSX (Watson+, 2006–2014, http://www.aavso.org/vsx).
Wilson, R. E. 1978, *Astrophys. J.*, **224**, 885.
Wilson, R. E., and Devinney, E. J. 1971, *Astrophys. J.*, **166**, 605.
Woźniak, P. R., *et al.* 2004, *Astron. J.*, **127**, 2436.
Zickgraf, J., Krautter, J., Reffert, S., Alcalá, J. M., Mujica, R., Covino, E., and Sterzik, M. F. 2005, *Astron. Astrophys.*, **433**, 151.

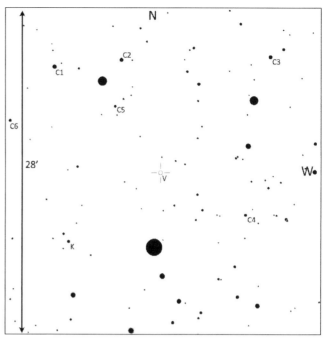

Figure 1. Finder chart for V338 Dra (V), comparison (C1–C6), and check (K) stars.

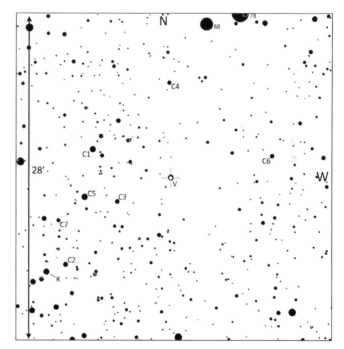

Figure 2. Finder chart for NSVS 6133550 (V), comparison (C1–C7), and check (K) stars.

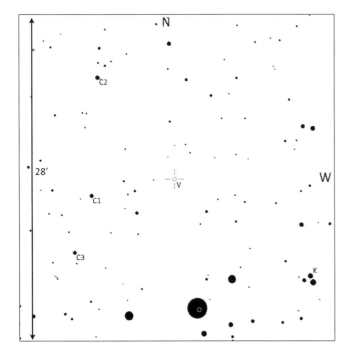

Figure 3. Finder chart for V1377 Tau (V), comparison (C1–C3), and check (K) stars.

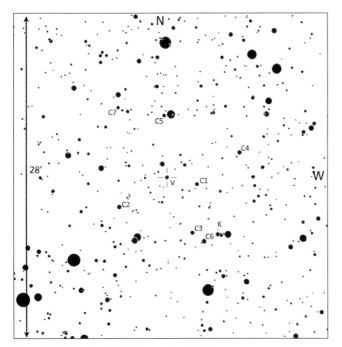

Figure 4. Finder chart for NSVS 3917713 (V), comparison (C1–C7), and check (K) stars.

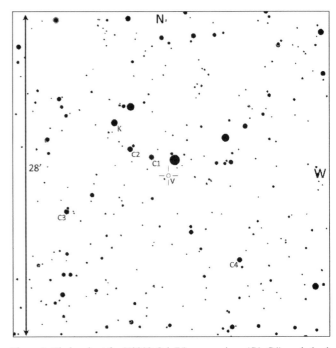

Figure 5. Finder chart for V2802 Ori (V), comparison (C1–C4), and check (K) stars.

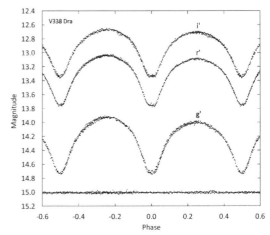

Figure 6. Observed light curves in standard magnitudes of V338 Dra (g' r' i' filters). The bottom curve shows the Sloan r' check star magnitudes (offset +1.6 magnitudes).

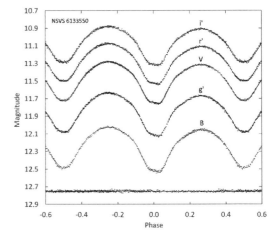

Figure 7. Observed light curves in standard magnitudes of NSVS 6133550 (B V g' r' i' filters). The bottom curve shows the Sloan r' check star magnitudes (offset +1.36 magnitudes).

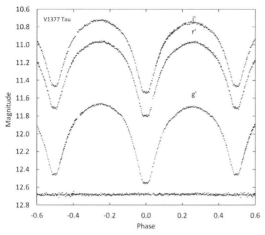

Figure 8. Observed light curves in standard magnitudes of V1377 Tau (g' r' i' filters). The bottom curve shows the Sloan r' check star magnitudes (offset +1.18 magnitudes).

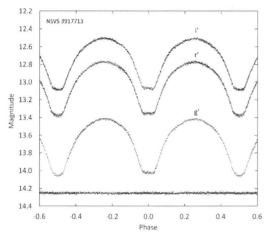

Figure 9. Observed light curves in standard magnitudes of NSVS 3917713 (g' r' i' filters). The bottom curve shows the Sloan r' check star magnitudes (offset +1.5 magnitudes).

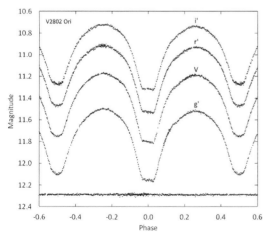

Figure 10. Observed light curves in standard magnitudes of V2802 Ori (V g' r' i' filters). The bottom curve shows the Sloan r' check star magnitudes (offset +1.5 magnitudes).

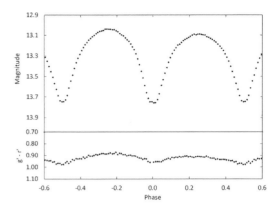

Figure 11. Light curve of all Sloan r'-band observations in standard magnitudes for V388 Dra (top panel). The observations were binned with a phase width of 0.01. The errors for each binned point are about the size of the plotted points. The g'- r' colors were calculated by subtracting the linearly interpolated binned g' magnitudes from the linearly interpolated binned r' magnitudes.

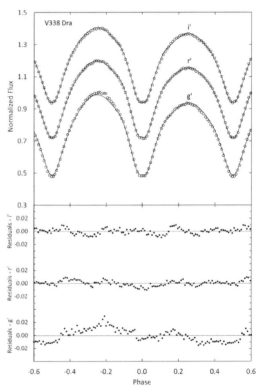

Figure 12. Comparison between the WD model fit (solid curve) and the observed normalized flux curves for V338 Dra (g' r' i' filters). Each curve is offset by 0.2 for this combined plot. The residuals for the best-fit model are shown in the bottom panel. Error bars are omitted from the points for clarity.

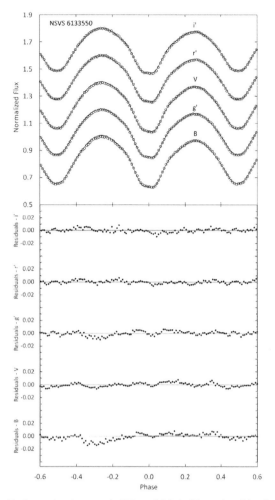

Figure 13. Comparison between the WD model fit (solid curve) and the observed normalized flux curves for NSVS 6133550 (B V g' r' i' filters). Each curve is offset by 0.2 for this combined plot. The residuals for the best-fit model are shown in the bottom panel. Error bars are omitted from the points for clarity.

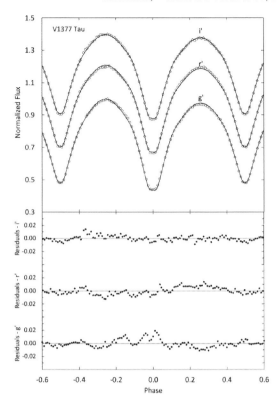

Figure 14. Comparison between the WD model fit (solid curve) and the observed normalized flux curves for V1377 Tau (g' r' i' filters). Each curve is offset by 0.2 for this combined plot. The residuals for the best-fit model are shown in the bottom panel. Error bars are omitted from the points for clarity.

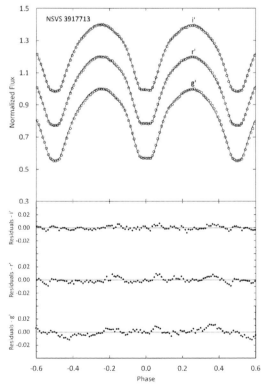

Figure 15. Comparison between the WD model fit (solid curve) and the observed normalized flux curves for NSVS 3917713 (g' r' i' filters). Each curve is offset by 0.2 for this combined plot. The residuals for the best-fit model are shown in the bottom panel. Error bars are omitted from the points for clarity.

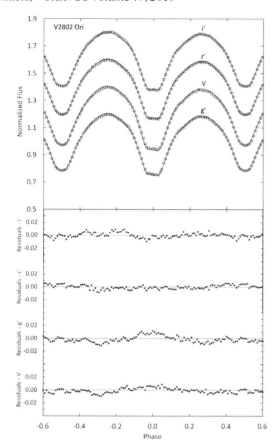

Figure 16. Comparison between the WD model fit (solid curve) and the observed normalized flux curves for V2802 Ori (V g' r' i' filters). Each curve is offset by 0.2 for this combined plot. The residuals for the best-fit model are shown in the bottom panel. Error bars are omitted from the points for clarity.

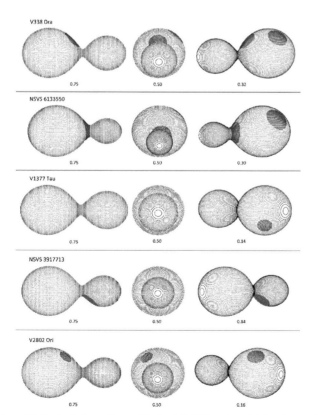

Figure 17. Roche Lobe surfaces of the best-fit WD spot model showing spot locations. The orbital phase is shown below each diagram.

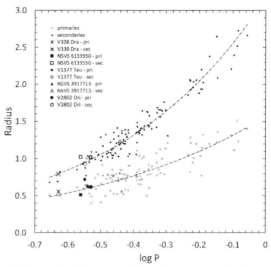

Figure 18. The radii of the stars in this study are compared with the radius distributions of 112 contact binaries with well determined geometrical and physical properties. The dashed lines are the least-square fits from the analysis of Gazeas and Stępień (2008).

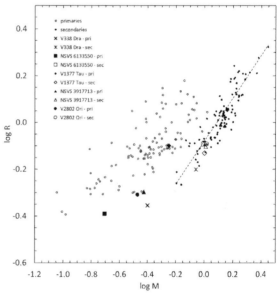

Figure 19. The radius distribution versus mass for the primary and secondary components of 112 contact binaries with well determined geometrical and physical properties. The stars in this study are shown for comparison. The dashed line is the mass-radius relation for the secondary components of contact binaries (Gazeas and Stępień 2008).

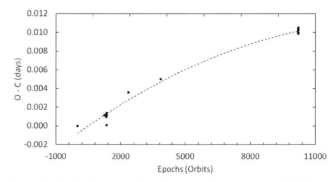

Figure 20. The O–C residuals (dots) from the initial ephemeris of V1377 Tau. The solid line is the quadratic ephemeris fit of Equation 7.

Medium-term Variation in Times of Minimum of Algol-type Binaries: XZ And, RZ Cas, U Cep, TW Dra, U Sge

Geoff B. Chaplin
Hokkaido, Kamikawa-gun, Biei-cho, Aza-omura Okubo-kyosei, 071-0216, Japan; geoff@geoffgallery.net

Received August 25, 2019; revised October 5, 2019; accepted October 6, 2019

Abstract We use methods of singular spectrum analysis to examine the observed minus expected times of minimum of five Algol-type binaries looking for significant periodicities. All the stars examined show long "waves" of 20 to 50 years but only XZ And and RZ Cas appear to have significant periodicities.

1. Introduction

In this paper we examine the Algol-type variables listed in Table 1 with physical data in Table 2. It is common to attempt to explain secular changes in times of minimum (of the eclipse) in terms of harmonics caused by third bodies in the system, for example Borkovits and Hegedus (1996), or additionally apsidal motion, Hoffman et al. (2006). A study by Li et al. (2018) of 542 short-period binaries showed that many such stars do indeed show the presence of third bodies. Other causes of periodicity (not necessarily harmonic) are interactions between electromagnetic fields (Hall 1989) or electromagnetic-gravitational interactions (Applegate 1992). A general review of Algol variables is given in Budding (1986) and references therein, and a table of such stars in Budding et al. (2004).

Our objective here is to determine if there are oscillations in the observed minus calculated (O–C) times of minimum with "periods" (i.e. non-secular changes) in ranges determined by the available length and density of data but broadly speaking in the range of a few years to 20 years—rather than the long-term variation, and without any presumption as to the cause of such variation. The upper limit of 20 years is driven by the desire to see a minimum of at least 6 or so complete periods in order to be confident of their real nature, and the lower period by the bucketing (binning) size chosen which in turn is driven by the density of data.

XZ And Demicran et al. (1995) examined the O–C history and postulated two or three other bodies with periods 137.5, 36.8, and 11.2 years as a possible cause of the variation, although they suggested the third period may be related to cyclic magnetic activity of the secondary star. Borkovits and Hegedus (1996) could not find a good fit for any third body orbit.

Manzoori (2016) quote periods of 34.8 ± 2.4 years and 23.3 ± 3.0 years.

RZ Cas This star's variability was discovered by Muller in 1906. The light-curve sometimes shows a flat bottom near primary minimum, which may be caused by the effect of short-term δ-Scuti type variability (Rodriguez et al. 2004) or hot and cool spots on the surface of the primary (Olson 1982). Ohshima et al. (2001) and Rodriguez et al. (2004) found very short-term periodicity of around 22 minutes. Mkrtichian et al. (2018) also confirmed the high frequency variability and in addition, based on a 1999–2009 data sample, a 6- to 9-year magnetically induced variation in times of minimum. Using a statistical analysis of times of minimum, Chaplin (2018) postulated a period of 22.5 to 24 years.

U Cep Ceraski (1880) first determined U Cep to be a variable binary star with a period of 2.49 days. It is amongst the most active Algols, and it is known to cycle through periods of extremely active mass exchange and times of relative quiescence and was found to exhibit a 9 ± 4 yr cycle of relative quiescence and high activity (Hall 1975) with period changes related to mass loss and non-conservation of angular momentum.

TW Dra This star forms a visual triple with ADS 9706, complicating both amateur and professional observations. Zejda et al. (2008) quote cycles of approximately 20 years caused by electromagnetic-gravitational interaction and 6.5 years caused by a third body.

U Sge Manzoori and Gozaliasl (2007) used polynomial fitting and Fourier analysis of times of minimum to conclude there were cyclic variations of lengths 15.8 and 9.5 years, whereas Simon (1997) reports a roughly defined period of 39 years.

Table 1. Stars analyzed.

Name	GSC (HD)	R.A. (J2000) h m s	Dec. (J2000)[1] ° ' "	Data History Start	Period[2] (days)
XZ And	02824-01360	01 53 48.76	+41 51 24.97	1891	1.35730911
RZ Cas	04317-01793	02 48 55.51	+69 38 03.44	1901	1.19525031
U Cep	04505-00519	01 02 18.44	+81 52 32.08	1880	2.4930911
TW Dra	(139319A)	15 33 51.06	+63 54 25.67	1898	2.80684701
U Sge	01607-00913	19 18 48.41	+19 36 37.72	1901	3.380619331

[1] *Wenger et al. (2000).* [2] *Frank and Lichtenknecker (1987).*

2. Methodology

2.1. Data

Data in the form of times of minimum (Tmin) are taken from the Lichtenknecker database (Frank and Lichtenknecker 1987), augmented in the case of RZ Cas by data from Mkrtichian *et al.* (2018). Data and the period (possibly after minor adjustment) are used to calculate observed minus expected (O–C) differences in times of minimum (expressed in days).

2.2. Data cleaning and weighting

Initial relatively isolated observations are ignored together with early very noisy data if any, and analysis is performed on the entire data series (visual and electronically based observations). Where more recent data include a long run of high quality observations (by electric photometer or CCD) we perform the analysis additionally using only the electronically based data.

It is to be noted that the accuracy of Tmin is the result of the combination of the accuracy of individual magnitude estimates and the number of such estimates (as well as a good analytical method to reduce these to a Tmin). In principle many unbiased visual observations can be as accurate as a small number of highly accurate CCD observations although in practice electronic data are far better; further theoretical discussion and analysis is given in Chaplin *et al.* (2018). While accuracy estimates are given for recent CCD results such explicit estimates are generally not available for other observation methods.

Various weighting (or unweighted) regimes were tested with relatively minor variation in results. In this paper we analyze using two regimes: weights according to the observational type: 10 (CCD), 3.25 (electric and wedge photometer), or 1 (others) ("fixed" weights), or weighted, by 1/observational-error squared ("inverse error" weighting). Observational error (for Tmin in days) is taken as given by Mkrtichian *et al.* (2018) (applicable to RZ Cas only), 0.0004 for other CCD data, 0.0004 for photometric data, 0.01 for a series of photographic data, and 0.02 for others (primarily visual). For the CCD data the error is a conservative estimate based on Samolyk (2011 and others), for photometric data it is based on a comparison of scatter compared to CCD, and for visual it is derived by assuming 0.1 accuracy in magnitude estimates and between 10 and 20 observations in the series—based on Chaplin *et al.* (2018)). Outlier rejection is performed by repeated local polynomial fitting rejecting outliers at the 4-sigma level.

2.3. Bucketing (binning)

SSA techniques require data at equally spaced time points whereas observations of Tmin are unevenly spaced. We use two alternative methods to obtain the required spacing after defining a bucket size (the time span—150-, 200-, 300-, and 500-day buckets were used). The first method ("date buckets") takes the (weighted) average of data within that interval, filling empty buckets by linear interpolation; the second method ("local poly buckets") fits a local polynomial and takes values from the fitted curve at evenly spaced time intervals. In the case of date bucketing a small bucket size means more empty buckets; where the percentage of empty buckets exceeds about one quarter the results of the analysis are ignored as unlikely to be reliable.

It should be noted that the bucketing process necessarily imposes a limit on the shortest periods that can be reliably detected. For example, with a binning of 500 days periods shorter than this are unlikely to make their presence known in the analysis (although it is theoretically possible if the period and bucketing interval are not simple multiples of each other) and detected periods are likely to be longer than a multiple of the bucketing interval.

2.4. SSA

We use techniques of Singular Spectrum Analysis ("SSA") as explained in detail by Chaplin (2018, 2019 and references therein) and implementations in the R language (R Found. 2018a), CRAN libraries (R Found. 2018b), using RStudio (2018) and in particular the "Rssa," and "simsalabim" (for significance testing, Gudmundsson 2017) libraries.

By way of background SSA can be thought of as a means of calculating averages (called EV series) from a data series— rather like (but much more complicated than) moving averages. These averages have certain properties in common with Fourier series—orthogonality between different series—although the EV series are generally neither periodic nor of constant amplitude. The R language primarily gives access to an extensive library of statistical, mathematical, graphical, and other routines, while RStudio provides an interface to the code, written output, all variable values, graphical output, libraries and help files, and more. Both are free and available for Windows, Mac and a range of Linux systems. Outline code for the analysis is given in Appendix A.

The EV series derived from the SSA analysis are ordered according to the magnitude of the associated eigen value (equivalent to ordering by the strength of the series in the total data), and we refer to them as series 1, 2, 3, etc.

Table 2. Stars' physical data, and sources.

Star	Spectral Type	Masses (solar) (primary, secondary)	Radii (solar)	Source
XZ And	A4IV-V, G5IV	3.2, 1.3	2.4, 2.6	Demircan *et al.* (1995)
RZ Cas	A3V[1], carbon[2] or K0IV[3]	2.2, 0.7	1.7, 1.9	Maxted *et al.* (1994)
U Cep	B 7/8 V, G 5/8 III–IV[4]	4.2, 2.8	2.8, 4.9	Batten (1974), Singh *et al.* (1995)
TW Dra	A5 V, K0 III	2.2, 0.9	2.6 (primary)	Tkachenko *et al.* (2010)
U Sge	B7III, K1III C	5.7, 1.9	4.2, 5.3	Dobias and Plavec (1985)

[1] *Duerbeck and Hänel (1979).* [2] *Abt and Morrell (1995).* [3] *Maxted* et al. *(1994). Rodriguez* et al. *(2004).* [4] *Tupa* et al. *(2013).*

The given orbital period is adjusted to give a best fit horizontal line to the O–C data. "Sequential SSA" is performed on the bucketed data to determine the trend in the signal, and removing this trend from the O–C data gives the residuals. Trends are defined as EV series which have no identified periodicity. The residuals are then analyzed using SSA again and the reconstructed signals are calculated after grouping residuals according to similarity of pattern and having high correlation. Periodicities of the reconstructed signals are then determined. Only signals which appear under a variety of bucketing time intervals and regimes ("consistency") are considered and tested for significance.

The length of the data series determines the longest period we can reasonable expect to be confident in identifying—our rule of thumb is 6 or more periods in the time series (so 20 years for a 120-year data series). The shortest period is determined by the density of data and how short a bucket size we can take without empty buckets exceeding around 25% of the total. With, for example, a 150-day bucket interval, periods of around 3 years or longer should be apparent from the analysis of the data.

SSA is a "data driven" method of analysis to be contrasted with a "model driven" approach where (as in Hoffman *et al.* 2006), for example, a parabola or sinusoid might be used to remove the trend component of the change in O–C. The strength of the argument for using one method rather than the other depends on the evidence from other observations. If, again for example, a distant companion star is observed, then the calculated orbital period can be used to determine the sinusoidal change (light time effect) in the O–C that its orbit will cause, and this can be used to remove (a component of) the trend. The argument for favoring the data-driven approach is stronger when no such external evidence in favor of a specific long-term change is known. Results regarding any discovered shorter period signals are to some extent dependent of what trend signal has been removed—in other words a model for the trend may lead to different shorter period signals being discovered compared to the purely data-driven approach.

2.5. Significance testing

We use the MCSSA routine from the simsalabim library as a tool to assign a level of significance to the results. Anything below a 95% confidence level is rejected. However, in stating that a result is significant we qualitatively take into account more factors than a single test number—the number of periods observed in the data, changes in amplitude of the signal, etc.

3. O–C charts and data types

Table 3 shows the symbols are used consistently in the O–C charts plotting raw data to identify different equipment used to make the observation leading to the Tmin calculation.

3.1. XZ And

After rejecting 18 outliers, a total of 1,086 times of minimum from year 1891 on are displayed in Figure 1. Analysis using an adjusted period of 1.357283 days showed a trend from signals 1, 2, and 5 across all bucketing and weighting regimes, illustrated in Figure 2.

Table 3. O–C charts and data types.

Symbol	Data Type
Cross	Visual observation
Tilted cross	Photographic observation
Diamond	Electric photometer
Square	CCD (amateur)
Filled dots	CCD (Mkrtichian *et al.* (2018), RZ Cas only)

Residuals showed a clear and consistent pattern across bucketing and weighting regimes with signals 1 and 2 being responsible for over 90% of the residual, giving rise to a period of approximately 38 years (36–40 years) and are shown in Figure 3. MCSSA testing showed the signal as better than 99% significant. An apparent period of approximately 23 years from signals 3 and 4 is both weak and largely present only when visual observations dominate the data, and does not show as significant.

Figure 4 shows the raw data minus trend value and we note the latest CCD data present a different pattern and indicate a significantly smaller magnitude of variation, casting doubt on possible interpretation as a (single) third body effect. There are indications that earlier turning points also showed more complicated behavior.

3.2. RZ Cas

In total, 4,930 observations were used, and bucketing into 150-day or longer intervals had less than 15% empty buckets with only one empty bucket at 300- and 500-day bucketing. Figure 5 shows the data after rejecting 46 outliers at the 4 standard deviation level.

Consistently across all bucketing intervals (data or polynomial buckets) and weighting regimes, the first two signals describe the trend pattern, and after adjusting the period to 1.1952500 days Figure 6 shows the 500-day bucketed data and the trend from the first two signals.

Decomposition of the residual is again the same for all bucketing periods and data weighting methods—the first two series constitute the signal with a period of 22.7 to 25 years, which is significant to better than 99%. Weaker signals are present in higher series and show (near) harmonics at 11.5 and 5.5 years although they are not significant. The reconstituted series from these two signals together with the residuals is shown in Figure 7 using 500-day bucketing.

CCD and electric photometer data amount to 377 observations after removing two outliers, and are shown in Figure 8.

After period adjustment (to 1.1952500 days again) and removal of the trend (from the first two EV series) signal decomposition of the residual series reveals no surprises; decomposition is very similar across different bucketing intervals and repeats signals at the same periods as found from the long-term analysis. The dominant signal remains the 24-year period and Figure 9 shows the 200-day bucketed data and the primary signal. The signal remains significant at better than 98%.

We note that the long-term variation shown in Figure 7 is of the order of 0.05 day, whereas the variation relative to the long-term trend is of the order of 0.01 day while the accuracy of the

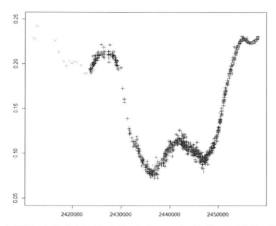

Figure 1. XZ And, O–C data (deviation in days against Julian Date). See Table 3 for key to the symbols.

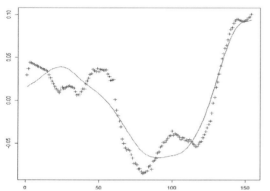

Figure 2. XZ And, 300-day bucketed data and fitted trend (O–C against bucket number).

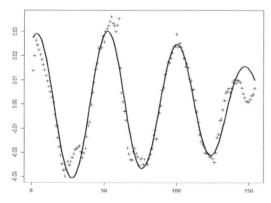

Figure 3. XZ And, residual data and reconstructed signal.

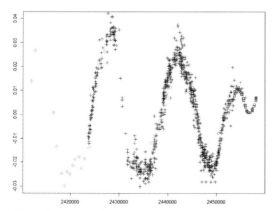

Figure 4. XZ And, raw data minus trend.

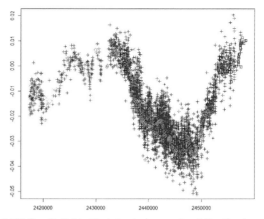

Figure 5. RZ Cas, O–C data (deviation in days against Julian Date).

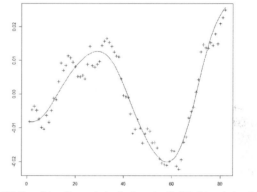

Figure 6. RZ Cas, data after period adjustment with fitted trend signal (O–C against bucket number).

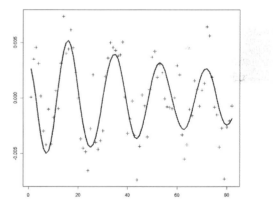

Figure 7. RZ Cas, residuals (500-day buckets) and reconstructed signal.

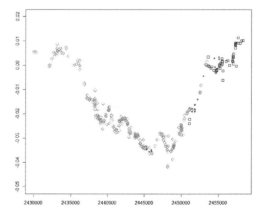

Figure 8. RZ Cas, electronic data (O–C against JD).

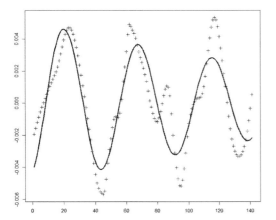

Figure 9. RZ Cas, 200-day bucketed electronic data and long-term trend (O–C vs bucket number).

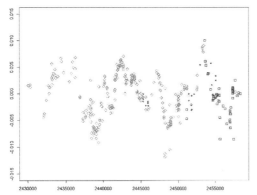

Figure 10. RZ Cas, electronic data minus the trend values.

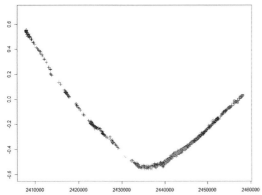

Figure 11. U Cep, O–C data (deviation in days against Julian Date).

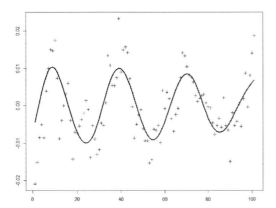

Figure 12. U Cep, residuals and reconstructed signal.

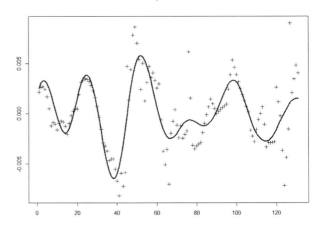

Figure 13. U Cep, 200-day bucketed data and reconstructed signal.

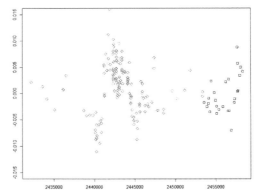

Figure 14. U Cep, electronic data minus the trend values.

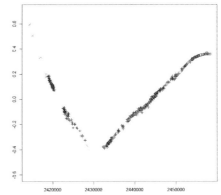

Figure 15. TW Dra, O–C data (deviation in days against Julian Date).

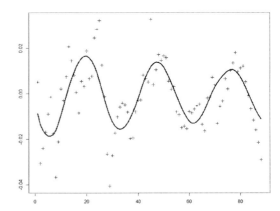

Figure 16. TW Dra, residuals and reconstructed signal.

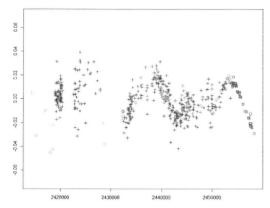

Figure 17. TW Dra, data minus trend.

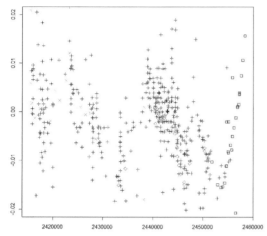

Figure 18. U Sge, data after eliminating outliers.

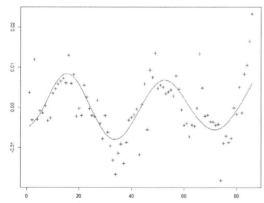

Figure 19. U Sge, 500-day bucket data and trend.

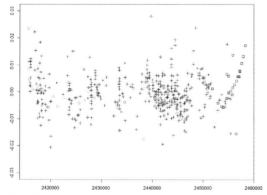

Figure 20. U Sge, data minus trend.

electronic data is generally much better than 0.001. Figure 10 shows data minus the trend values. It can be seen that RZ Cas regularly shows very short-term changes in period of the order of 0.01 day in between 3 and 6 years.

3.3. U Cep

1,242 observed minus calculated times of minimum after rejecting 37 outliers from 1,880 to 2,018 using a period of P = 2.493009 are shown in Figure 11.

Between 4 and 7 signals constitute trends (period shown as infinity), depending on bucketing method and length. After calculating residuals the reconstructed signal repeatedly shows periodicities of 11.7–12.1 years, 20.5–24 years, with a longer period of 40 years occasionally appearing—however, none of these appear to be significant. Only at the 500-day interval does the signals analysis agree for both bucketing methods (and weighting regimes). Figure 12 shows the residuals under 500-day bucketing with the strongest signal of 40 years from EV series 1 and 2.

There are 190 electronic (CCD and electric photometer) data after removal of 5 outliers. Trend and residual signal analysis is clearer and more consistent than with the entire data series, although the strongest periodicity in the residual signal is identified as 17–19 years and several other signals with periods as short as 12 years also appear, but none of these appear to be significant. Figure 13 shows the 200-day bucketing residual series together with the reconstructed signal from vectors 1 to 4.

At 200-day bucketing there are approximately 30% empty buckets, while at 500 days there are still 20% empty buckets and the data points are much sparser. Although testing indicates the period is significant (over 95%) the lack of data with the resulting interpolated points, together with the lack of a match to results found from the longer data series, indicates not much confidence can be placed in this result.

Figure 14 shows the data minus the trend values and shows the clumping of observations around the middle of the series and to a lesser extent at the end. The accuracy of the observations is considerably smaller than the range of variation so we can conclude that U Cep undergoes irregular changes in period of the order of 0.01 to 0.02 day—increasing and decreasing—taking place within a timespan of the order of 7 years.

3.4. TW Dra

After rejecting 10 outliers 534 times of minimum from year 1901 on were bucketed into 150-day or longer buckets, with at best 20% of the buckets being empty using a 500-day bucket. The data are shown in Figure 15.

Four EV series constituted tends and after adjusting the period to 2.806772 days analysis of the residuals showed periodicity 36.5–37.5 years (signals 1 and 2 accounting for 40%), see Figure 16. In addition there is a weak signal of 13 to 17 years but the amplitude of the signal is far greater during the time of visual observations and virtually disappearing during the latter interval of electronic observations. Neither of the periodicities appears to be significant.

Figure 17 shows the deviations of the observations from the trend and we see the above trend pattern repeated and period changes of the order of 0.04 day occurring over 14 years or so.

3.5. U Sge

After rejecting 16 outliers, 480 times of minimum data from year 1901 on were analyzed; the data are shown in Figure 18 using a period of 3.380619 days. The raw data show the evidence of changes in the period of the order of 0.03 day—both increasing and decreasing—taking place over a timespan varying in the range of 10 to 25 years, in particular evidenced by the recent series of CCD observations. Unlike previous examples U Sge does not show a clear secular pattern.

Initial analysis presented a confusing picture with periodicities of 42 to 68 years occurring but with no regular repetition or pattern and, unlike the previous stars, no obvious long-term pattern. After including EV signals with periods of 40 years or longer with signals having undefined periods, Figure 19 using 500-day bucketing shows a wave of roughly 48 years, but with only just over two such cycles occurring in the data series we feel it more appropriate to regard this as a trend rather than genuine periodicity.

Analysis of the residuals using 500-day bucketing produced a dominant signal of 22.5 to 24 years although this was not confirmed with shorter bucketing intervals. Periods around half this occurred both with 500-day bucketing and with shorter bucketing intervals. However, testing showed no significant signals. Figure 20 shows the data minus the trend and confirms the sharp changes of 0.02 day in 9 years or so.

4. Summary and conclusions

It should be born in mind that a data-driven method of analysis may produce periodicities different from those which would arise if a model-driven approach for removing the trend had been used.

XZ And has a secular pattern spanning about 0.02 day (change in Tmin) and shows a clear and consistent residual signal across the range of bucketing intervals, methods, and weighting regimes with a highly significant signal of approximately 38 years but with much lower amplitude in the recent cycle and evidence of more complicated behavior at turning points.

RZ Cas has a secular pattern spanning about 0.06 day and shows a fairly clear and consistent residual signal, with a significant signal of approximately 24 years repeating 5 cycles throughout the data history. Analyzed separately, electronic data reveal the same signal. RZ Cas exhibits rapid shifts in Tmin of around 0.01 day over periods of 3 to 6 years.

U Cep has a secular pattern spanning about 1.1 days does but not show well defined or significant signals of any periodicity, although a long wave of 40 years is often revealed (as are 12- and 24-year periods) and is visually present in Figure 12. Electronic data suggest a different period; they also shows changes in Tmin of the order of 0.02 day occurring over a period of 7 years or so.

TW Dra has a secular pattern spanning about 1.0 day and shows a clear and consistent breakdown of the residual signal with a dominant wave of 37 years, although this does not show as significant. Electronic data show shifts in Tmin taking place over a timespan of 14 years or so.

U Sge exhibits no long-term trend, unlike the other stars discussed here, with no clear breakdown of the signals. The only period revealed is a long wave of approximately 48 years. There is evidence of changes in the period of the order of 0.02 day taking place over a timespan varying in the range of 9 years and longer, in particular evidenced by the recent series of CCD observations.

All the above stars exhibit long waves from 24 to 48 years but with only 2 to 5 cycles appearing in the data history we can not feel completely confident that these are stable periods despite what statistical significance tests might say. The strongest candidates for such stable periods (and which show as strongly significant) are XZ And, additionally because of the clarity of the signal breakdown and the magnitude of the wave, and RZ Cas, with 5 cycles present in the data and the clarity of the breakdown of the signal.

The value of amateur CCD observations is apparent from the analysis of these stars, and continued observations with accurate reduction of the results to a time of minimum are strongly encouraged for these and other close binaries.

5. Acknowledgements

This research has made use of the Lichtenknecker-Database of the BAV, operated by the Bundesdeutsche Arbeitsgemeinschaft für Veränderliche Sterne e.V. (BAV). This paper has benefited in terms of readability and content through the constructive comments of an anonymous referee.

References

Abt, H. A., and Morrell, N. 1995, *Astrophys. J., Suppl. Ser.*, **99**, 135.

Applegate, J. H. 1992, *Astrophys. J.*, **385**, 621.

Batten, A. H. 1974, *Publ. Dom. Astrophys. Obs. Victoria*, **14**, 191.

Borkovits, T., and Hegedus, T. 1996, *Astron. Astrophys., Suppl. Ser.*, **120**, 63.

Budding, E. 1986, *Astrophys. Space Sci.*, **118**, 241.

Budding, E., Erdem, A., Çiçek, C., Bulut, I., Soydugan, F., Soydugan, E., Bakiş, V., and Demircan, O. 2004, *Astron. Astrophys.*, **417**, 263.

Ceraski, W. 1880, *Astron. Nachr.*, **97**, 319.

Chaplin, G. B. 2018, *J. Amer. Assoc. Var. Star Obs.*, **46**, 157.

Chaplin, G. B. 2019, *J. Amer. Assoc. Var. Star Obs.*, **47**, 17.

Chaplin, G. B., Samolyk, G., and Screech, J. T. 2018, *J. Br. Astron. Assoc.*, **128**, 167.

Demircan, O., Akalin, A., Selam, S., Derman, E., and Mueyesseroglu, Z. 1995, *Astron. Astrophys., Suppl. Ser.*, **114**, 167.

Dobias, J. J., and Plavec, M. J. 1985, *Publ. Astron. Soc. Pacific*, **97**, 138.

Duerbeck, H. W., and Hänel, A., 1979, *Astron. Astrophys., Suppl. Ser.*, **38**, 155.

Frank, P. and Lichtenknecker, D. 1987, *BAV Mitt.*, No. 47, 1 (Lichtenknecker database http://www.bav-astro.eu/index.php/veroeffentlichungen/service-for-scientists/lkdb-engl).

Gudmundsson, L. 2017, Singular System/Spectrum Analysis (SSA; https://r-forge.r-project.org/projccts/simsalabim).

Hall, D. S. 1975, *Acta Astron.*, **25**, 1.

Hall, D. S. 1989, *Space Sci. Rev.*, **50**, 219.

Hoffman, D. I., Harrison, T. E., McNamara, B. J., Vestrand, W. T., Holtzman, J. A., and Barker, T. 2006, *Astron. J.*, **132**, 2260.
Li, M. C. A., *et al.* 2018, *Mon. Not. Roy. Astron. Soc.*, **480**, 4557.
Manzoori, D., and Gozaliasl, G. 2007, *Astron. J.*, **133**, 1302.
Manzoori, D. 2016, *Astron. Lett.*, **42**, 329.
Maxted, P. F. L., Hill, G., and Hilditch, R. W. 1994, *Astron. Astrophys.*, **282**, 821.
Mkrtichian, D. E., *et al.* 2018, *Mon. Not. Roy. Astron. Soc.*, **475**, 4745.
Muller, G. 1906, *Astron. Nachr.*, **171**, 357.
Ohshima, O., *et al.* 2001, *Astron. J.*, **122**, 418.
Olson, E. C. 1982, *Astrophys. J.*, **259**, 702.
The R Foundation for Statistical Computing. 2018a, R: A language and environment for statistical computing (https://www.R-project.org).
The R Foundation for Statistical Computing. 2018b, CRAN: The Comprehensive R Archive Network (https://cran.r-project.org).
RStudio. 2018, RStudio software (https://www.rstudio.com).
Rodriguez, E., *et al.* 2004, *Mon. Not. Roy. Astron. Soc.*, **347**, 1317.
Samolyk, G. 2011, *J. Amer. Assoc. Var. Star Obs.*, **39**, 177, and numerous other similar reports.
Simon, V. 1997, *Astron. Astorphys.*, **327**, 1087.
Singh, K. P., Drake, S. A., and White, N. E. 1995, *Astrophys. J.*, **445**, 840.
Tkachenko, A., Lehmann, H., and Mkrtichian, D. 2010, *Astron. J.*, **139**, 1327.
Tupa, P. R., DeLeo, G. G., McCluskey, G. E., Kondo, Y., Sahade, J., Giménez, A., and Caton, D. B. 2013, *Astrophys. J.*, **775**, 46.
Wenger, M., *et al.* 2000, *Astron. Astrophys., Suppl. Ser.*, **143**, 9.
Zejda, M., Mikulášek, Z., and Wolf, M. 2008, *Astron. Astrophys.*, **489**, 321.

Appendix A: Outline code for SSA analysis and significance testing

This code outline shows the steps in the analysis but for a single bucketing period, method, and weighting regime, with user entering the trend and signal components after viewing the relevant SSA analysis output. Comments are in *italics*

```
rm(list=ls(all=TRUE))
#Load User-Defined Functions
setwd("D:/Documents/R/GBC defined functions")
source("astro_udf.R")
#load Rssa R library from Install Packages
library(Rssa)
library(simsalabim) # used for MCSSA

# USER INPUT
run = "findTrends" # check O–C" "findTrends" "findSignals" "final"
star = "TW Dra"
allData = "ALL" # choose ALL or CCD for electronic only
dataMethodForSSA = "bucket" # "localPoly" or "bucket"
weightsMethod = "coded" # "inverseError" or "coded"
ylimRange = c(-0.6,0.7) # "y" range for some charts - initially a guess
periodAdjustment = -0.00008 # sometimes needed because epoch changes, try 0 at start
bucketSize = 500 # (days,) for example

# END USER INPUT

# STEP 1: Read in data and calculate OminusC

setwd("D:/Documents/EBdata")
tsIn<-read.csv(paste0(star,".csv"))
# data is named "TW Dra.csv", csv column data with header line as follows
# Tmin,N,period,OminusC,Visual,error
# N is not used; period column not used except first line gives stated period and second the epoch
# Visual is a flag as below; error is the data accuracy
# X Mkrkitchian data (RZ Cas only)
# C CCD
# E electric photometer
# K wedge photometer
# P single density photoplate
# F series of exposures

period = tsIn$period[1]; period = period + periodAdjustment
epoch = tsIn$period[2]
n1 = nrow(tsIn)
tsIn = na.omit(tsIn) # eliminate data rows where NA values exist
cat(paste0(n1-nrow(tsIn)," incomplete data rows eliminated"))
if (allData == "CCD") tsIn = tsIn[(tsIn[,5] == "C ") | (tsIn[,5] == "X ") | (tsIn[,5] == "E "),] # for electronic data only
ts = tsIn[-c(1:dropFirstNRows),]
ndata = nrow(ts)

# calculate expected Tmin and OminusC, and plot

expectedTmin = expectedTmin_udf(ts$Tmin,period,epoch) # library code calculates expected Tmin
ts$OminusC = ts$Tmin - expectedTmin
ts = cbind(ts,Weights=seq(1,1,length.out=ndata)) # add bad data flag column
if(weightsMethod == "inverseError") ts$Weights=1/ts$error else if (weightsMethod == "coded") {
 ts$Weights = ifelse(ts$Visual == "X " | ts$Visual == "C ",10,
 ifelse(ts$Visual == "E " | ts$Visual == "K ",3.25,1))
}

ts = cbind(ts,badFlag=seq(0,0,length.out=ndata)) # set bad data flag
ts = cbind(ts,bucketNumber=seq(0,0,length.out=ndata)) # add column to say which bucket data is
ts <- within(ts,badFlag <- ifelse(!is.na(OminusC),0,1)) # bad flag for missing values in OminusC
ts$OminusC[is.na(ts$OminusC)] <- 0 # convert NAs to zero
xAxis = ts$Tmin
xlimRange = c(min(xAxis),max(xAxis))
# the following chart allows the user to check whether O–C data is based on two or more epochs
# user adjusts period to get a continuous set of data
plot(y=ts$OminusC,x=ts$Tmin,xlim=xlimRange,ylim=ylimRange,xlab="",ylab="",type="p",col=" grey20",pch=3, main=paste(star,": O–C raw data"))
if (run == "check O–C") stop()

# STEP 2: do a Local Poly Regression and reject outliers

ndataOld = ndata
if (allData == "CCD") {
 sigLevels = c(4,4,4,4,4,4)
 polySpan = 0.2
} else if (allData == "ALL") {
 sigLevels = c(4,4,4,4)
 polySpan = 0.03
}
ts = badRawDataMethodLP_udf(ts,polySpan,sigLevels,ylimRange) # library code eliminates outliers
```

```
ndata = nrow(ts)

# STEP 3: calculate bucketed data and find trends

# STEP 3: bucket data
if (dataMethodForSSA == "localPoly"){
 maxBuckets = floor((ts$Tmin[ndata]-ts$Tmin[1])/bucketSize)
 bucketOmC = seq(0,0,length.out=maxBuckets)
 cat("maxBuckets ",maxBuckets,"\n")
 bucketOmC = localPolyBuckets_udf(ts,bucketSize,maxBuckets,ylimRange,
xlimRange) # library code

} else if (dataMethodForSSA == "bucket") {
 tmp = bucketDataAndFlag_udf(bucketSize,ts,drawPlot=FALSE) # library code
 maxBuckets = unlist(tmp[2])
 emptyBuckets = unlist(tmp[3])
 avgFilledBucketCount = unlist(tmp[4])
 bucketOmC = unlist(tmp[1])
 bucketOmC[is.na(bucketOmC)] <- 0
 ts$bucketNumber = unlist(tmp[6])
 flag = unlist(tmp[5])
}

# adjust period so best fit line is horizontal
Tmin <- ts$Tmin
x = c(1:maxBuckets)
fit = lm(bucketOmC~x)
slope = fit$coefficients[["x"]]
newPeriod = period + slope / bucketSize
bucketOmC = bucketOmC - slope * x
mu = mean(bucketOmC)
bucketOmC = bucketOmC - mu
plot(bucketOmC,xlim=c(1,maxBuckets),xlab="",  ylab=paste(""),
type="p",col="black",
 lwd=2, pch=19, main=paste(star,"bucket O–C adjusted period"))

# find trends
L = floor(maxBuckets / 2)
s<-ssa(bucketOmC,L,kind="1d-ssa") #run Rssa
outputVecCount = 10
plot(s,type="vectors",idx=1:outputVecCount,xlim=c(1,L),col="black",lwd=2)
# vector data plots
plot(w<-wcor(s,groups=c(1:outputVecCount)),title=paste(star,"1d-ssa
correlation matrix"))
if (run == "findTrends") stop()
trendEV = c(1,2,5) # for example after inspection of the above charts
trend <- reconstruct(s, groups = list(EV = trendEV))

trend = unlist(trend[1])
plot(bucketOmC,xlim=c(1,maxBuckets),xlab="",  ylab=paste(""),
type="p",col="black",
 lwd=1, pch=3, main=paste(star,"bucket O–C and trend"))
lines(xAxis=c(1,maxBuckets),trend,type="l",col="blue",lwd=1,lty=1)

# STEP 4: find signals

residual = bucketOmC - trend
s<-ssa(residual,L,kind="1d-ssa") #run Rssa
outputVecCount = 10
plot(s,type="vectors",idx=1:outputVecCount,xlim=c(1,L),col="black",lwd=2)
plot(w<-wcor(s,groups=c(1:outputVecCount)),title=paste(star,"1d-ssa
correlation matrix"))
if (run == "findSignals") stop()

# STEP 5: get reconstructed signals and frequencies

SSSAsignalEV = c(1:2) # for example after review of the above charts
r2 <- reconstruct(s, groups = list(EV = SSSAsignalEV))
signal = unlist(r2[1])
plot(signal,xlim=c(1,length(signal)),xlab="",  ylab=paste(""),
type="l",col="black",
 lwd=2, main=paste(star,"signal1 SSSA"))
spectrum = spectral_udf(signal,drawPlot=TRUE,paste(star,"signal1 SSSA
spectrum"),smoothing="ar")
SSSAspecPeaks1 = unlist(spectrum[1])*bucketSize/365.25
spectrum2 = spectral_udf(signal,drawPlot=TRUE,paste(star,"signal1 SSSA
spectrum"),smoothing="pgram")
plot(residual,xlim=c(1,length(trend)),xlab="bucket",  ylab=paste("O–C"),
type="p",col="black",
 pch=3, main=paste(star,"residual series"))
lines(signal,col="black",lwd=3,lty=1)

# STEP 6: MCSSA analysis

y = MCSSA(s, residual, 1000, conf = 0.99, keepSurr = FALSE, ar.method="mle")
plot(y, by = "freq", normalize = FALSE, asFreq = TRUE,
 lam.pch = 1, lam.col = "black", lam.cex = 1, sig.col = "black",
 sig.pch = 19, sig.cex = 1, conf.col = "darkgray", log = "xy",
 ann = TRUE, legend = TRUE, axes = TRUE)
```

CCD Photometry, Period Analysis, and Evolutionary Status of the Pulsating Variable V544 Andromedae

Kevin B. Alton
UnderOak and Desert Bloom Observatories, 70 Summit Ave, Cedar Knolls, NJ 07927; kbalton@optonline.net

Received October 7, 2019; revised November 15, 2019; accepted November 20, 2019

Abstract Multi-color (BVR_c) CCD-derived photometric data were acquired from V544 And, a pulsating variable presently classified as an SX Phe-type subdwarf system. Deconvolution of time-series light curve (LC) data was accomplished using discrete Fourier transformation and revealed a mean fundamental mode (f_0) of oscillation at 9.352 d^{-1} along with five other partial harmonics ($2f_0$–$6f_0$). No other statistically significant frequency shared by all bandpasses was resolved. Potential secular period changes were evaluated using four new times-of-maximum (ToMax) light produced from the present study along with other values reported in the literature. In addition, photometric data collected during the SuperWASP (2006–2007) survey combined with CCD-derived V-mag data acquired from the AAVSO archives produced twenty-nine more ToMax measurements. Corresponding residuals from the observed minus calculated values indicate very little change in the primary pulsation period since 2004. The evolutionary status, age, and physical nature of V544 And were investigated using the PAdova and TRieste Stellar Evolution Code (PARSEC) for generating stellar tracks and isochrones. At this time, the weight of evidence points to a HADS rather than an SX Phe-type pulsating star.

1. Introduction

High amplitude δ Scuti stars, hereafter HADS, represent less than 1% of all δ Sct variables (Lee *et al.* 2008). They commonly oscillate via low-order single or double radial pulsation modes (Poretti 2003a, 2003b; Niu *et al.* 2013, 2017). Pulsations are driven by the κ-mechanism resulting from partial ionization of He II (Pamyatnykh *et al.* 2004). Many (~40%) are double pulsators showing simultaneous pulsations in the fundamental and the first overtone mode with amplitudes generally higher in the fundamental mode (McNamara 2000). Non-radial pulsations have also been detected with the HADS variable V974 Oph (Poretti 2003a, 2003b). HADS variables have historically been divided according to metallicity relative to the Sun ([Fe/H]=0). The metal-poor ([Fe/H]<<0) group has been traditionally classified as SX Phe-type variables based on the prototype SX Phoencis. Purportedly they have shorter periods (0.02 < P < 0.125 d) and lower masses (~1.0–1.3 $M_\odot$) than their sibling HADS variables possessing near solar metal abundance (McNamara 2011). SX Phe stars frequently reside in globular clusters (GC) which are ancient collections of Population II stars. Therein, the majority of SX Phe variables are classified as blue straggler stars, paradoxically appearing much younger than their GC cohorts. Balona and Nemec (2012) proposed that it is not possible to differentiate between δ Sct and field SX Phe variables based on pulsation amplitude, the number of pulsation modes, period, or even metallicity (Garg *et al.* 2010). Much more sensitive space telescopes like Kepler (Gilliland *et al.* 2010), CoRoT (Baglin 2003), and MOST (Walker *et al.* 2003) have found many examples that contradict these historically accepted definitions. They further contend that the evolutionary status of each star is the only way to distinguish between these two classes.

The variability of V544 Andromedae (GSC 02815-00790) was first identified from an evaluation of unfiltered photometric data collected during the ROTSE-I Survey (Akerlof *et al.* 2000; Wozniak *et al.* 2004). Khruslov (2008) reported that V544 And (NSVS 3844113, NSVS 6459386, and NSVS 6472319) was an SX Phe-type variable with a period of 0.10694 d. Photometric (V-mag) data from V544 And were downloaded from the AAVSO International Database (Kafka 2019). Herein, an assessment of these and other LC data from the SuperWASP Survey (Butters *et al.* 2010) further confirmed the fundamental period reported by Khruslov (2008). This report marks the first multi-color photometric study on V544 And which also critically assesses its previous classification as an SX Phe-type pulsator.

2. Observations and data reduction

2.1. Photometry

Time-series images were acquired at Desert Blooms Observatory (DBO; 110.257 W, 31.941 N) with an SBIG STT-1603ME CCD camera mounted at the Cassegrain focus of a 0.4-m ACF-Cassegrain telescope. This telecompressed (0.62×) f/6.8 instrument produced an image scale of 1.36 arcsec/pixel (bin=2×2) and a field-of-view (FOV) of 17.2'×11.5'. Image acquisition (75-s) was performed using THESKYX PRO Version 10.5.0 (Software Bisque 2019). The CCD-camera is equipped with B, V, and R_c filters manufactured to match the Johnson-Cousins Bessell specification. Dark subtraction, flat correction, and registration of all images collected at DBO were performed with AIP4WIN v2.4.0 (Berry and Burnell 2005). Instrumental readings were reduced to catalog-based magnitudes using the MPOSC3 star fields (Warner 2007) built into MPO CANOPUS v10.7.1.3 (Minor Planet Observer 2019). Light curves (LC) for V544 And were generated using an ensemble of five non-varying comparison stars. The identity, J2000 coordinates, and MPOSC3 color indices (B–V) for these stars are provided in Table 1. A representative image containing the target (T) and matching ensemble of comparison stars (1–5) is also provided (Figure 1). Data from images taken below 30° altitude (airmass > 2.0) were excluded; considering the proximity of all program stars, differential atmospheric extinction was ignored. During each imaging session comparison stars typically stayed within ±0.009 mag for V and R_c filters and ±0.015 mag for B passband.

Table 1. Astrometric coordinates (J2000), V-mag, and color indices (B–V) for V544 And and five comparison stars (1–5) used during this photometric study.

FOV ID	Star Identification	R.A. h m s	Dec. ° m s	MPOSC3[a] V-mag	MPOSC3[a] (B–V)
T	V544 And[b]	01 44 27.968	+37 58 53.697	12.790	0.231
1	GSC 02815-01525	01 44 44.500	+38 02 06.345	11.082	0.385
2	GSC 02815-01590	01 44 49.205	+38 02 27.708	13.298	0.531
3	GSC 02815-01205	01 45 00.547	+38 05 22.596	11.709	0.623
4	GSC 02815-00872	01 44 07.721	+38 01 57.396	13.066	0.572
5	GSC 02815-01467	01 44 44.597	+38 02 54.960	13.543	0.683

[a] V-mag and (B–V) for comparison stars derived from MPOSC3 database described by Warner (2007).
[b] Mean V-mag and (B–V) determined during this study.

Table 2. Times of maximum (ToMax), measurement uncertainty, filter type, epoch, and fundamental pulsation timing differences (PTD) for V544 And used to calculate a new linear ephemeris.

ToMax (HJD–2400000)	Uncertainty	Filter[a]	Cycle No.	PTD	Ref.	ToMax (HJD–2400000)	Uncertainty	Filter[a]	Cycle No.	PTD	Ref.
53201.6971	0.0011	T	–49048	–0.0011	1	56587.3408	0.0005	C	–17388	–0.0005	6
53220.6268	0.0014	T	–48871	0.0007	1	56587.4482	0.0004	C	–17387	0.0000	6
53228.6461	0.0007	T	–48796	–0.0004	1	56587.5549	0.0003	C	–17386	–0.0002	6
53229.6101	0.0016	T	–48787	0.0011	1	56920.3443	0.0035	TG	–14274	–0.0005	7
53232.6039	0.0021	T	–48759	0.0007	1	56920.3443	0.0035	TR	–14274	–0.0005	7
53235.5976	0.0013	T	–48731	0.0002	1	56920.3445	0.0035	TB	–14274	–0.0003	7
53238.5923	0.0013	T	–48703	0.0006	1	56920.4507	0.0035	TG	–14273	–0.0010	7
53241.5851	0.0015	T	–48675	–0.0008	1	56920.4509	0.0035	TR	–14273	–0.0008	7
53241.6918	0.0014	T	–48674	–0.0011	1	56920.4513	0.0035	TB	–14273	–0.0004	7
53245.6504	0.0016	T	–48637	0.0008	1	56928.3646	0.0010	C	–14199	–0.0005	8
53256.5579	0.0018	T	–48535	0.0007	1	56928.4713	0.0005	C	–14198	–0.0008	8
53256.6635	0.0011	T	–48534	–0.0006	1	56970.2481	0.0015	0	–13807.5[b]	0.0169	7
54049.3928	0.0016	T	–41121	0.0006	1	56970.3565	0.0013	0	–13806.5[b]	0.0184	7
54049.6064	0.0012	T	–41119	0.0002	1	57004.2907	0.0006	V	–13489	–0.0001	8
54103.3948	0.0016	T	–40616	–0.0009	1	57004.3981	0.0009	V	–13488	0.0004	2
55117.3794	0.0014	V	–31134	0.0018	2	57004.3983	0.0008	V	–13488	0.0006	8
55117.4861	0.0016	V	–31133	0.0016	2	57294.4125	0.0009	V	–10776	0.0001	2
55117.5923	0.0012	V	–31132	0.0008	2	57294.5182	0.0008	V	–10775	–0.0011	2
55227.3092	0.0009	V	–30106	–0.0002	2	57667.3050	0.0008	V	–7289	0.0014	2
55227.4153	0.0008	V	–30105	–0.0010	2	57727.2963	0.0008	V	–6728	0.0007	2
55452.4119	0.0007	V	–28001	–0.0011	3	58104.3571	0.0007	V	–3202	–0.0004	2
55452.5192	0.0006	V	–28000	–0.0007	3	58177.2880	0.0007	V	–2520	–0.0008	2
55452.6263	0.0005	V	–27999	–0.0005	3	58384.6407	0.0007	V	–581	–0.0001	2
55531.4399	0.0005	V	–27262	0.0001	3	58443.6707	0.0010	R_c	–29	0.0004	9
55851.2895	0.0004	C	–24271	–0.0006	4	58443.6711	0.0012	B	–29	0.0008	9
55870.2183	0.0006	V	–24094	0.0003	4	58443.6715	0.0011	V	–29	0.0012	9
55870.3251	0.0004	V	–24093	0.0002	4	58443.7765	0.0005	R_c	–28	–0.0007	9
55870.4321	0.0004	V	–24092	0.0002	4	58443.7775	0.0004	V	–28	0.0002	9
55870.5389	0.0005	V	–24091	0.0001	4	58443.7785	0.0016	B	–28	0.0013	9
56228.3511	0.0007	V	–20745	–0.0008	5	58446.6635	0.0011	R_c	–1	–0.0010	9
56228.3521	0.0010	V	–20745	0.0002	2	58446.6647	0.001	V	–1	0.0001	9
56235.3026	0.0014	C	–20680	–0.0002	5	58446.6656	0.0014	B	–1	0.0011	9
56581.2459	0.0012	C	–17445	0.0001	6	58446.7703	0.0013	R_c	0	–0.0012	9
56581.3541	0.0008	C	–17444	0.0014	6	58446.7712	0.0014	V	0	–0.0003	9
56587.2341	0.0005	C	–17389	–0.0002	6	58446.7724	0.0018	B	0	0.0009	9

[a] T = TAMMAG2 from SuperWASP; C = clear; 0 = no filter; TG = tricolor green channel; TR = tricolor red channnel; TB = tricolor blue channel; V, B, and R_c = photometric Johnson-Cousins.

[b] Anomalous values not used for determination of linear ephemeris.

Reference note: 1. SuperWASP: Butters et al. (2010); 2. AAVSO International Variable Star Index: Watson et al. (2014); 3. Wils et al. (2011); 4. Wils et al. (2012); 5. Wils et al. (2013); 6. Wils et al. (2014); 7. Hübscher (2015); 8. Wils et al. (2015); 9. This study at DBO.

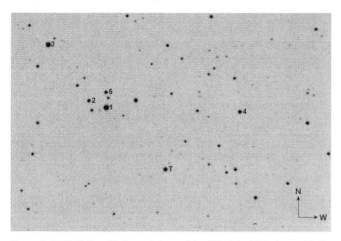

Figure 1. FOV (17.2 × 11.5 arcmin) containing V544 And (T) along with the five comparison stars (1–5) used to reduce time-series images to MPOSC3-catalog based magnitudes.

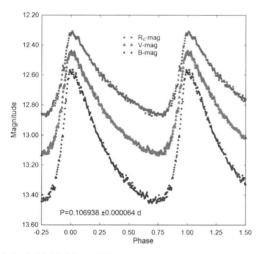

Figure 2. Period folded (0.106938 ± 0.000064 d) LCs for V544 And produced from photometric data obtained between 21 Nov and 24 Nov 2018, at DBO. LCs shown at the top (R_c), middle (V_c) and bottom (B) represent catalog-based (MPOSC3) magnitudes determined using MPO CANOPUS.

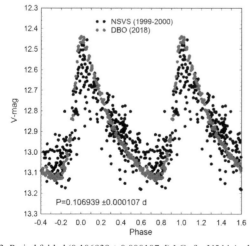

Figure 3. Period folded (0.106939 ± 0.000107 d) LCs for V544 And produced with sparsely sampled data from the ROTSE-I (1999–2000) Survey. Precise time-series V-mag data acquired at DBO (2018) are superimposed for direct comparison with ROTSE-I magnitudes which have been offset to conform to the DBO-derived V-mag values.

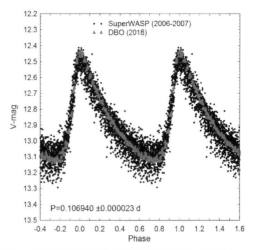

Figure 4. Period folded (0.106940 ± 0.000023 d) LCs for V544 And produced with broad-band (400–700 nm) filtered data from the SuperWASP survey (2006–2007). Precise time-series V-mag data acquired at DBO (2018) are superimposed for direct comparison with SuperWASP magnitudes which have been offset to conform to the DBO-derived V-mag values.

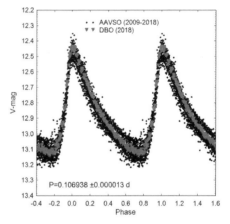

Figure 5. Period folded (0.106938 ± 0.000013 d) LCs for V544 And produced with V-mag data from the AAVSO archives (2009–2018). Precise time-series V-mag data acquired at DBO (2018) are superimposed for direct comparison with the AAVSO data which have been offset to conform to the DBO-derived V-mag values.

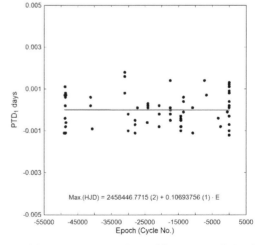

Figure 6. Straight line fit of pulsation timing difference vs. period cycle number indicating overall little or no change to the fundamental pulsation period of V544 And had occurred between 2004 and 2018.

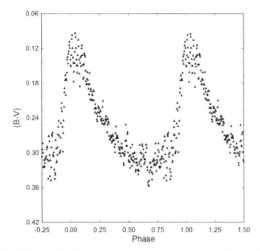

Figure 7. V544 And LC illustrating significant increase in reddening (0.136 < (B–V) < 0.326) as maximum light slowly descends to minimum light. This effect is most closely associated with a decrease in the effective surface temperature during minimum light.

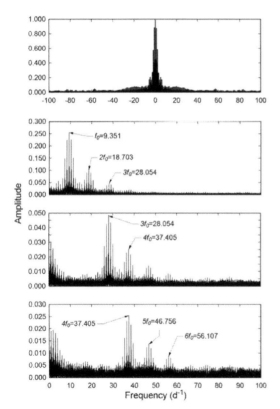

Figure 8. Amplitude spectra showing spectral window (top) and all significant pulsation frequencies following DFT analysis of photometric data from V544 And acquired by the SuperWASP survey between 21 Aug 2006 and 03 Jan 2007. This includes the fundamental f_0 frequency through its highest partial harmonic $6f_0$ detected (S/N ≥ 6) following prewhitening.

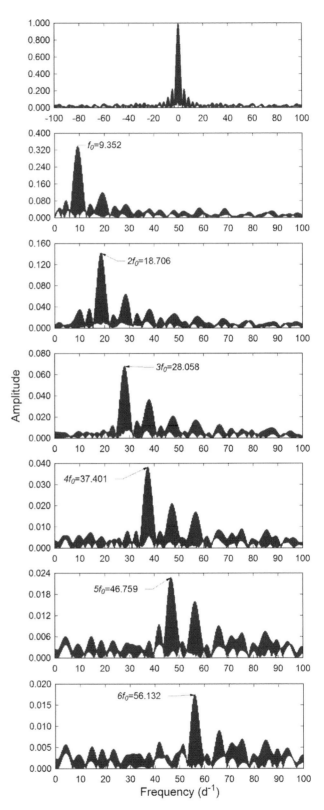

Figure 9. Amplitude spectra showing all significant pulsation frequencies following DFT analysis of B-mag photometric data from V544 And acquired at DBO between 21 Nov and 24 Nov 2018. This includes the spectral window (top) and the fundamental f_0 frequency through its highest partial harmonic $6f_0$ detected (S/N ≥ 6) following prewhitening.

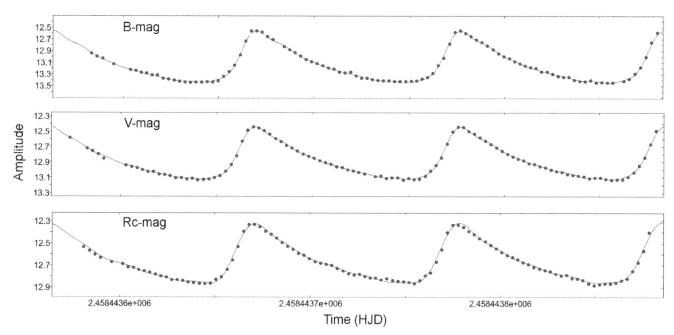

Figure 10. Representative DFT fit of B- (top), V- (middle), and R_c-mag (bottom) time series data based on elements derived from DFT. These data were acquired on 21 Nov 2018 at DBO.

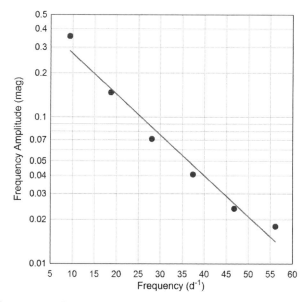

Figure 11. Amplitude decay of the fundamental (f_0) pulsation period and its corresponding partial harmonics ($2f_0$–$6f_0$) observed in the B-passband from LCs acquired in 2018 (DBO).

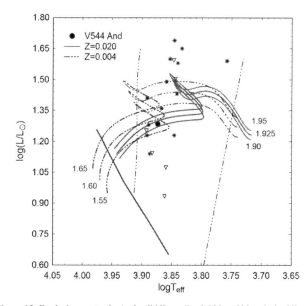

Figure 12. Evolutionary tracks (red solid lines; Z = 0.020 and blue dashed lines; Z = 0.004) derived from PARSEC models (Bressan *et al.* 2012). The position of V544 And (solid-black circle) is shown relative to ZAMS (thick maroon line) and within the theoretical instability strip (dashed lines) for low-order radial mode δ Scuti pulsators. The position of other HADS (*) and SX Phe (▽) variables reported by Balona (2018) are included for comparison.

3. Results and discussion

3.1. Photometry and ephemerides

Photometric values in B (n = 194), V (n = 196), and R_c (n = 198) passbands were each processed to produce LCs that spanned three days between 21 Nov and 24 Nov, 2018 (Figure 2). Period determinations were performed using PERANSO v2.6 (Paunzen and Vanmunster 2016) by applying periodic orthogonals (Schwarzenberg-Czerny 1996) to fit observations and analysis of variance to assess fit quality. In this case a mean period solution for all passbands (0.106938 ± 0.000064 d) was obtained. Folding together the sparsely sampled ROTSE-I survey data revealed a period at 0.106939 ± 0.000107 d (Figure 3). The SuperWASP survey (Butters et al. 2010) provided a wealth of photometric data taken (30-s exposures) at modest cadence that repeats every 9 to 12 min. These data acquired in 2006–2007 were period folded and reached superimposition when P = 0.106940 ± 0.000023 d (Figure 4). Finally, V-mag data mined from the AAVSO International Variable Star Index (2009–2018) were similarly evaluated (Figure 5) to produce a fundamental period where P = 0.106938 ± 0.000013 d. ToMax estimates were derived using the polynomial extremum fit utility featured in PERANSO 2.6 (Paunzen and Vanmunster 2016). New data derived from DBO (4), AAVSO (14), and SuperWASP (15) LCs along with published ToMax values (Table 2) were used to produce the following linear ephemeris (Equation 1):

$$\text{Max (HJD)} = 2458446.7714(2) + 0.1069376(1)\,E. \quad (1)$$

Secular changes in the fundamental period can potentially be uncovered by plotting the difference between the observed ToMax values and those predicted by the reference epoch against cycle number (Figure 6). The time difference values vs. cycle number (Table 2) are best described by a straight line relationship (Equation 1), suggesting no apparent long-term change to the fundamental period has occurred since 2004. These results along with the period folded LCs from DBO (Figure 2), ROTSE-I (Figure 3), SuperWASP (Figure 4), and AAVSO (Figure 5) also reveal that the V-mag amplitude has not changed significantly over the same period of time.

3.2. Light curve behavior

Morphologically, light curves from HADS variables are asymmetrical with rapid brightening to produce a sharply defined maximum peak. Thereafter a slower decline in magnitude results in a broad minimum. The largest difference between maximum and minimum light is observed in the blue passband (ΔB-mag = 0.88), followed by V (ΔV-mag = 0.69), and finally the smallest difference detected in infrared (ΔR_c-mag = 0.55). This behavior is characteristic of pulsating F- to A-type stars. Plotting (B–V) vs. phase reveals significant reddening during minimum light (Figure 7). Interstellar extinction (https://irsa.ipac.caltech.edu/applications/DUST) was obtained according to Schlafly and Finkbeiner (2011). The reddening value (E(B–V) = 0.0454 ± 0.0009 mag) corresponds to an intrinsic color index (B–V)$_0$ for V544 And that varies between 0.091 ± 0.036 at maximum light and 0.281 ± 0.017 mag at minimum brightness. Based on the polynomial transformation equations derived by Flower (1996), the average effective temperature (T_{eff}) was estimated to be 7853 ± 1524 K. These results based strictly on (B–V) photometry at DBO are higher than findings for V544 And (T_{eff} = 7398$^{+133}_{+104}$ K) included in the Gaia DR2 release of stellar parameters (Andrae et al. 2018). The latest data release (DR5) from the LAMOST telescope (Zhao et al. 2012) reports that T_{eff} = 7190 ± 18 K, wherein V544 And is assigned spectral type A7 based upon low resolution (R ≈ 1000) spectra (http://dr5.lamost.org/spectrum/view?obsid=371004085) (Rui et al. 2019). Parameter estimates (T_{eff}, [Fe/H], log g, and radial velocity) were derived from LAMOST DR5 using Generative Spectrum Networks following training on PHOENIX spectra (Husser et al. 2013). The final T_{eff} value (7480 ± 339 K) adopted for this study represents a mean value from LAMOST DR5 (7190 K), Gaia DR2 (7398 K), and DBO (7853 K).

3.3. Light curve analysis by Discrete Fourier Transformation

Discrete Fourier Transformation (DFT) was used to extract the fundamental pulsating frequency (spectral window = 100 d^{-1}) using PERIOD04 (Lenz and Breger 2005). Pre-whitening steps which successively remove the previous most intense signals were employed to tease out other potential oscillations from the residuals. Only those frequencies with a S/N ≥ 6 (Baran et al. 2015) in each passband are presented in Table 3. In all cases, uncertainties in frequency, amplitude, and phase were estimated by the Monte Carlo simulation (n = 400) routine featured in PERIOD04. Minor differences in the fundamental and associated harmonic frequencies between data sources (DBO vs. SuperWASP) may be attributed to decreases in the S/N ratio with harmonic order, varying filter bandpass, and/or differences in the data acquisition cadence and duration.

The DFT results demonstrate that V544 And is a monoperiodic radial pulsator; changes in stellar size during each pulsation cycle are therefore symmetrical. The amplitude spectra derived from a subset (2006–2007) of the SuperWASP data are illustrated in Figure 8. Others from DBO (B-passband) are shown in Figure 9 but do not include V- and R_c-passbands since they are essentially redundant. As would be expected, the fundamental pulsation period (f_0 = 9.352 ± 0.0001 d^{-1}) has the greatest amplitude. Sequential pre-whitening of LCs from DBO uncovered partial harmonics out as far as 6f_0, except for the R_c-passband where detection was limited to 5f_0. Representative DFT fits to B-, V-, and R_c-mag LC data collected at DBO (21 Nov 2018) are illustrated in Figure 10. The amplitude decay appears to be exponential as a function of harmonic order (Figure 11). This behavior has been observed with other HADS variables such as VX Hya (Templeton et al. 2009), RR Gem (Jurcsik et al. 2005), V460 And (Alton and Stępień 2019a), and V524 And (Alton and Stępień 2019b). No other independent pulsation modes were consistently detected during the short campaign at DBO or when the SuperWASP data were analyzed by PERIOD04.

3.4. Global parameters

Pulsating stars have long endured as standard candles for estimating cosmic distances ever since Henrietta Leavitt discovered a period-luminosity (P-L) relationship with

Table 3. Fundamental frequency (f_0) and corresponding partial harmonics ($2f_0$–$6f_0$) detected following DFT analysis of time-series photometric data (BVR_c) collected from V544 And at DBO (2018) along with TAMMAG2 (T) results acquired from SuperWASP (2006–2007).

	Freq. (d^{-1})	Freq. Error	Amp. (mag.)	Amp. Error	Phase	Phase Error	Amp. S/N
f_0–B	9.352	0.001	0.357	0.002	0.534	0.173	154
f_0–V	9.352	0.001	0.274	0.001	0.801	0.171	130
f_0–R_c	9.352	0.001	0.217	0.002	0.549	0.140	122
f_0–T	9.351	0.001	0.256	0.001	0.538	0.340	140
$2f_0$–B	18.706	0.001	0.148	0.002	0.252	0.172	70
$2f_0$–V	18.706	0.001	0.116	0.001	0.434	0.187	45
$2f_0$–R_c	18.707	0.001	0.094	0.001	0.738	0.159	72
$2f_0$–T	18.703	0.001	0.104	0.002	0.274	0.382	61
$3f_0$–B	28.058	0.002	0.056	0.002	0.433	0.173	41
$3f_0$–V	28.064	0.001	0.056	0.001	0.075	0.184	24
$3f_0$–R_c	28.064	0.002	0.045	0.002	0.509	0.156	32
$3f_0$–T	28.054	0.001	0.049	0.001	0.503	0.352	38
$4f_0$–B	37.401	0.003	0.041	0.002	0.030	0.182	22
$4f_0$–V	37.415	0.003	0.031	0.001	0.166	0.186	19
$4f_0$–R_c	37.072	0.004	0.023	0.002	0.057	0.151	14
$4f_0$–T	37.405	0.002	0.026	0.002	0.231	0.379	19
$5f_0$–B	46.759	0.005	0.024	0.002	0.663	0.163	10
$5f_0$–V	46.754	0.005	0.018	0.001	0.359	0.176	9
$5f_0$–R_c	46.421	0.007	0.013	0.002	0.773	0.149	11
$5f_0$–T	46.756	0.001	0.014	0.001	0.749	0.394	10
$6f_0$–B	56.132	0.007	0.018	0.002	0.329	0.179	12
$6f_0$–V	55.767	0.007	0.011	0.001	0.293	0.164	8
$6f_0$–T	56.107	0.001	0.009	0.001	0.709	0.373	7

Table 4. Global stellar parameters for V544 And using Gaia DR2 derived values and those determined directly from observations at DBO.

Parameter	Gaia DR2	DBO
Mean T_{eff} [K]	7398^{+104}_{-133}	7480 ± 339[a]
Mass [$M_\odot$]	2.00 ± 0.15	1.99 ± 0.14
Radius [$R_\odot$]	2.69 ± 0.09	2.61 ± 0.09
Luminosity [$L_\odot$]	19.56 ± 4.68	19.27 ± 5.88
$\rho_\odot$ [g/cm^3]	0.145 ± 0.018	0.157 ± 0.019
log g [cgs]	3.879 ± 0.044	3.903 ± 0.042
Q [d]	0.034 ± 0.003	0.036 ± 0.012

[a] Value adopted is mean T_{eff} from DBO, Gaia DR2, and LAMOST DR5.

Cepheid variables in the Small Magellanic Cloud (Leavitt and Pickering 1912). Later on, this P-L relationship was refined to accommodate differences between metal rich (Population I) and metal-poor (Population II) Cepheids (Baade 1956). Similar to Cepheids, other variable stars that pulsate via the κ-mechanism were found to obey distinct P-L relationships. P-L relationships for RR Lyrae-type variables in the mid-infrared (Neeley et al. 2015) and near infrared (Longmore et al. 1986) have been empirically-derived to estimate distances to globular clusters. A refinement of the P-L relationship for δ Sct variables was reported by McNamara (2011), albeit with now outdated Hipparcos parallaxes. A new P-L relationship using, for the most part, more accurate values determined by the Gaia Mission (Lindegren et al. 2016; Brown et al. 2018) was recently published (Ziaali et al. 2019). Accordingly, this empirically-derived expression (Equation 2):

$$M_V = (-2.94 \pm 0.06) \log(P) - (1.34 \pm 0.06), \quad (2)$$

is similar to the equation published by McNamara (2011) but with improved precision. The corresponding value for absolute V_{mag} ($M_V = 1.47 \pm 0.08$) when substituted into the reddening corrected distance modulus produced an estimated distance (1723 ± 78 pc) to V544 And. This result is slightly higher than the Gaia-DR2 distance (1695^{+294}_{-222} pc) but well within the confidence intervals calculated from parallax using the Bailer-Jones bias correction (Bailer-Jones 2015).

Alternatively, McNamara (2011) describes an empirical relationship (Equation 3) for metal poor stars ([Fe/H] < –2.0) where the absolute magnitude can be estimated according to:

$$M_V = (-2.90 \pm 0.05) \log(P) - (0.89 \pm 0.05). \quad (3)$$

In this case $M_V = 1.93 \pm 0.07$, so the reddening corrected distance resulted in a much lower value (1395 ± 53 pc) compared to that calculated (1723 ± 78 pc) from Equation 2. Given the uncertainty about the Gaia DR2 value (1695^{+294}_{-222} pc), which is presently considered the gold standard for distance, it is unlikely that these values are statistically different.

The location of V544 And, which is ~682 pc below the Galactic plane, suggests residence in the thick disc where the scale height is ~0.9 kpc (Li and Zhao 2017) rather than the halo where many metal poor ([Fe/H] < –1.6) stars reside (Carollo et al. 2010). The proper motion ($\mu_\alpha = 17.873 \pm 0.101$; $\mu_\delta = -8.304 \pm 0.174$ km·s^{-1}) and distance (1695 pc) to V544 And are known from Gaia DR2 (Lindegren et al. 2016; Brown et al. 2018) while the radial velocity ($v_R = -15.3 \pm 7.1$ km·s^{-1}) has been reported in LAMOST DR5 (Rui et al. 2019). Space velocities (($U_{LSR} = -81.7$, $V_{LSR} = -122.4$, and $W_{LSR} = -16.9$ km·s^{-1}) were determined (http://kinematics.bdnyc.org/query) relative to the local standard of rest assuming the solar motion reported by Dehnen and Binney (1998) where U = 10, V = 5.3, and W = 7.2 km·s^{-1}. More recent (Coşkunoğlu et al. 2011) estimates for the solar motion standard of rest (U = –8.5, V = 13.38, and W = 6.49 km·s^{-1}) lead to somewhat different space velocities for V544 And ($U_{LSR} = 93.02$, $V_{LSR} = -104.32$, and $W_{LSR} = -23.75$ km·s^{-1}) (https://rdrr.io/cran/astrolibR/man/gal_uvw.html). However, other than a change in convention where U_{LSR} is positive outward toward the Galactic anticenter, total spatial velocity

$$\left(\sqrt{U^2_{LSR} + V^2_{LSR} + W^2_{LSR}} \right)$$

by either local standard of rest is too small (≤ 148.1 km·s^{-1}) for a halo object (> 180 km·s^{-1}) but convincingly within the range for a thick disk resident (Yan et al. 2019).

Gaia DR2 also includes estimates for solar radius and luminosity (Andrae et al. 2018). In this regard it is worth noting that there are some differences in the equations used to produce the Gaia DR2 values reported for V544 And and those otherwise determined herein (Table 4). The corresponding data reduction has been discussed in more detail for two other HADS variables V460 And (Alton and Stępień 2019a) and V524 And (Alton and Stępień 2019b).

The mass of a single isolated field star is very difficult to determine directly. Instead, a mass-luminosity (M-L) relationship ($1.05 < M/M_\odot \leq 2.40$) developed by Eker *et al.* (2018) for main sequence (MS) stars in detached binary systems was used as the best estimate. This expression (Equation 4):

$$\log(L) = 4.329(\pm 0.087) \cdot \log(M) - 0.010(\pm 0.019), \quad (4)$$

leads to a mass (2.00 ± 0.15) in solar units as derived from the Gaia DR2 stellar parameters where $L = 19.56 \pm 4.68\ L_\odot$. All values summarized in Table 4 fall well within expectations for a HADS variable. It should be emphasized that these fundamental physical parameters were derived by assuming a null value for the Gaia extinction ($A_G = 0$). V544 And resides in a region of the Milky Way (Gal. coord. (J2000): $l = 134.3183$; $b = -23.7149$) where the interstellar extinction ($A_V = 0.1417 \pm 0.0028$) is measureably different from zero (Schlafly and Finkbeiner 2011). Therefore, the same equations (Equations 6–8 in Alton and Stępień 2019a) were applied but this time using the data obtained at DBO where $V_{avg} = 12.790 \pm 0.023$, $A_V = 0.1417 \pm 0.0028$, $M_V = 1.528$, $T_{eff} = 7480$ K, and $BC_V = 0.0243$. The results summarized in Table 4 indicate that the Gaia DR2 reported values for luminosity and radius appear to be comparable even with different assumptions about interstellar extinction. The luminosity ($19.27 \pm 5.88\ L_\odot$) produced from the DBO data translates into a radius ($R_* = 2.61 \pm 0.09$) and a mass of 1.99 ± 0.141. Additionally, stellar radius in solar units was independently determined from an empirically derived period-radius (P-R) relationship (Equation 5) reported by Laney *et al.* (2003) for HADS and classical Cepheids:

$$\log(R_*) = a + b \cdot \log(P) + c, \quad (5)$$

where $a = 1.106 \pm 0.012$, $b = 0.725 \pm 0.010$, and $c = 0.029 \pm 0.024$. In this case the value for R_* ($2.70 \pm 0.40\ R_\odot$) was closer to the value obtained from the Gaia DR2 estimates ($2.69 \pm 0.09\ R_\odot$). Khruslov (2008) suggested the "large distances from the Galactic plane make it possible to consider" classifying V544 And as an SX Phe-type pulsator. However, as mentioned earlier, more recent distance and kinematic evidence suggests that V544 And does not inhabit the halo where metal deficient ([Fe/H] <-1.6) stars typically reside. Moreover, the mass threshold of an SX Phe-type variable is generally regarded to be less than 1.3 $M_\odot$ (McNamara 2011). This significant distinction challenges the notion that V544 And is an SX Phe-type variable.

Derived values for density (ρ_*), surface gravity ($\log g$), and pulsation constant (Q) are also included in Table 4. Stellar density (ρ_*) in solar units (g/cm^3) was calculated according to Equation 6:

$$\rho_* = 3 \cdot G \cdot M_* \cdot m_\odot / (4\pi (R_* \cdot r_\odot)^3). \quad (6)$$

where G is the cgs gravitational constant, $m_\odot =$ solar mass (g), $r_\odot =$ solar radius (cm), M_* is the mass, and R_* the radius of V544 And, both in solar units. Using the same algebraic assignments, surface gravity ($\log g$) was determined by the following expression (Equation 7):

$$\log g = \log (M_* \cdot m_\odot / (R_* \cdot r_\odot)^2). \quad (7)$$

The dynamical time that it takes a p-mode acoustic wave to internally traverse a star is related to its size but more strongly correlates to the stellar mean density. The pulsation constant (Q) is defined by the period-density relationship (Equation 8):

$$Q = P \sqrt{\bar{\rho}_* / \bar{\rho}_\odot}\ . \quad (8)$$

where P is the pulsation period (d) and $\bar{\rho}_*$ and $\bar{\rho}_\odot$ are the mean densities of the target star and Sun, respectively. The mean density can be expressed (Equation 9) in terms of other measurable stellar parameters where:

$$\log(Q) = -6.545 + \log(P) + 0.5 \log(g) + 0.1 M_{bol} + \log(T_{eff}). \quad (9)$$

The full derivation of this expression can be found in Breger (1990). The resulting Q value (Table 4) derived from observations at DBO is consistent with theory and the distribution of Q-values (0.03–0.04 d) from fundamental radial pulsations observed with other δ Sct variables (Breger 1979; Joshi and Joschi 2015; Antonello and Pastori 1981).

3.5. Evolutionary status of V544 And

The evolutionary status of V544 And was evaluated (Figure 12) using PARSEC-derived stellar tracks and isochrones (Bressan *et al.* 2012) and then plotted ($\log T_{eff}$ vs. $\log(L/L_\odot)$ as a theoretical Hertzsprung-Russell diagram (HRD). The thick solid maroon-colored line defines the zero-age main sequence (ZAMS) position for stars with metallicity $Z = 0.014$. The two broken lines nearly perpendicular to the ZAMS delimit the blue (left) and red (right) edges of the theoretical instability strip for radial low-p modes (Xiong *et al.* 2016). Included are the positions of several known HADS and SX Phe-type variables (Balona 2018). The solid-black circle indicates the position of V544 And using the DBO derived parameters and corresponding error estimates provided in Table 4. Over the last few decades, the reference metallicity values used by several authors for computing stellar models have ranged between 0.012 and 0.020 (Amard *et al.* 2019). Investigations focused on a definitive value for $Z_\odot$ have been recently reported (von Steiger and Zurbuchen 2016; Serenelli *et al.* 2016; Vagnozzi *et al.* 2017); ironically a single undisputed value for solar metallicity remains elusive. Nonetheless, a metal abundance (Z) value is required in order to determine the mass, radius, and age of V544 And from theoretical evolutionary tracks. A Z-value can be estimated indirectly from its Galactic coordinates and kinematic properties. As mentioned earlier, the total spatial velocity (148.1 km·s^{-1}) suggests affiliation with the galactic thick disc (Yan *et al.* 2019). Its distance from the galactic plane (-682 pc) also favors a thick disc membership rather than residence in the halo. It can therefore be assumed that V544 And approaches solar metallicity, or at most a few times lower.

Two separate PARSEC evolutionary models (Bressan *et al.* 2012) each ranging in age between 1×10^8 and 1.7×10^9 y are illustrated in Figure 12. The red solid lines show the models ($M_\odot = 1.90$, 1.925, and 1.95) when $Z = 0.020$, while the blue, dash-dotted lines define the models ($M_\odot = 1.55$, 1.60, and 1.65) where $Z = 0.004$. The latter simulations correspond to a decrease in metallicity by a factor of 3 to 5 depending on the reference solar metallicity. Assuming $Z = 0.020$, it can be seen that V544 And would have a mass of 1.91 ± 0.02 $M_\odot$ and a radius of 2.69 ± 0.13 $R_\odot$. The position of this intrinsic variable along the $M_\odot = 1.90$ evolutionary track corresponds to an age of 1.09 Gyr suggesting it is a moderately evolved MS object lying amongst other HADs close to the blue edge of the instability strip.

By comparison, if V544 And is a metal poor ($Z = 0.004$) star, then it would have a similar radius (2.51 ± 0.20), but would be less massive (1.58 $M_\odot$). Its position lies just prior to the HRD region where evolutionary tracks of low metallicity stars begin stellar contraction near the end of hydrogen burning in the core. This star would still be a MS object but with an age approaching 1.50 Gyr.

The theoretical mass (1.91 $M_\odot$) where $Z = 0.020$ favors the higher metallicity of V544 And and is in good agreement with results independently determined (Table 4) using an empirical M-L relationship. In this regard, the LAMOST DR5 metallicity value ([Fe/H] = -0.571) may be underestimated or does not scale with its total metal abundance (Z). By any measure, V544 And is very likely not a metal poor ([Fe/H] < -1) star traditionally associated with SX Phe-type variables. Should high resolution spectroscopic data become available in the future, uncertainty about the mass of V544 And will likely improve.

4. Conclusions

This first multi-color (BVR_c) CCD study of V544 And has produced 4 new times-of-maximum along with another 29 values culled from the SuperWASP survey and the AAVSO archives. These results and other published values resulted in a new linear ephemeris. An assessment using the observed and predicted times-of-maximum suggests that since 2004 very little change in the fundamental pulsation period has occurred and that the V-mag amplitude has not changed significantly over the same period of time. Deconvolution of time-series photometric data by discrete Fourier transformation shows that V544 And is a monoperiodic radial pulsator ($f_0 = 9.352$ d^{-1}) which also oscillates in at least five other partial harmonics ($2f_0 - 6f_0$). A mean effective temperature for V544 And (7480 ± 339 K) was estimated from LAMOST DR5, Gaia DR2, and DBO results and corresponds to spectral type A7. These results, along with the distance (1724 ± 78 pc) estimated from the distance modulus, agreed reasonably well with the same findings (1695^{+294}_{-222} pc) provided in Gaia DR2. The pulsation period (~0.10694 d), radial oscillation mode, V_{mag} amplitude (0.69 mag), and LC morphology are all consistent with the defining characteristics of a HADS variable. However, these three parameters do not necessarily exclude the possibility that V544 And is a field SX Phe-type pulsator. Nonetheless in this case, the estimated mass of V544 And (~1.9–2 $M_\odot$) far exceeds the generally accepted threshold (M < 1.3 $M_\odot$) for SX Phe stars (McNamara 2011). Moreover, evolutionary tracks from the PARSEC model which assume near solar abundance ($Z = 0.020$) are best matched by a MS star with a mass of 1.91 $M_\odot$ and radius of 2.69 $R_\odot$. Finally there is a high probability that V544 And resides in the thick disk based on total spatial velocity. Given these results, the weight of evidence indicates that V544 And should be classified as a HADS rather than an SX Phe variable.

5. Acknowledgements

This research has made use of the SIMBAD database operated at Centre de Données astronomiques de Strasbourg, France. Time-of-maximum light data from the Information Bulletin on Variable Stars (IBVS) website proved invaluable to the assessment of potential period changes experienced by this variable star. In addition, the Northern Sky Variability Survey hosted by the Los Alamos National Laboratory and the International Variable Star Index archives maintained by the AAVSO were mined for critical information. This work also presents results from the European Space Agency (ESA) space mission Gaia. Gaia data are being processed by the Gaia Data Processing and Analysis Consortium (DPAC). Funding for the DPAC is provided by national institutions, in particular the institutions participating in the Gaia MultiLateral Agreement (MLA). The Gaia mission website is https://www.cosmos.esa.int/gaia. The Gaia archive website is https://archives.esac.esa.int/gaia. This paper makes use of data from the first public release of the WASP data as provided by the WASP consortium and services at the NASA Exoplanet Archive, which is operated by the California Institute of Technology, under contract with the National Aeronautics and Space Administration under the Exoplanet Exploration Program. The use of public data from LAMOST is also acknowledged. Guoshoujing Telescope (the Large Sky Area Multi-Object Fiber Spectroscopic Telescope LAMOST) is a National Major Scientific Project built by the Chinese Academy of Sciences. Funding for the project has been provided by the National Development and Reform Commission. LAMOST is operated and managed by the National Astronomical Observatories, Chinese Academy of Sciences. The thoughtful guidance provided by Professor K. Stępień during is much appreciated while the careful review and helpful commentary provided by an anonymous referee is gratefully acknowledged.

References

Akerlof, C., et al. 2000, Astron. J., **119**, 1901.
Alton, K. B., and Stępień, K. 2019a, J. Amer. Assoc. Var Star Obs., **47**, 53.
Alton, K. B., and Stępień, K. 2019b, Acta Astron., **69**, 283.
Amard, L., Palacios, A., Charbonnel, C., Gallet, F., Georgy, C., Lagarde, N., and Siess, L. 2019, Astron. Astrophys., **631A**, 77.
Andrae, R., et al. 2018, Astron. Astrophys., **616A**, 8.
Antonello, E., and Pastori, L. 1981, Publ. Astron. Soc. Pacifc, **93**, 237.
Baade, W. 1956, Publ. Astron. Soc. Pacifc, **68**, 5.
Baglin, A. 2003, Adv. Space Res., **31**, 345.
Bailer-Jones, C. A. L. 2015, Publ. Astron. Soc. Pacifc, **127**, 994.
Balona, L. A. 2018, Mon. Not. Roy. Astron. Soc., **479**, 183.
Balona, L. A., and Nemec, J. M. 2012, Mon. Not. Roy. Astron. Soc., **426**, 2413.
Baran, A. S., Koen, C., and Pokrzywka, B. 2015, Mon. Not. Roy. Astron. Soc., **448**, L16.
Berry, R., and Burnell, J. 2005, The Handbook of Astronomical Image Processing, 2nd ed., Willmann-Bell, Richmond VA.

Breger, M. 1979, *Publ. Astron. Soc. Pacifc*, **91**, 5.

Breger, M. 1990, *Delta Scuti Star Newsl.*, **2**, 13.

Bressan, A., Marigo, P., Girardi, L., Salasnich, B., Dal Cero, C., Rubele, S., and Nanni, A. 2012, *Mon. Not. Roy. Astron. Soc.*, **427**, 127.

Brown, A. G. A., et al. 2018, *Astron. Astrophys.*, **616A**, 1.

Butters, O. W., et al. 2010, *Astron. Astrophys.*, **520**, L10.

Carollo, D., et al. 2010, *Astrophys. J.*, **712**, 692.

Coşkunoğlu, et al. 2011, *Mon. Not. Roy. Astron. Soc.*, **412**, 1237.

Dehnen, W., and Binney, J. J. 1998, *Mon. Not. Roy. Astron. Soc.*, **298**, 387.

Eker, Z., et al. 2018, *Mon. Not. Roy. Astron. Soc.*, **479**, 5491.

Flower, P. J. 1996, *Astrophys. J.*, **469**, 355.

Garg, A., et al. 2010, *Astron. J.*, **140**, 328.

Gilliland, R. L., et al. 2010, *Publ. Astron. Soc. Pacifc*, **122**, 131.

Hübscher, J. 2015, *Inf. Bull. Var. Stars*, No. 6152, 1.

Husser, T.-O., Wende-von Berg, S., Dreizler, S., Homeier, D., Reiners, A., Barman, T., and Hauschildt, P. H. 2013, *Astron. Astrophys.*, **553A**, 6.

Joshi, S., and Joshi, Y. C. 2015, *J. Astrophys. Astron.*, **36**, 33.

Jurcsik, J., et al. 2005, *Astron. Astrophys.*, **430**, 1049.

Kafka, S. 2019, variable star observations from the AAVSO International Database (https://www.aavso.org/aavso-international-database).

Khruslov, A. V. 2008, *Peremennye Zvezdy Prilozhenie*, **8**, 5.

Laney, C. D., Joner, M., and Rodriguez, E. 2003, in *Interplay of Periodic, Cyclic and Stochastic Variability in Selected Areas of the H-R Diagram*, ed. C. Sterken, ASP Conf. Ser. 292, Astronomical Society of the Pacific, San Francisco, 203.

Leavitt, H. S., and Pickering, E. C. 1912, *Harvard Coll. Obs. Circ.*, No. 173, 1.

Lee, Y.-H., Kim, S. S., Shin, J., Lee, J., and Jin, H. 2008, *Pub. Astron. Soc. Japan*, **60**, 551.

Lenz, P., and Breger, M. 2005, *Commun. Asteroseismology*, **146**, 53.

Li, C. and Zhou, G. 2017, *Astrophys. J.*, 850, 25.

Lindegren, L., et al. 2016, *Astron. Astrophys.*, **595A**, 4.

Longmore, A. J., Fernley, J. A., and Jameson, R. F. 1986, *Mon. Not. Roy. Astron. Soc.*, **220**, 279.

McNamara, D. H. 2000, in *Delta Scuti and Related Stars, Reference Handbook and Proceedings of the 6th Vienna Workshop in Astrophysics*, ed. M. Breger, M. Montgomery, ASP Conf. Ser. 210, Astronomical Society of the Pacific, San Francisco, 373.

McNamara, D. H. 2011, *Astron. J.*, **142**, 110.

Minor Planet Observer. 2019, MPO Software Suite (http://www.minorplanetobserver.com), BDW Publishing, Colorado Springs.

Neeley, J. R., et al. 2015, *Astrophys. J.*, **808**, 11.

Niu, J.-S., Fu, J.-N., and Zong, W.-K. 2013, *Res. Astron. Astrophys.*, **13**, 1181.

Niu, J.-S., et al. 2017, *Mon. Not. Roy. Astron. Soc.*, **467**, 3122.

Pamyatnykh, A. A., Handler, G., and Dziembowski, W. A. 2004, *Mon. Not. Roy. Astron. Soc.*, **350**, 1022.

Paunzen, E., and Vanmunster, T. 2016, *Astron. Nachr.*, **337**, 239.

Poretti, E. 2003a, *Astron. Astrophys.*, **409**, 1031.

Poretti, E. 2003b, in *Interplay of Periodic, Cyclic and Stochastic Variability in Selected Areas of the H-R Diagram*, ed. C. Sterken, ASP Conf. Ser. 292, Astronomical Society of the Pacific, San Francisco, 145.

Rui, W., et al. 2019, *Pub. Astron. Soc. Pacifc*, **131**, 024505.

Schlay, E. F., and Finkbeiner, D. P. 2011, *Astrophys. J.*, **737**, 103.

Schwarzenberg-Czerny, A. 1996, *Astrophys. J., Lett.*, **460**, L107.

Serenelli, A., Scott, P., Villante, F. L., Vincent, A. C., Asplund, M., Basu, S., Grevesse, N., and Peña-Garay, C. 2016, *Mon. Not. Roy. Astron. Soc.*, **463**, 2.

Software Bisque. 2019, THESKYX PRO imaging software (http://www.bisque.com).

Templeton, M. R., Samoly, G., and Dvorak, S. 2009, *Pub. Astron. Soc. Pacifc*, **121**, 1076.

Vagnozzi, S., Freese, K., and Zurbuchen, T. H. 2017, *Astrophys. J.*, **839**, 55.

von Steiger, R., and Zurbuchen, T. H. 2016, *Astrophys. J.*, **816**, 13.

Walker, G., et al. 2003, *Publ. Astron. Soc. Pacifc*, **115**, 1023.

Warner, B. D. 2007, *Minor Planet Bull.*, **34**, 113.

Watson, C., Henden, A. A., and Price, C. A. 2014, AAVSO International Variable Star Index VSX (Watson+, 2006–2014; http://www.aavso.org/vsx).

Wils, P., et al. 2011, *Inf. Bull. Var. Stars*, No. 5977, 1.

Wils, P., et al. 2012, *Inf. Bull. Var. Stars*, No. 6015, 1.

Wils, P., et al. 2013, *Inf. Bull. Var. Stars*, No. 6049, 1.

Wils, P., et al. 2014, *Inf. Bull. Var. Stars*, No. 6122, 1.

Wils, P., et al. 2015, *Inf. Bull. Var. Stars*, No. 6150, 1.

Wozniak, P. R., et al. 2004, *Astron. J.*, **127**, 2436.

Xiong, D. R., Deng, L., Zhang, C., and Wang, K. 2016, *Mon. Not. Roy. Astron. Soc.*, **457**, 3163.

Yan, Y., Du, C., Liu, S., Li, H., Shi, J., Chen, Y., Ma, J., and Wu, Z. 2019, *Astrophys. J.*, **880**, 36.

Zhao, G., Zhao, Y.-H., Chu, Y.-Q., Jing, Y.-P, and Deng, L.-C. 2012, *Res. Astron. Astrophys.*, **12**, 723.

Ziaali, E., Bedding, T. R., Murphy, S. J., Van Reeth, T., and Hey, D. R. 2019, *Mon. Not. Roy. Astron. Soc.*, **486**, 4348.

Sky Brightness Measurements and Ways to Mitigate Light Pollution in Kirksville, Missouri

Vayujeet Gokhale
Jordan Goins
Ashley Herdmann
Eric Hilker
Emily Wren
Truman State University, 100 E Normal Street, Kirksville, MO 63501; gokhale@truman.edu

David Caples
James Tompkins
Moberly Area Community College, Kirksville, MO 63501

Received July 23, 2019; revised July 29, July 30, 2019; accepted August 2, 2019

Abstract We describe the level of light pollution in and around Kirksville, Missouri, and at Anderson Mesa near Flagstaff, Arizona, by measuring the sky brightness using Unihedron sky quality meters. We report that, on average, the Anderson Mesa site is approximately 1.3 mag/arcsec^2 darker than the Truman State Observatory site, and approximately 2.5 mag/arcsec^2 darker than the roof of the science building at Truman State University in Kirksville. We also show that at the Truman observatory site, the North and East skies have significantly high sky brightness (by about 1 mag/arcsec^2) as compared to the South and West skies. Similarly, the sky brightness varies significantly with azimuth on the top of the science building at Truman State—the west direction being as much as 3 mag/arcsec^2 brighter than the south direction. The sky brightness at Anderson Mesa is much more uniform, varying by less than 0.4 mag/arcsec^2 at most along the azimuthal direction. Finally, we describe the steps we are taking in the Kirksville area to mitigate the nuisance of light pollution by installing fully shielded outdoor light fixtures and improved outdoor lights on Truman State University's campus.

1. Introduction

Light pollution is the introduction of artificial light, either directly or indirectly, into the natural environment. It refers to wasted light that performs no useful function or task and leads to light trespass, and increased sky glow and glare. While having some level of outdoor lighting is prudent for aesthetic reasons and for public safety, badly designed light fixtures are wasteful and lead to decrease in human, animal, and plant well-being. Most of the wasted light comes from outdoor lighting such as residential lights, street lights, and business lights. While the problem of light pollution is of particular importance to the astronomy community, the safety, health, ecological, and environmental costs of light pollution are gaining increasing attention (Gaston *et al.* 2012; Chepesiuk 2009). The problem of light pollution is expected to become even more acute with the growing use of increasingly affordable and efficient LED lighting, which will not only affect the quantity of light pollution (brightness), but also the quality of the light (color). In addition to disrupting the sleep cycles of humans, light pollution also disrupts ecological systems (Sanchez *et al.* 2017; Aube *et al.* 2013). Natural patterns of light determine wildlife behavior in a variety of ways, including predation, reproduction, and fatigue, which are drastically altered when subjected to light pollution. Furthermore, light pollution comes at the price of unnecessary energy costs and carbon emissions, impacting both the consumer and the environment (Gaston *et al.* 2012).

In the town of Kirksville, Missouri, the University itself is a major source of light pollution, and poses several problems to students living on or close to campus. Fiscally, this can be seen in the electricity usage of the University, which is incorporated into the prices of room and board, as well as tuition. Light trespass from unshielded light shines directly into the windows of residence halls, disrupting the sleep patterns of students. An additional problem inflicted by light pollution is glare, which can impair the vision of drivers passing through campus, putting pedestrians at risk. Commonly used outdoor lights emit a significant amount of light in wavelengths shorter than ≈ 500 nm, toward the blue-end of the electromagnetic spectrum. Blue light scatters more than red light, and hence using outdoor lights which emit more energy towards the blue-end (wavelength < 500 nm) of the electromagnetic spectrum causes greater sky glow than lights emitting most of their energy towards the red end (wavelength ≥ 500 nm). Additionally, it is known that blue light causes more glare than red light (Intl. Dark Sky Assoc. 2010) and so it is prudent to use outdoor lights which emit less light towards the blue-end of the electromagnetic spectrum. Thus, using lights which emit most of their energy at wavelengths greater than 500 nm effectively reduces the sky brightness and is comparatively soothing to the eye (Intl. Dark Sky Assoc. 2010; Luginbuhl *et al.* 2010; Mace *et al.* 2001).

There are several effective methods of light pollution reduction. These include retrofitting existing fixtures with fully shielded light shields which direct light towards the ground, and installing outdoor lights with color temperature $T \leq 3000$ K, with lesser emission at wavelengths < 500 nm than the outdoor lights currently in use. As mentioned, a significant contribution of outdoor lighting is from street and business lights, and any change in the quality and quantity of outdoor lighting will have to involve the active involvement of local, regional, and federal government authorities. One way to garner public support, and to convince authorities to make the necessary changes in policies, is providing evidence for light pollution via long-term monitoring of the sky brightness at several locations. This will ensure mitigation of any biases introduced by a particular location (proximity to playground lights, for example) as well as due to temporal factors (moon phase, cloud cover, decorative lights during holidays, and so on). With this in mind, we have devised a three-point plan to quantify and mitigate the nuisance of light pollution in the Kirksville (Missouri, USA) community that involves:

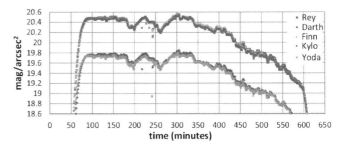

Figure 1. A typical plot showing the sky quality measurements using five sensors placed at the same location (Truman State Observatory). The newer sensors (*Rey, Finn, Kylo*) give darker measurements than do the older sensors *Yoda* and *Darth*. See text for details.

1. Quantifying light pollution in and around the town of Kirksville, Missouri, using Unihedron SQMs,
2. Increasing awareness about the nuisance and dangers of light pollution, and
3. Working with city and school authorities to transition to outdoor lights with color temperature less than 3000 K, and install light shields and light friendly fixtures to reduce light pollution.

In this paper, we describe the current light pollution level in and around Kirksville, Missouri (population approximately 17,000), and at Anderson Mesa, about 15 miles southeast of Flagstaff, Arizona (population approximately 72,000). Kirksville presents a semi-rural setting in the north of the state of Missouri, while Flagstaff is a designated dark-sky location with light ordinances and zoning codes. We compare the night sky brightness levels in Kirksville with similar measurements made near Flagstaff by using Unihedron (http://unihedron.com/index.php) sky quality meters (SQMs) installed at various locations. Unihedron manufactures different types of SQM sensors though, for practical reasons, we favor the handheld version for Alt-Az measurements and the datalogging SQM-LU-DL for continuous sky brightness measurements. We describe some of the properties of the SQMs we use and our set-up in the following section. In section 3, we present our results and analyses of SQM measurements made at various sites over the past few years. In section 4 we discuss our ongoing efforts and future plans regarding the quantification and mitigation of light pollution. In particular, we describe how we have involved students and student organizations and used them as leverage to push administrators towards installing light shields and improved outdoor lighting on our campus and downtown area.

2. Sensor properties and set-up

For this project, our main concern was to ensure that the sensors are mutually consistent, so that we could compare the sky brightness at various locations on a given night. In order to test this, we periodically set up the sensors right next to each other and compared the SQM readings from these sensors. In all we have used five sensors named *Yoda, Darth, Rey, Finn,* and *Kylo*. All of these are datalogging SQMs (SQM-LU-DL) which allow for continuous monitoring of the sky. These sensors have a full-width-at-half-maximum of about 20°. The SQM sky brightness is given in units of magnitudes per square arcsecond (mags/arcsec^2), which means that a difference in 5 units is equivalent to a ratio of 100 in luminance (for details, please see Kyba *et al.* (2011).

From the simultaneous runs of these sensors at the same location, we noticed that the two older sensors, *Darth* and *Yoda*, were consistently giving us values between 0.65 to 0.75 "higher" than the newer sensors (*Rey, Finn,* and *Kylo*) when the sky brightness is about 19 mags/arcsec^2 or darker (Figure 1). The variations between the three new sensors are minimal—at most 0.1 unit when the sky brightness is about 19 mags/arcsec^2 or darker. Consequently, for the sake of consistency, we added an offset of 0.7 mag/arcsec^2 to all our readings obtained from *Darth* and *Yoda* while comparing sky brightness at various locations. Note that we found that the offset is not constant at different levels of darkness and may have a temperature-dependence as well. (We have informed Unihedron about this discrepancy. Tekatch (2019) informs us that the most common issue is of darker readings caused by a frosted IR sensor, though it is unclear if this is the issue with our sensors. We intend to take up Unihedron's offer to recalibrate and clean the optics (free of charge) in order to identify and correct the problem.) For future studies, we intend to discard the older sensors and closely monitor and calibrate the sensors to ensure they give consistent results to avoid this problem.

Data were collected at numerous sites in two cities - Kirksville, Missouri, and Flagstaff, Arizona. Kirksville is a small town (population ≈ 17,000, elevation 300 m) in northeast Missouri, while Flagstaff is a "dark sky city" in northern Arizona (population ≈ 72,000, elevation 2,130 m). The different sites (see Table 1 and Figure 2) in Kirksville included the roof

Table 1. Geographical characteristics of sites used in this work.

Site	Nearest City	Geographic Coordinates (°)		Elevation	Comments
AM	Flagstaff, Arizona (North-East)	35.0553 N	111.4404 W	2163 m, 7096 ft	dark site, usually low humidity
MG	Kirksville, Missouri	40.1866 N	92.5809 W	299 m, 981 ft	urban site, usually high humidity
TSO	Kirksville, Missouri (North-East)	40.177 N	92.6010 W	299 m, 981 ft	semi-rural site, usually high humidity
VG Roof	Kirksville, Missouri (South-East)	40.2110 N	92.6305 W	299 m, 981 ft	semi-rural site, usually high humidity

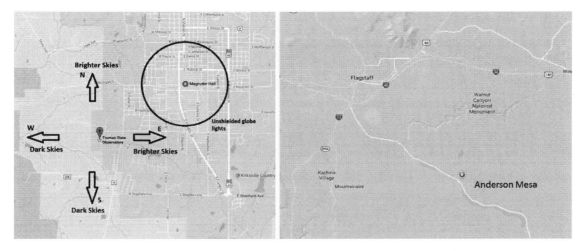

Figure 2. Geography of the Kirksville sites (left panel) and the Anderson Mesa site near Flagstaff, Arizona (right panel).

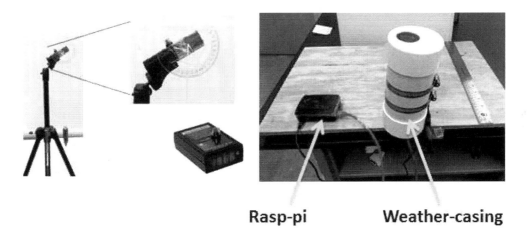

Figure 3. Left Panel: Set up for sky brightness measurements as a function of altitude-azimuth using hand-held SQMs Birriel and Adkins (2010). Right Panel: SQM-LE sensor inside the weather casing attached to a raspberry-pi via an Ethernet cable for continuous measurements. The SQM-LU-DL USB-connected sensor has a similar setup.

of the science building (Magruder Hall, MG from here on), the roof of the authors' residence (VG-roof) located 5 miles northwest of campus, and the Truman State Observatory (TSO), located about 2 miles south west of campus. The MG site is surrounded on all sides with unshielded "globe" lights (Figures 2 and 10) which are a significant source of light pollution. The Anderson Mesa (AM) site, the location of Lowell Observatory research telescopes, is about 12 miles southeast of Flagstaff and is considered a rural "dark site."

3. Quantifying sky brightness

In this paper, we describe two ways in which we quantified the sky brightness. In the first method, we use the manually operated Unihedron SQM light sensor (Half Width Half Maximum (HWHM) of the angular sensitivity is $\approx 42°$) and the SQM-L sensor (HWHM of the angular sensitivity is $\approx 10°$) to measure the sky brightness as a function of the altitude in four directions (east, west, north, and south). (Two different versions of the manually operated SQM were used since in 2017 we only had access to the SQM. The SQM-L sensor became available to us only after 2018.) In the second method, we use the "automatic" datalogging SQM-LU-DL sensor housed in a weather-proof case to monitor the sky brightness. These datalogging sensors can be powered by batteries which allow us to measure sky brightness at remote locations over several nights.

3.1. Angle dependence

We measured the sky brightness as a function of the altitude and azimuthal angle following the procedure outlined by Birriel and Adkins (2010). The manually operated SQM is mounted on a tripod with a clear protractor attached horizontally to it. A plumb bob is suspended using a string to enable accurate measurements of the zenith angle (see Figure 3). As much as possible, measurements were made on moonless, cloudless nights away from trees, tall walls, and buildings. At each angle, we recorded the sky brightness five times, and calculated the average before moving on to a different angle.

The angles were changed by either 10 or 15 degrees. We carried out these measurements at three different locations: the roof of the science building (MG hall), the Truman State Observatory, and at Anderson Mesa near Flagstaff, Arizona. The differences in the three sites are striking (Figures 4, 5, and 6). As shown in Figure 2, the science building is surrounded on all sides by dark-sky unfriendly "globe" lights and consequently the sky brightness is significant at low altitude (this is partly due

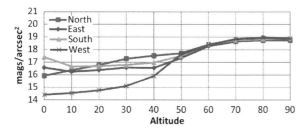

Figure 4. Typical plot of the measurements of sky brightness as a function of altitude and direction at the MG site using the SQM sensor. The sky is darkest close to the zenith, and brightest along the horizon. The sky brightness variation in a given direction is between 4.5 mag/arcsec^2 (west) and 1.5 mag/arcsec^2 (south). The sky brightness is the greatest toward the west due to the presence of a large number of unshielded "globe" lights in that direction.

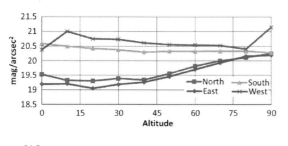

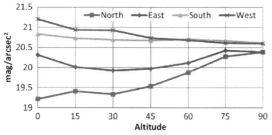

Figure 5. Sky brightness measurements as a function of altitude and direction at the TSO site from data collected approximately two years apart using the SQM (upper panel) and SQM-L (lower panel) sensors. Note the high sky brightness levels in the east and north, the direction of Kirksville town. The presence of a state park and wilderness towards the west and south directions results in relatively darker skies in these directions

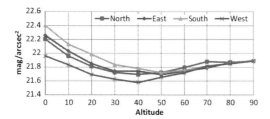

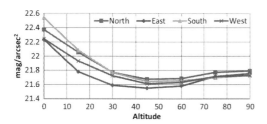

Figure 6. Same as Figure 5, but at the AM site near Flagstaff, Arizona on two clear night two years apart. Note the relatively short range of variation in the sky brightness as compared to the MG location. On average, the sky brightness levels at AM are about 3 mag/arcsec^2 better at the zenith, and as much as 8 mag/arcsec^2 better at the horizon than the MG location (see Figure 4).

to reflection from buildings and trees, see Figure 4). The TSO location has high sky brightness in the direction of Kirksville, which lies to its north-east (Figure 2). The sky brightness does not appear to depend strongly on altitude toward the south and west due to lack of any street lights or residences in those directions (Figure 5). The data from the two epochs are largely consistent, except close to the horizon where local land features (trees, bushes, street lights) might affect the measurements. At the TSO site (Figure 5), there is considerable discrepancy in the "east" data from the two epochs, of the order of 1 mag/arcsec^2 near the horizon. This can be attributed to street lights on a road that runs north-south to the east of the TSO site. Also, note that the 2017 data were collected using the SQM meter which has a much greater angular sensitivity than the SQM-L meter used for obtaining the 2019 data, which explains the "brighter" readings on the SQM meter when pointed in the direction of the town, though a similar effect is not seen in the north direction.

The altitudinal dependence of sky brightness in Flagstaff is somewhat different than the city and semi-rural locations in Kirksville. At the Anderson Mesa location near Flagstaff, absence of street lights in all directions results in a significantly dark horizon, with gradually increasing sky brightness to about 45° in altitude. This increase in brightness may be a result of natural or artificial sky glow, or from near or far field luminance and illumination (Schaefer 2019; Walker 1977). The sky brightness then decreases toward the zenith. The AM-site is south-east of Flagstaff, resulting in slightly poorer skies towards the north and west (the Moon was rising in the east during the 2019 measurements, biasing our results in that direction).

Our results qualitatively match that of Birriel and Adkins (2010), though there are quantitative differences due to the geography vis-a-vis terrain and the location of town with respect to the measurement site. In particular, the variation in the sky brightness in the azimuthal direction is much greater closer to town (≈ 2 mag/arcsec^2, their Figure 3a) as against at a location farther from town (≤ 1 mag/arcsec^2, their Figure 2a). This is comparable to the results presented here in Figures 4, 5, and 6: at the most light polluted location, MG roof, the azimuthal variation is the largest at the horizon (≈ 2.7 mag/arcsec^2), at the semi-rural location (TSO) the azimuthal variation is about 2 mag/arcsec^2, while at the rural, "dark sky" location, the azimuthal variation is ≤ 0.5 mag/arcsec^2. In terms of altitudinal variation, Birriel and Adkins (2010) do not observe a brightening at around 45° at the "dark sky" location (their Figure 3) as we do at the AM-site (Figure 6). This could be because the AM-site is much darker (21.9 mag/arcsec^2 at zenith) than the Cave Run Lake (21.2 mag/arcsec^2 at zenith) site.

3.2. Continuous monitoring

The second method we are using to measure sky brightness is by using the "datalogging" mode of the SQM-LU-DL sensor provided by Unihedron. This sensor can be powered by a battery and is capable of storing data for several nights. Encased in the weather-proof casing, this set up is ideal for sky-brightness measurements in remote locations without power and/or wireless internet. We use this sensor in two ways: one is battery operated, and in the other mode, we connect the SQM-LU-DL to a Raspberry-pi microcomputer, which then transmits data via wireless to a webserver (Figure 3). We set up the SQM-LU-DL in its weather-proof case, pointing toward the zenith, at a location away from any trees, buildings, and overhead lights. Within city limits, the sensor is always mounted on the roof of a convenient, accessible building above the level of the street lights, while in semi-rural and rural areas the sensor is mounted on a laboratory-stand placed on an even surface, usually the ground.

Figures 7 and 8 show plots of some of our "continuous monitoring" runs. These data were taken over the past few years at various locations in varying weather conditions. Results from the AM-site are fairly consistent over the measurements made two years apart. The sky brightness is about 21.9 mag/arcsec^2 in the absence of clouds and the moon. On the other hand, the data from the different locations in Kirksville show a significant level of light pollution. The TSO location, about two miles south west of campus, is the darkest site, with a sky brightness of about 20.7 mag/arcsec^2, while the VG-roof location is about 20.4 mag/arcsec^2. The MG-site, located on the Truman State University campus, has a sky brightness of about 19.4 mag/arcsec^2, approximately 2.5 units worse than the AM-site. Again, the data are fairly consistent several months apart, which is reassuring.

Note that passing clouds are evident in the plots shown below (Figures 7, 8, and 9)—light is reflected back toward the Earth from the bottom of the clouds, resulting in an uneven sky brightness curve as measured by the SQMs (Kyba *et al.* 2011). In general, the sky brightness increases as a consequence of clouds—especially so in light polluted areas since there is more ambient light to reflect back from the bottom of the clouds. This is evident in 2019 data shown in Figure 8. On this night, clouds cleared up by timestamp 300 leading to a leveling out of the SQM readings at the three different sites around Kirksville. Note, however, that at the AM-site, the sky brightness actually decreased (right panel, Figure 7) on 27 May 2019, a significantly cloudy night. This may be due to the lack of light projected upward due to the absence of outdoor light sources near the AM-site. It is also possible that the sky brightness is affected differently by different kinds of clouds (high cirrus, stratus, cumulus, etc.), something that we are investigating presently.

Figure 9 shows a comparison of the sky-brightness measured at various locations in Kirksville and at Anderson Mesa near Flagstaff. It is clear that the AM-sky is darker by about 1.3 mags/arcsec^2 as compared to the semi-rural sites in Kirksville. Also, the AM-site is, on average, about 2.5 mags/arcsec^2 darker than MG-hall on the Truman State campus.

4. Discussion and future work

We are currently in the process of installing the SQM-LU-DL sensors at various locations across the town of Kirksville, Missouri. We intend to have three permanent SQM-LU-DL sensors attached to Raspberry-pi micro-computers that can automatically transmit data via wireless internet to a webserver, two of which are currently operational at the TSO and MG-hall sites. These long-term measurements can serve as a baseline for comparison in the sky brightness levels at different locations, as well as for comparison between measurements made several years apart at the same location. In turn, these can then be used

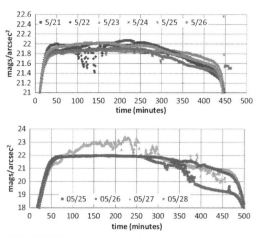

Figure 7. SQM-LU-DL measurements at Anderson Mesa for six nights in May 2017 (upper panel) and four nights in May 2019 (lower panel) as a function of time. Some of nights were cloudy, leading to several "spikes" in the sky brightness measurements. For 2017, moonrise is around timestamp 300 (2:40 AM MST) on 21 May (dark blue curve), while moonrise is around timestamp 250 for 25 May 2019.

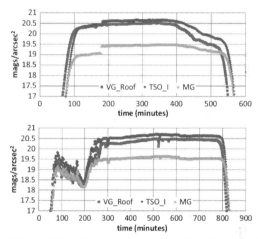

Figure 8. Different SQM-LU-DL measurements in Kirksville at four different locations on 8 May 2018 and 7 January 2019. The urban location (MG) has the greatest sky brightness, while the semi-rural location at the Truman Observatory (red) is comparably dark. The light blue curve is the sky brightness from the roof of the author's residence, at a location 5 miles west/north-west of Kirksville. On the upper panel, an abrupt jump can be seen in the sky brightness around timestamp 175 at locations close to the Truman campus (most prominent in the TSO-curve, but can also be seen in the MG-curve) corresponding to the switching "off" of the lights on the football stadium on campus. The lower panel shows the effects of clouds at the beginning of the night, until approximately timestamp 300 after which the sky was reasonably clear and the SQM curves flatten out.

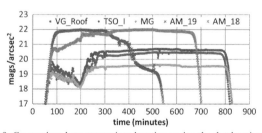

Figure 9. Comparison between various locations using the datalogging SQM-LU-DL sensors. The Anderson Mesa site is clearly the darkest of the sites we monitored. The green dots represent data from the roof of Magruder Hall, the science building on the Truman State University campus, which is by far the most light-polluted site. The data were staggered to align the onset of darkness after sunset—the different durations of darkness represent the differing duration of night. The Truman data were taken in early January, while the Anderson Mesa data are from March 2018 and May 2019.

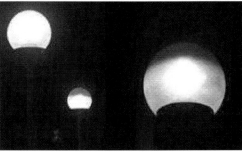

Figure 10. Upper Panel: Unshielded "globe" lights galore on Franklin Street. One of the student dorm buildings can be seen toward the right of the image. Lower Panel: An example of an unshielded light fixture ("globe lights") housing a $T = 5000$ K (blue/white color) light bulb and another fixture fitted with a "dark sky reflector" housing a $T = 3000$ K (off-white) light bulb. $T \leq 3000$ K light is known to cause less glare and is less detrimental to the environment and plant and animal health (Intl. Dark Sky Assoc. 2010; Luginbuhl et al. 2010).

to convince authorities to improve outdoor lighting and to provide objective evidence for the effectiveness of using fully shielded light fixtures, once they are implemented.

We have recently set up a weather station, cloud sensor, and an all-sky-camera at the Truman State Observatory. We are currently analyzing data from the SQMs installed at the Observatory and correlating these data with weather parameters such as humidity, temperature, and cloud cover. The all-sky-camera images provide a visual check to the cloud cover measurements made by the cloud sensor. Preliminary results show a strong dependence of the sky brightness on cloud cover (Kyba et al. 2011}. We are particularly interested in monitoring the effects of different kinds of clouds (cirrus, stratus, cumulus, etc.) on sky brightness.

In addition to the quantitative and qualitative elements of the research, we have devoted a significant amount of time and effort toward increasing awareness about light pollution. This has involved engaging and educating both the Kirksville and Truman populations about the harmful effects of light pollution. Students from the Truman State "Light Pollution Group" make presentations about the detrimental effects of light pollution to supplement shows at the Del and Norma Robison Planetarium at Truman State University. IDA brochures (https://www.darksky.org/our-work/grassroots-advocacy/resources/public-outreach-materials/) are distributed to the audience after a viewing of the IDA documentary "Losing the Dark" (https://www.darksky.org/our-work/grassroots-advocacy/resources/losing-the-dark/).

We are in the process of acquiring night-sky friendly light shields (Figure 10) to cover some of the unshielded light fixtures on campus. In collaboration with the "Stargazers" student astronomy group on campus, we obtained funding for these light shields from the Environmental Sustainability Fee Committee (EFC) and the Funds Allotment Council (FAC) at Truman State University. Both these funds are generated via a nominal ($\approx$ \$5 per semester) fee imposed on each student attending Truman State. The proposals for these grants were written by participating students. The level of funding from the EFC and the FAC is \$6,000 and \$2,500, respectively. These funds were used to purchase IDA-approved dark-sky shields to retrofit the "globe" lights on campus to reduce skyglow. Starting fall 2019, fifty such shields are being installed in a selected area (Franklin Street, Figure 10) to test the effectiveness of the shields in terms of student approval, reduction of skyglow, and maintenance factors. Franklin Street was chosen due to the presence of several unshielded lights close to on-campus housing halls. If the light-shields are successful in combating light pollution and glare, we plan on applying for additional funds to purchase and install more shields. In addition, we are transitioning to improved outdoor lighting (The lights being installed (color-temperature $T \approx 3000$ K) emit about 15% of their light at wavelengths below 500 nm (Maa 2019) by replacing the blue/white lamps currently in use on most outdoor light fixtures. These actions present us with an opportunity to do before-and after-studies to investigate the impact of the dark-sky reflectors and the changed lighting on the sky brightness.

As outlined in the Introduction (section 1) we have made significant progress in our three-step program to address the issue of wayward outdoor lighting. We have successfully set up several SQM meters at various locations to quantify sky brightness and have engaged in creating awareness. We have used our data and analyses to convince authorities to implement dark-sky friendly light fixtures. In collaboration with the Missouri chapter of the International Dark Sky Association (https://darkskymissouri.org/), we are working on establishing a network of SQM sensors across several cities, parks, and recreational areas across the state of Missouri in the near future. We hope that in the next few years, we can establish several "dark-sky friendly" parks and recreational areas in the state of Missouri by significantly reducing light pollution. Though it exists world-wide, light pollution remains a local problem with local solutions. While the data presented in this paper are specific to measurements made at the particular locations under consideration, we believe that the overall rationale, methodology, and analyses presented here can be duplicated at other locations. We hope that concerned citizens, and amateur and professional astronomers across the nation and the world will follow suit and work towards creating safe, night-sky friendly environments in their communities. Human beings have evolved and grown up with an unadulterated view of the beautiful night sky for millennia. We owe it to ourselves, and to future generations, to be not deprived of this beauty.

5. Acknowledgements

This work is supported by funds from the Missouri Space Grant Consortium and the Provost's office at Truman State University. The authors are thankful to the support from the Environmental Sustainability Fee Accountability Committee and the Funds Allotment Council at Truman State University for funding to purchase the dark sky reflector light-shields and new $T = 3000$ K outdoor lights. VG would like to thank the Physical Plant personnel at Truman State University for their willingness to work with us on this project. The authors would also like to thank Andrew Neugarten for setting up the Raspberry-pi microcomputers. The authors would also like to thank the anonymous referee for useful comments and suggestions, which greatly improved the manuscript.

References

Aube, M., Roby, J., and Kocifaj, M. 2013, *PLoS ONE*, **8**, e67798.

Birriel, J., and Adkins, J. K. 2010, *J. Amer. Assoc. Var. Star Obs.*, **38**, 221.

Chepesiuk, R. 2009, *Environ. Health Prospect.*, **117**, A20.

Gaston, K. J., *et al.*. 2012, *J. Appl. Ecology*, **49**, 1256.

International Dark Sky Association. 2010, "Visibility, Environmental, and Astronomical Issues Associated with Blue-Rich White Outdoor Lighting" (https://www.darksky.org/our-work/grassroots-advocacy/resources/ida-publications/).

Kyba C. C. M., Ruhtz, T., Fischer, J., and Holker, F. 2011, *PLoS ONE*, **6**, e17307.

Luginbuhl, C., Moore, C., and McGovern, T., eds. 2010, *Nightscape*, No. 80, 8 (https://www.darksky.org/wp-content/uploads/bsk-pdf-manager/3_SEEINGBLUE.PDF).

Maa, M. 2019, private communication.

Mace, D., Garvey, P., Porter, R. J., Schwab, R., and Adrian, W. 2001, "Countermeasures for Reducing the Effects of Headlight Glare" (https://trid.trb.org/view/707950), AAA Foundation for Traffic Safety, Washington, DC.

Sanchez de Miguel, A., Aubé, M., Zamorano, J., Kocifaj, M., Roby, J., and Tapia, C. 2017, *Mon. Not. Roy. Astron. Soc.*, **467**, 2966.

Schaefer, B. 2019, private communication.

Tekatch, A. 2019, private communication.

Walker, M. F., 1977, *Publ. Astron. Soc. Pacific*, **89**, 405.

Automated Data Reduction at a Small College Observatory

Donald A. Smith
Physics Department, Guilford College, 5800 W. Friendly Avenue, Greensboro, NC 27410; dsmith4@guilford.edu

Hollis B. Akins
Physics Department, Grinnell College, 1116 Eighth Avenue, Grinnell, IA 50112

Received July 15, 2019; revised August 14, 2019; accepted August 30, 2019

Abstract We report on our success in automating the data analysis processes for a small (half-meter class) reflecting telescope at Guilford College, a small liberal arts college in an urban location. We give a detailed description of the hardware that makes up the instrument. We use commercial software to run the telescope, and we have written Python scripts to automatically carry out the daily tasks of image reduction, source extraction, and photometric calibration of the instrumental magnitudes to the AAVSO APASS catalogue. We characterize the effect of light pollution in terms of sky brightness and typical limiting magnitudes ($V \sim 15$ for a 60-s image), and we report that our RMS scatter in light curves for constant sources approaches 0.5% for bright ($V < 10$) sources. As a test of our system, we monitored two known variable stars, and we confirm the predicted ephemerides as given by the AAVSO web site target tool. We hope that this paper will provide a context in which future observations reported from this site can be interpreted.

1. Introduction

The Guilford College Cline Observatory is a primarily educational resource for students interested in learning about observational astronomy. An array of small reflecting telescopes can be set up on clear nights to allow beginners, photographers, and the public to view the night sky. With solar filters and an Hα scope, we enable the observations of sunspots and eclipses. Two 2.4-m radio telescopes extend our vision to the 21-cm neutral hydrogen line. In recent years, however, we have endeavored to bring the primary optical telescope into a reliably automated state such that it can more robustly contribute to scientific research.

In this paper, we report on the technical specifications of the 16-inch optical telescope and our automated data processing pipeline. We detail the structure and characterize the performance of the pipeline; every morning it reduces the images recorded by the telescope the previous night. This software also extracts a list of sources from every image and calibrates their instrumental magnitudes to the AAVSO APASS catalogue.

The result is an ever-updating database of brightness measurements that can be used to construct light curves, perform brightness distribution studies, and monitor stars that have been flagged as of interest by the AAVSO. We intend this paper to characterize the specifics of our instrument so that future reports and updates from our site can be understood. In a separate paper, we plan to report on our efforts to set up an accompanying spectroscopic telescope that can be used in conjunction, simultaneously, with the automated photometric telescope. This will be useful, as both telescopes will share the same observing conditions.

In section 2, we lay out the technical specifications of the hardware and software that make up the telescope. We follow, in section 3, with a short description of the observing conditions that our urban environment affords us. Section 4 describes the design and operation of the data reduction and analysis pipeline, and section 5 presents two examples of the results our observatory can provide. We close with a short description of what we hope to offer in future work. It is our hope that this paper will provide the reader the context to evaluate the strengths and challenges underlying future science reports.

2. Telescope components and control

The centerpiece of the observatory is a 6-m diameter Observadome. The dome opens a 1.6-m wide slit using two laterally sliding gate doors. Power is conveyed to the dome from below using seven circular rails with sliding conductive contacts, enabling unrestricted rotation in either direction. The hemispherical dome rests on eight spring-loaded tires, and rotation is driven by two friction-drive wheels. An optical sensor counts wheel rotations and allows the calculation of angular displacement. A magnetic switch defines the home position. A wall-mounted enclosure contains the two circuitboards of a MAXDOME II automation system (Diffraction Limited 2019a) and four relays wired in parallel with the manual buttons. We have also installed wireless communication between the MAXDOME boards and the slit window motors, as the unrestricted rotation of the dome disallows any attempt to run a wire across the boundary. This system enables computer control of dome rotation and slit operation. The dome also has a wind shield that can move up and down, but we have not automated this feature and generally leave the shield down.

At the center of the dome is a 16-inch RCOS reflecting telescope (f/8.3). The focal length of 3.4 m yields an image scale of 61 arcsec/mm at the focal plane. This closed truss optical tube is mounted on a Paramount ME robotic telescope mount (Software Bisque 2019a). The entire instrument rests on a sand-filled pier, 30 cm in diameter. This pier sits on a steel I-beam, which extends 70 cm to the North (center to center) from a 80-cm diameter cylindrical concrete base. The base is acoustically isolated from the metal floor, and rests on a skeletal structure that is isolated from the rest of the building. The result is that the rotation axes of the telescope cross roughly three meters

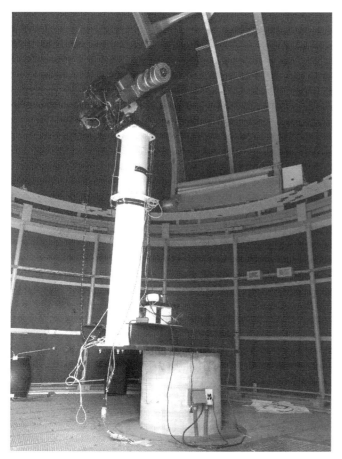

Figure 1. Image of the inside of the dome. The cantilevered pier is visible at the bottom of the image—the concrete acoustically isolated base, the black I-beam, and the white sand-filled pier. The Paramount ME and the RCOS telescope are on top, with the CCD camera to the left. The dome slit is visible in the background. The orange curves are the rails that carry AC power to the dome from below.

above the floor, to put the telescope at the geometric center of the dome. Figure 1 shows the interior of the dome, from the concrete base up to the RCOS OTA.

The cantilevered offset of the pier was the result of a compromise between scientific functionality and the aesthetics of how the dome would look from a distance. Although we worried that the cantilever structure might introduce harmful vibrations into the observing process, an undergraduate thesis project revealed that in the absence of humans, neither the dome's rotation nor the mount's slewing introduced any measurable jitter into the images (Corbett 2011). The same research found that vibrations created by striking the pier directly faded with an e-folding time of less than 30-s. Automated operations have never revealed significant degradation due to pier oscillations.

Images are captured with an SBIG STF-8300M CCD camera (Diffraction Limited 2019b). This instrument has 3326 × 2504 pixels at 5.4 microns, yielding an image scale of 0.33 arcsec/pix at the chip. The shutter is designed to afford even illumination over the entire chip, and the full-frame readout time is less than one second. The camera is outfitted with an FW8 filter wheel. We have chosen to use Baader Red, Green, and Blue filters for astrophotography, and Bessel R, V, and B filters for photometry. The final two slots contain an Hα filter and an open, "Lum" filter. Focus is driven by a motor that moves the secondary mirror along the optical axis. This motor is controlled by the RC Optical Systems Telescope Command Center (TCC-I; Deep Sky Instr. 2019). The STF-8300 has the capability to add a secondary camera for autoguiding, but we have not yet implemented this feature.

We have two devices to monitor the weather. An Aurora Cloud Sensor (Aurora Eurotech 2019) measures the difference between the sky temperature and the ambient air temperature to estimate cloud cover, and an exposed sensor reports the presence of rain drops. A Davis Vantage Vue weather station (Davis Instr. 2019) records humidity, pressure, wind speed and direction, and temperature. These data are recorded and stored on the central computer.

All the features of this telescope system are controlled via computer. As of May 2019, the control computer is a Hewlett-Packard HPE-470F Pavillion Elite, running 64-bit Windows 10. Software Bisque's THESKYX (2019b) anchors the system, connecting to the mount, focuser, camera, and dome. Aurora and Davis provide their own software to connect to the weather stations. We have augmented THESKYX with the TPoint and All Sky ImageLink features. The mount, camera, and focuser are plugged into a IP-enabled power strip so that sockets may be turned on and off individually through a web browser interface, and we have a wall-mounted IR-enabled Foscam security camera. These features allow us to operate the facility remotely using the TeamViewer program as a remote desktop. The use of TeamViewer's App makes it possible to run the whole operation from the screen of a mobile phone.

THESKYX can receive scripted commands over the Internet. We have written programs to schedule a night's observations without human intervention. The python programs are stored on a Linux computer in another room in the building. Each script starts with targets as a list of strings with associated R.A. and Dec. coordinates (J2000.0). Exposure time, filter choice, and delays between observations are also specified. The script updates the coordinates to the current day's epoch and instructs THESKYX to slew the telescope, wait a desired time to allow the dome and the telescope to both reach the target, and then take an image. Each image is automatically passed to THESKYX's All Sky Image Link process for source identification and the derivation of world coordinate system (WCS) astrometric parameters. This ensures that our images can be seamlessly passed to the analysis pipeline (section 4).

The script includes definitions for procedures to turn all the pieces of the telescope system on at the start of the night as well as code to shut everything down at a pre-determined time. After startup is complete, the script will loop through the indicated observations until the shutdown time is reached. However, we have not yet scripted a connection to the weather monitor, so the system is not yet capable of shutting itself down in the case of unanticipated weather developments. We plan to implement this feature in the future.

3. Urban sky brightness

The Guilford College Cline Observatory (Lat. +36° 5' 42", Lon. –79° 53' 24", altitude 280 m) is situated on the roof of a three-story building that houses the natural sciences offices, labs, and classrooms. Although the College has committed to replacing outdoor light fixtures with downward facing baffles, and the campus contains more than 100 acres of undeveloped forest, we are still in an urban environment. As of 2019, Greensboro proper houses more than 250,000 inhabitants, and is one of three major cities in an urban area called "The Triad" (also including High Point and Winston-Salem). The center of Greensboro is ~9 km southeast of our location, and the Piedmont Triad International airport is ~3 km to the west.

Although the northern skies are slightly darker than the southern, this cannot be considered a dark sky site. A 2001 global light pollution map, based on satellite observations, atmospheric modeling, and other statistics, predicts a range in sky background brightness values from 17.80 to 18.95 magnitudes per square arcsecond of subtended area at Guilford College's location (Cinzano et al. 2001). A 2016 student senior thesis project recorded roughly 100 images of two fields with 10 s and 30 s of exposure in each of the Baader Green and Blue filters and found that the actual sky brightness at our location was consistent with approximately 17 mag per sq. arcsec (Seitz 2016).

A follow-up senior thesis project attempted to measure the contribution to sky brightness due to the addition of lights to our primary football field, roughly 500 m southeast of the science building. Turning the field lights on and off had to be coordinated with a company in Iowa, but Oulette (2016) observed multiple fields at an elevation of approximately 30° above the stadium and found that the presence or absence of the field's lights had no significant impact on the sky brightness or limiting magnitude of the images.

Our own observations of open cluster NGC~6811 indicate that for a 60-s exposure, we can expect to reach roughly 15th magnitude. Figure 2 shows a histogram of the number of sources extracted from the field with our automated pipeline. For a 60-s image, there is a steep decline in the number of sources at $V \sim 15$, and no sources at all below $V \sim 15.5$.

Another way to characterize the stability of the analysis is to measure magnitude fluctuations over time. Tracking M29 for the night of 14 July 2019 yielded 35 measurements of 101 stars. The pipeline analysis reported median values of the FWHM for each image that ranged between 1 and 2 arcseconds. Figure 3 shows a plot of the RMS variability in the resulting light curves as a function of the mean magnitude. For stars brighter than $V \sim 11$, the variation is consistent with fluctuations at the ~0.005-mag level. Below this brightness, the stability decays, rising to a maximum of 0.07 mag. The statistical uncertainty of the dimmest stars in these 60-s images is ± 0.03 mag.

These fluctuations can be considered an estimate of the systematic uncertainty in the magnitude, and for dim stars they dominate over the statistical error associated with the photon counts. In the next section, we describe the pipeline that produced these measurements.

4. Data analysis pipeline

We have developed software to reduce and analyze all images captured with the 16-inch telescope, based on the pipeline used in the ROTSE-III telescope system (Akerlof

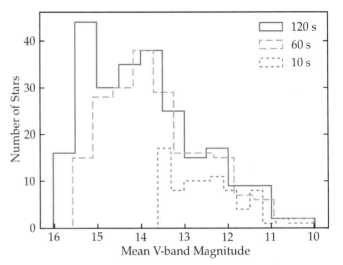

Figure 2. Histogram of number of sources as a function of V-band magnitude for a 10-s exposure (81 stars in green/dotted), a 60-s exposure (203 stars in orange/dashed), and a 120-s exposure (241 stars in blue/solid) of open cluster NGC 6811. The limiting magnitudes can be estimated by seeing where the histogram drops off: $V \sim 13.5$, 15.0, and 15.5, respectively.

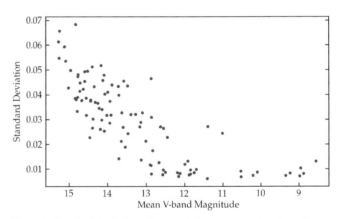

Figure 3. Standard deviation of source brightness as a function of mean magnitude. This graph is derived from 35 observations of 101 sources in open cluster M29 over the course of the night of 14 July 2019. Each image was exposed for 60 s. Fluctuation levels approach 0.5% for sources brighter than $V \sim 11$, but degrade to 0.07 mag at a magnitude of $V \sim 15.5$. These RMS values can be interpreted as an estimate of the systematic uncertainty in our ensemble photometry method.

et al. 2003; Smith et al. 2003). The pipeline consists of python scripts housed on a Linux computer. The data directories on the telescope control computer are remote-mounted on the Linux computer for easy access. All images are recorded in the FITS format (Wells et al. 1981).

The primary script is launched via a cron utility every morning. If the script ascertains that new images have been recorded in the last night, it copies them to the Linux computer. It then applies dark, bias, and flat-field corrections to create a new archive of calibrated images, using the ASTROPY package (Astropy Collaboration et al. 2013). The script also checks to see if new calibration files have been recorded. If possible, new calibration masters are made from individual files to ensure up-to-date correction images.

Any further pipeline processing requires the image to have WCS conversion equations recorded in the FITS headers. If the images were not solved at the time they were recorded

Table 1. Observing targets.

Star	R.A. (J2000) h m s	Dec. (J2000) ° ′ ″	Date	Type	T^a (days)	Number of Images	Cadence (seconds)	$(N)^b$
RR Lyr	19 25 27.9	+42 47 03.7	2018-08-15	RR	0.57	73	160	86
V457 Lac	22 36 23.0	+38 06 18.4	2018-08-16	E	6.16	29	800	101

[a] From The AAVSO International Variable Star Index (VSX; Watson et al. 2006-2014). [b] Average number of sources in field.

with the ImageLink process within THESKYX (as programmed in our observing PYTHON scripts), a plate solution using nova.astrometry.net, or equivalent, is also sufficient.

Following calibration, in the presence of astrometric data, we use the SExtractor PYTHON wrapper (Barbary 2016) to perform source extraction, annulus subtraction, and aperture photometry. We use a fixed aperture radius of 12 pixels (scaled down linearly at higher pixel binning) and a surrounding annulus with inner and outer radii set to 1.5 to 2.0 times the aperture radius, respectively. Sources with pixel values less than four times the background RMS are discarded. We use an outlier-removal algorithm to account for source contamination, and all stars flagged as saturated or truncated by an image boundary are ignored. This aperture photometry method yields uncalibrated instrumental magnitudes for every source identified in the image.

Next, measured celestial locations of all unsaturated sources are then matched to entries in the AAVSO APASS catalogue (Henden et al. 2015), with a tolerance of 3 arcseconds, to retrieve catalogue magnitudes. The median offset between instrumental and catalogue magnitudes is used to correct all the instrumental magnitudes and thereby derive calibrated magnitudes for all sources in the field. This approach is called ensemble photometry (Honeycutt 1992; Everett and Howell 2001). The resulting calibrated magnitudes are saved to a master database along with other identifying information for each source. From this database we can extract light curves.

5. Example results

To test the reliability of the pipeline reduction and analysis process, we chose two variable stars to monitor throughout a night. We used the AAVSO Target Tool (AAVSO 2017) to identify two sources that were predicted to engage in interesting behavior on the nights we would be observing. RR Lyrae, the eponym for the RR Lyrae class of variable stars, was predicted to reach a maximum at approximately 06:30 UT on the morning of 16 Aug 2018, so we expected to see the brightness begin to rise approximately 90 minutes before then. The next night, we monitored V457 Lac, an eclipsing binary that was predicted to have an eclipse event at 03:00 UT.

Table 1 presents details of our two observing programs. It gives the target coordinates and the dates of the observations, the type of the target star, and the period of its variation. The number of images taken each night and the typical time between images are also reported. Both fields held close to 100 sources (last column), more than enough for the ensemble photometry technique, and we obtained over 25 images of each field. We used a Bessel V-band filter for all images.

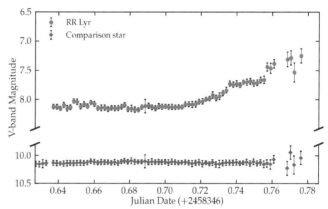

Figure 4. Light curve for RR Lyr from 15 Aug 2018. The light curve for a (constant intensity) comparison star at tenth magnitude is also included to indicate the level of systematic fluctuations in the analysis. Error bars indicate formal statistical uncertainty derived from random counting. The gaps and increased uncertainty towards the end of the observing run are the result of incoming clouds.

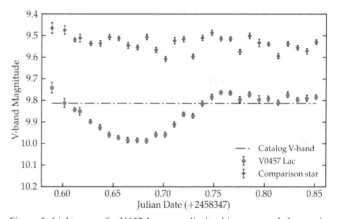

Figure 5. Light curve for V457 Lac, an eclipsing binary, recorded over six hours on the night of 17 Aug 2018. The light curve for a (constant intensity) comparison star is also included to indicate the level of systematic fluctuations in the analysis. Error bars indicate formal statistical uncertainty derived from random counting. A horizontal dashed line indicates the magnitude for V457 Lac given in the UCAC catalogue (Zacharias et al. 2013). The timing of the eclipse is consistent with the ephemeris reported via the AAVSO web site.

Figure 4 shows a light curve for RR Lyr. We observe the star rise from $V \sim 8.1$ to ~ 7.3 before clouds obscured the image around 06:30 UT. The comparison star (UCAC 664-074503) shows no systematic trends (nor does any other star in the field—this star is included as representative). The RMS scatter in the brightness of the comparison star (before the clouds arrived) is ~ 0.01 mag, consistent with the results reported in Figure 3. Since the rise is clearly unique to RR Lyr, greatly exceeds the systematic variability, and coincides with the predicted rise of this target, we are confident that we have robustly detected the variability of this star.

Figure 4 shows a light curve for eclipsing binary V457 Lac. This curve shows a significant dip in the brightness of the system, from an average magnitude of $V \sim 9.8$ to a minimum of ~ 10.0. The comparison star shows no such systematic deviation from a constant brightness. Note that this magnitude range for V457 Lac is in line with the AAVSO reported magnitude range from 9.84 to 10.04. The time of onset for this eclipse is consistent with the 03:00 (UT) time predicted by the ephemeris in the AAVSO Target Tool.

These observations give us confidence in the reliability and accuracy of our tools.

6. Future work

Now that the images from the 16-inch telescope are automatically and reliably reduced and analyzed, we hope to use the facility for scientific studies. Being a small observatory in an urban setting, the most likely venue for us to make meaningful contributions is in the study of bright variables and/or transients. With a typical limiting magnitude of $V \sim 15$ for a 60-s image, many variable stars are within our reach (e.g. Woźniak *et al.* 2004; Paczyński 2006). A few supernovae a year should be visible (e.g. Lae *et al.* 2015; Bellm *et al.* 2019), and occasionally even a Gamma-Ray Burst (e.g. Smith *et al.* 2016) or gravitational wave merger/kilonova event (Abbott *et al.* 2017). Asteroids are also potential targets.

We hope to bring our companion spectroscopic telescope on line soon, and with the two together we hope to monitor spectroscopic binaries for doppler shifts. The $\sim 1\%$ variations at the bright end of our observations means that detecting exoplanet transits will be a challenge. We can improve the statistics with longer exposure times, of course, at the cost of temporal precision in determining the ingress and egress times of the transits.

On the technical side, in addition to the autoguiding capability mentioned previously, we plan to integrate the weather monitoring stations into the operating software so that if the stations detect poor observing conditions, they will automatically trigger a shutdown. Although we have scripting software that can execute a planned set of observations, the observing queue is not dynamical nor does it take elevation into account. We hope to improve this in the future.

Our preliminary analysis of the limiting magnitudes of our images suggests that perhaps our source extraction cutoff of four times the background RMS might be too conservative. For the dimmest stars in the images (around $V \sim 16$), the statistical uncertainty is about 0.03 mag, and the RMS fluctuation over 35 images in one night is only 0.07 mag. This suggests we are not even reaching as far down as a SNR of 10, and we could be missing dimmer stars. For the purposes of the *automated* pipeline, we wanted to err on the conservative side, but we could in the future explore whether that cutoff could be reduced without introducing too many false positives.

In any case, we hope that this report will give future readers confidence in understanding the conditions and limitations under which our observations were taken.

7. Acknowledgements

The authors would like to thank J. Donald Cline for his continuing support of the observatory, as well as colleagues and students at Guilford College for help and discussions. In particular, Prof. T. Espinola, C. Potts, S. Kirwan, N. Cleckler, E. Ruprecht, and A. Jordan have been of great help along the way. K. Iverson, of Plane Wave Instruments, helped with the automation of the dome. K. Sturrock's PYTHON library was critical in enabling us to automate observing in THESKYX, and J. Lynch donated the telescope control computer. Thanks also to Prof. D. Reichart for support in the early phases of this work. We would also like to thank the anonymous referree, whose suggestions were very helpful in improving both this paper and the analysis on which it is based. This research was made possible through the use of the AAVSO Photometric All-Sky Survey (APASS), funded by the Robert Martin Ayers Sciences Fund and NSF AST-1412587.

References

AAVSO. 2017, AAVSO Target Tool (https://www.aavso.org/aavso-target-tool).
Abbott, B. P., *et al.* 2017, *Astrophys. J., Lett.*, **848**, L12.
Akerlof, C. W., *et al.* 2003, *Publ. Astron. Soc. Pacific*, **115**, 132.
Astropy Collaboration, *et al.* 2013, *Astron. Astrophys.*, **558A**, 33.
Aurora Eurotech. 2019, Aurora cloud sensor (http://www.auroraeurotech.com/CloudSensor.php).
Barbary, K. 2016, *J. Open Source Software*, **1**, 58.
Bellm, E. C., *et al.* 2019, *Publ. Astron. Soc. Pacific*, **131**, 078002.
Cinzano, P., Falchi, F., and Elvidge, C. D. 2001, *Mem. Roy. Astron. Soc.*, **328**, 689.
Corbett, M. 2011, Bachelor's thesis, Guilford College.
Davis Instruments. 2019, Vantage Vue weather station (https://www.davisinstruments.com/solution/vantage-vue).
Deep Sky Instruments. 2019, RC optical systems telescope command center (TCC-I; https://www.rcopticalsystems.com/accessories/tcc.html).
Diffraction Limited. 2019a, MAXDOME II observatory dome control system (https://diffractionlimited.com).
Diffraction Limited. 2019b, SBIG STF-8300M CCD camera (https://diffractionlimited.com/product/stf-8300).
Everett, M. E., and Howell, S. B. 2001, *Publ. Astron. Soc. Pacific*, **113**, 1428.
Henden, A. A., Welch, D. L., Terrell, D., and Levine, S. E. 2009, *Bull. Amer. Astron. Soc.*, **41**, 669.
Honeycutt, R. K. 1992, *Publ. Astron. Soc. Pacific*, **104**, 435.
Law, N. M., *et al.* 2015, *Publ. Astron. Soc. Pacific*, **127**, 234.
Ouellette, J. 2016, Bachelor's thesis, Guilford College.
Paczyński, B. 2006, *Publ. Astron. Soc. Pacific*, **118**, 1621.
Seitz, T. 2016, Bachelor's thesis, Guilford College.
Smith, A. B., Caton, D. B., and Hawkins, R. L. 2016, *Publ. Astron. Soc. Pacific*, **128**, 055002.
Smith, D., *et al.* 2003, in *Gamma-Ray Burst and Afterglow Astronomy 2001*, AIP Conf. Proc. 662, American Institute of Physics, Melville, NY, 514.

Software Bisque. 2019a, Paramount ME robotic telescope mount (http://www.bisque.com/sc/media/p/28169.aspx).

Software Bisque. 2019b, THESKYX professional edition (http://www.bisque.com).

Watson, C., Henden, A. A., and Price, C. A. 2006–2014, AAVSO International Variable Star Index VSX (http://www.aavso.org/vsx).

Wells, D. C., Greisen, E. W., and Harten, R. H. 1981, *Astron. Astrophys., Suppl. Ser.*, **44**, 363.

Woźniak, P. R., *et al.* 2004, *Astron. J.*, **127**, 2436.

Zacharias, N., Finch, C. T., Girard, T. M., Henden, A., Bartlett, J. L., Monet, D. G., and Zacharias, M. I. 2013, *Astron. J.*, **145**, 44.

The Contribution of A. W. Roberts' Observations to the AAVSO International Database

Tim Cooper
Astronomical Society of Southern Africa; tpcoope@mweb.co.za

Received June 29, 2019; revised August 1, 2019; accepted August 2, 2019

Abstract Alexander William Roberts observed around one hundred variable stars mainly during the period 1891–1912. In 2004 we succeeded in digitizing around 70,000 of his visual observations which were added to the AAVSO International Database (AID), ensuring these are available for further study. These observations, many of which pre-date the existence of the AAVSO, extend the period of observation of many variable stars by one or two decades earlier. This paper summarizes the observations made by Roberts and added to the AID, and gives examples of some apparent discrepancies which warrant further investigation.

1. Introduction

Alexander William Roberts (AAVSO observer code RAE) was born in Farr, Scotland, on December 4, 1857. He emigrated to South Africa in 1883 aged twenty-five to take up a teaching post at Lovedale Missionary Institution, and later became Principal of Lovedale College. It was in the field of native affairs that he achieved his greatest endeavors, and in 1920 he became a Senator responsible for the Native Affairs Commission (Snedegar 2015). Roberts had a passion for astronomy, and under the influence of his friend Sir David Gill, Astronomer Royal at the Cape, he took an increasing interest in visual observation of variable stars. He was familiar with the work of Gould in preparation of his "Uranometria Argentina" (1879) and in his own words:

> Gould's work opened up a very great field. My own thoughts, since early youth, had turned to astronomy, and so in 1888, after considerable correspondence with Gould and Pickering and in consultation with Gill, I determined to erect a small observatory at Lovedale for the single purpose of observing southern variable stars and other allied phenomena. I spent two years in getting my hand into the work, becoming skillful in determining differences of magnitude, acquainting myself with the labours of other observers. (McIntyre 1938)

Roberts duly set up his observatory at Lovedale and began observing in 1891, when only 35 variable stars were catalogued south of declination –30° (Roberts 1891). His first observation was of the delta Cepheid variable l Car on 1891 April 7 (JD 2411829). In the following years Roberts concentrated on variable stars generally south of declination –30°, the most northerly object being the long period variable RZ Sco at declination –23°. The star which he observed on the largest number of occasions is the eclipsing binary RR Cen, with 2,289 observations in the AID. Roberts himself discovered the variability of this star.

Roberts remained active as a variable star observer until 1922, when, apart from κ Pav, he ceased to observe due to work commitments. He wrote:

> During the year 1922 work has been interrupted at this observatory. The Director was called upon by the Union Government to take over the post of Senior Native Commissioner for South Africa.... The delicate conditions of many of the native questions and difficulties made it impossible for the Director to give any part of his time to astronomical work. (Roberts 1923)

After observing for three decades, Roberts' contribution to variable star observation had virtually ended, but the legacy he left behind was enormous. Apart from the large number of observations, Roberts was responsible for the formation of the Variable Star Section of the old Cape Astronomical Association, which amalgamated in 1922 with the Johannesburg Astronomical Association to form the Astronomical Society of South Africa (ASSA, today Astronomical Society of Southern Africa). In his Presidential Address he said:

> In our own southern land this Society, over which I have the honour to be president, has put variable star work in the forefront of its endeavours, and as an Association we may point with no mean gratification to the achievements of Watson, Long, Cousins, Houghton, Skjellerup, Smith, Ensor, and others, who in this wonderful, sundrenched, star-lit land of ours have found success and happiness in their scientific pursuits. (McIntyre 1938)

Members of ASSA are active still today in observing stars which were observed by Roberts, and we are fortunate to have had the opportunity to finally add the observations he made to the AID, and open these up for study more than a century later.

2. Roberts' observations, instrumentation, and methodology

Roberts died on January 27, 1938, at Alice, South Africa. His extensive records, including raw observations, lists of magnitude estimates, hand-drawn charts and sequences, and manuscripts on each star were stored, and were subsequently found, in four cupboards in the library of Boyden Observatory (van Zyl 2003). These records consisted of around 140 packets wrapped in brown paper, which when stacked on top of each other would have formed a stack four metres high! Following a suggestion by Dr. Janet Mattei in 2002, a team succeeded in digitizing the observations, with the result that more than 70,000 visual observations were added to the AID (Cooper *et al.* 2004). More recently a team from the Centre for Astronomical Heritage (CfAH), funded by the Endangered Archives Programme of the British Library, has embarked on a program to digitize the entire archive held at Boyden, including all of the Roberts documents.

Roberts' observations were made with rather modest equipment by today's standards. His early observations were made with a 1-inch theodolite by Troughton and Simms. From 1900 he used a 2-inch refractor made by Messrs. Thomas Cooke and Sons of York, and provided by Sir John Usher, Norton, Edinburgh (Roberts 1900). This instrument had a rotating prism at the front of the objective, which allowed Roberts to adjust the position angle of stars being compared in order to prevent errors due to "position error" (Roberts 1896a, 1896b). With its field of view of nearly 3°, Roberts was able to measure star magnitudes down to about magnitude 10. He also had access to a 3¼-inch Ross telescope, which was loaned to him by the RAS, but as Roberts commented (1902): "it is a matter of regret to me that I am unable to follow stars below magnitude 11.2, the limit of the Ross glass." Thus Roberts could not follow the many long period variables he studied when they were fainter than about magnitude 11, a fact which stands out when we plot his observations in the AID (see Figures 1–3 for examples). By 1904 he was using the 1-inch and 3¼-inch telescopes for short and long period variables, and the 2-inch Cooke prismatic telescope for eclipsing variables (Roberts 1905).

When he began observing in 1891, suitable variable star charts were not available for most of the stars he wished to observe and Roberts had to prepare his own. In order to do this he would plot the positions of all stars in the vicinity of a known variable star on a sheet of plain paper, using the positions given in the "Cape Photographic Durchmusterung" when that catalogue became available (Gill and Kapteyn 1896–1900). The next step was to "fix upon certain stars as starting points from which to give relative magnitudes to all the other stars in the zone." He generally used Gould's "Uranometria Argentina" (1879) for his early magnitudes, but later used Harvard College Observatory magnitudes where these became available, and in 1912 he acquired on loan "through the very great kindness of Professor Pickering" the 4-inch Harvard meridian photometer, which he used to determine accurate comparison star magnitudes, especially for observing Algol variables (Roberts 1913).

In this way Roberts continued to observe variable stars on a regular basis between 1891 and 1907, except for 1897, when he made no observations. There was a further lull in 1907, when Roberts spent much of the time away overseas, and used the opportunity to reduce the observations made so far. In 1909 he "stopped observing short period variables and Algol variables." mainly due to weather (Roberts 1910), and the following year paid special attention to the variability of S Ara and κ Pav (Roberts 1911a) as he stated: "these stars show evidence of possessing characteristics both of short period variation and Algol variation; that is, they exhibit variation due to eclipse, superimposed upon the ordinary Cepheid type of continuous light change."

After about 1912 his observations of many stars became infrequent, occupying his time rather with reducing the observations made over the preceding twenty years, and making observations in order to qualify some specific aspect of a specific star's period. The process of reducing his observations was largely completed by 1914, and appears to take the form of hand-written manuscripts on each star, which we found in the wrapped packs at Boyden Observatory. Results from these manuscripts were used to prepare numerous papers, but the actual manuscripts were never published. The digitization currently conducted by the CfAH will finally go a long way to enabling publication of his manuscripts.

3. Summary of observations and types of stars observed

Roberts' various reports mention having observed 105 different variable stars (Roberts 1906). In the process of digitizing his observations, we found observations of 98 separate stars, which were processed and added to the AID, listed in Appendix A by type, and summarized in Table 1. Appendix A also lists the Julian Date of the first observation by Roberts, the first date of observation not made by Roberts, and the number of days his observations predate the first AAVSO observation for each star. From this it can be seen that Roberts extended the observations for all stars he observed except for three (Beck 2019), being R Car (observed by Tebbutt), R Ret (by Pogson), and S Ara (by Innes). Note also that although Roberts is said to have observed RS Car (= Nova Car 1896), no observations were found. The total number of observations processed and added to the database was 70,034, whereas some, including no less than Sir David Gill, credit Roberts with over 280,000 observations. We ascribed this discrepancy to a difference in what constitutes "an observation" in the AID. A single observation is the result of estimating the brightness of a variable star and reporting that magnitude as a single-line entry for the time of observation. But this observation is never the result of a single estimation, and the proficient observer always makes several estimates in order to improve accuracy, before averaging the individual estimates to arrive at the final observation. Roberts himself commented on this procedure (Roberts 1896a), and so, for example, his total observations reported for 1895 was 2,893, but he noted:

Table 1. Types of stars for which observations were added to the AAVSO International Database (observer RAE).

Type of Variable	Classification	No. of Stars
Long period	M	55
Semi-regular	SR	10
Eclipsing	E	7
Cepheid	DCEP	20
Novae	N	1
R CrB	RCB	1
Irregular, poorly studied	I	3
Constant	CST	1
RR Lyr	RRAB/BL	1

Each of the observations is the mean of two, one direct and the other reverse. This mode of observation has been adopted to eliminate "position error." As each observation also means, on the average, the determination of five comparison stars, the individual determinations of magnitude throughout the year are considerably over 30,000.

For this reason we believe Roberts' contribution is correctly around 70,000 observations, and all these have been captured and entered into the AID.

4. Details of some individual stars

Despite Roberts being best known for his observation of eclipsing binaries, the largest number of observations was of long period variables, which account for more than half of the number. In fact, eclipsing variables (types E and EW) make up less than ten percent of the stars observed. The author has prepared light curve plots for all 98 stars observed by Roberts and for which observations were submitted to the AID. Examination of these light curves is useful to understand where Roberts' data can add to the knowledge of certain stars by extending their light curves by around two decades earlier than previous, as well as highlighting some opportunities which require further investigation. The following light curves are presented as examples.

4.1. The long period variables: R Cen, S Hor, and U CrA

Observations of long period variable stars makes up the majority of stars observed, numbering 55 stars. The light curve for R Cen (Figure 1) is typical of Roberts' observations of long period variables. Observations by Roberts are shown in blue, while all other observers are in black.

The absence of observations fainter than magnitude 11 is already evident, which was the limit of the Ross 3¼-inch refractor used when stars were faint, although R Cen at minimum was only just below his grasp. His first observation was on August 29, 1892 (JD 2411972), and his last was on January 25, 1917 (JD 2421253), making 999 observations during this interval. Following this Roberts produced a manuscript for R Cen, describing the discovery as a variable star by Gould in 1871, followed by a rigorous description of the derivation of comparison star magnitudes he used, a list of his observations, and his analysis of dates of maxima and minima. He concluded: "The Lovedale observations are sufficient in number and in range to indicate with certainty that during the time R Centauri has been observed at Lovedale its period has been steadily decreasing, falling from 567 days in 1891 to 552 days in 1916." Using these data Ramoshebi (2006) was able to model the changing period of R Cen, which has decreased further to the current value around 501.8 days (Samus *et al.* 2017).

Another southern long period variable which demonstrates the limitations of his equipment more clearly is S Hor (Figure 2). The star was confirmed as being variable by Roberts, following which it was entered in his catalogue "Southern Variable Stars" (see Roberts 1901a). From 420 observations made between JD 2414612 and 2421608 (November 1898 to January 1918) he

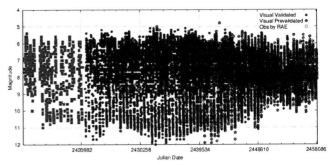

Figure 1. Light curve for R Cen from AAVSO data.

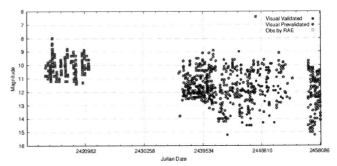

Figure 2. Light curve for S Hor from AAVSO data.

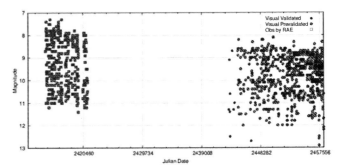

Figure 3. Light curve for U CrA from AAVSO data.

determined the period to be 330 days, close to the current value of 335.8 days. The cutoff in visibility of the star whenever it was fainter than about magnitude 11.2 is evident in the light curve. The light curve also shows that the first AAVSO observation of this star was submitted by Frank Bateson on March 17, 1957, so that the addition of Roberts' data permits study of the variability of S Hor extended by nearly sixty years.

The light curve for U CrA (Figure 3) similarly shows the cutoff at around magnitude 11.2, so that like many other long period variables Roberts was unable to determine the date of minima with any accuracy. The first AAVSO observation after Roberts was by Thomas Cragg on April 23, 1977, so that Roberts' data permits investigation of the variability extended by almost eighty years!

Initial inspection of the light curve for U CrA might indicate that the magnitude at maximum is slowly becoming fainter, and with that in mind the author visited Boyden during April 2019 to investigate Roberts' notes, including his charts and comparison sequences, more closely. A number of pages were found labelled "Comparison stars," with differing values over the years, and the table in Appendix B gives the values for the first and last of these pages. These show that while the comparison magnitudes were similar at the brighter end, the sequence is around 0.6 magnitude brighter in later years at the fainter end. More importantly, comparing with current values (AAVSO and GSC), his sequence is found to be around one magnitude brighter across the entire range, and probably explains the differences at maximum shown in Figure 3.

With the extension provided by his data in mind, all of the light curves of long period variables observed by Roberts would probably benefit from further scrutiny to determine whether amplitudes and periods have changed. A comprehensive review is required to compare Roberts' sequences to accepted modern magnitudes for all stars he observed. In regard to periods, Eddington and Plakidis (1929) referred previously to the complications in determining periodicity in long period variables by superposed irregularities, so that periods of these stars may appear to increase or decrease by several percent on time scales of decades. This would explain the differences found, for example, in S Hor, whereas the period in R Cen appears to be decreasing secularly (Anon. 2019). It should be remembered that in many cases there are large gaps between Roberts' and later AAVSO observations, which fact needs to be taken into account when investigating periods.

4.2. The semi-regular variables: Z Hya and L2 Pup

In all, Roberts observed ten semi-regular variables. An example is the variable star Z Hya, which was well observed by Roberts, but the star seems to have been neglected afterwards, with virtually no observations in the AAVSO database, as Figure 4 shows. The GCVS gives the type as SRB, with a V range of 8.8 to 9.8 and period 75 days. The star is mentioned by Ashbrook (1942) as being previously published as irregular, but she concludes it to be semi-regular with a period of about 75 days, from which, no doubt, the GCVS value is derived. Roberts found a visual range of 9.2 to 10.0 and comments:

> Light variations irregular. The rise to a maximum is rapid, but not continuous. Innes finds minima more distinctly marked than maxima; from observations of minima he also deduces a period of 53 days. In order to connect his observations with those made at Lovedale I have taken the probable period 52.5 days. A longer period, 62 days, would satisfy the Lovedale observations alone. (Roberts 1901a)

Very clearly Z Hya needs more attention before comparing the current classification with the behavior found by Roberts.

Another star for which Roberts data extends the period of observation by around two decades earlier is the semi-regular star L2 Pup (Figure 5).

He derived a mean period of 140.15 days and a visual range of 3.4–4.6 at maximum to 5.8–6.2 at minimum, but also commented:

> Variation subject to irregularities both as regards to limits and period. The form of the light-curve is also dissimilar for different periods, the star sometimes taking longer to rise to a maximum than to fall to a minimum. (Roberts 1901a)

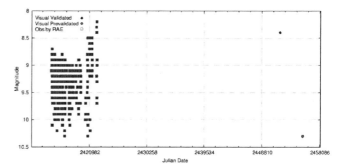

Figure 4. Light curve for Z Hya from AAVSO data.

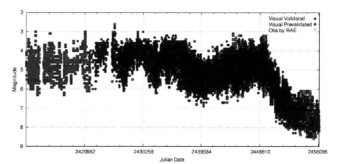

Figure 5. Light curve for L2 Pup from AAVSO data

In recent years the range of variability has become ever fainter, a process that started ca. 1992. Prior to that the light curve shows a broad dip, preceded by two or three dips of shorter duration. Bedding *et al.* (2005) analyzed the light curve for 1927–2005, and concluded the existence of two independent mechanisms: gradual dimming due to circumstellar dust, and pulsations within the star. Others have indicated that the pulsation amplitude varies cyclically on a time scale of thousands of days (Anon. 2019). The addition of Roberts' observations extends the data by nearly eight thousand days earlier, and may contribute to the further understanding of the evolution of this star.

4.3. The eclipsing variables: V Pup, RR Cen, and X Car

Roberts was well known for his observations of eclipsing stars, which make up 20% of the stars he observed, and the following are three examples. Roberts (1905) prepared his own charts and determined magnitudes of comparison stars for all three using the Oxford wedge photometer.

V Pup is classified as EB/SD, with a period of 1.4544859 days, V range from maximum 4.35 to 4.92 at primary minimum and 4.82 at secondary minimum (Samus *et al.* 2017). It was well observed by Roberts, who derived a period of 1.454475 days (1901a), but then appears to have been neglected until more recent times. Examination of the light curve in Figure 6 would indicate that Roberts observed both the maxima and minima to be fainter than listed in the GCVS, and also appears to show a peculiar decrease in the maximum magnitude during the period of Roberts' observations.

In an effort to explain these discrepancies the author investigated Roberts' notes. His chart and sequence are shown in Appendix C and indicate his sequence for V Pup was generally 0.2–0.5 magnitude fainter than the current AAVSO sequence, which explains the differences seen in Figure 6. What is not explained is the difference in magnitude when the star was at maximum light. For V Pup there is a single set of determinations of comparison magnitudes, which Roberts appears to have used invariably, unlike many other stars where he appears to have made ongoing determinations over a number of years. Therefore the apparent variation in maximum magnitude for V Pup remains to be explained.

The discovery of RR Cen as a variable star by Roberts was announced in *The Astronomical Journal*, No. 378 (Roberts 1896c), originally referred to as LAC. 5861, and is the star which he observed on the most occasions, with 2,289 observations in the AID. The GCVS gives its type as EW, period 0.6056845 day and V range 7.27 at maximum to 7.68/7.63 at minimum.

Roberts (1896d) published his results, deriving a period of 0.3028 day, and commented:

> The new short period variable in Centaurus recently announced in the *A.J.* no. 378, is one of considerable interest, as its type of variation is quite distinct from that of the other short-period variables discovered here. Its ascending period is almost exactly of the same duration as the descending period, a fact which relates the star indirectly to those of the *Algol-type*, and directly to *U Pegasi*. This is of extreme importance, as it materially strengthens the opinion expressed by Dr. Chandler that *U Pegasi* probably belongs to a new type of short period variable. (Roberts 1896d)

While Roberts' period appears to be exactly half that of the GCVS period, inspection of the light curve also indicates some apparent discrepancies between Roberts' and recent observations, and both historical and current data would benefit from closer scrutiny (Figure 7).

Another eclipsing star for which there are apparent discrepancies between Roberts and more recent AAVSO data is X Car (Figure 8).

The star is classified as EB, varying from 7.80 to 8.67 V with a period of 1.0826311 days. The range is very close to that observed by Roberts, but the star has been poorly observed since, apart from two sparse campaigns, with one set of data clearly discordant. A search of the archives produced Roberts' hand-drawn chart, but no indication of the magnitude sequence he used. X Car would benefit from closer scrutiny in order to clarify its current behavior.

4.4. Two stars paid particular attention: S Ara and κ Pav

Roberts was particularly interested in these two variables, believing that eclipses were somehow involved in the variability, and continued observation long after he had ceased observing other variable stars. Regarding S Ara (type RRAB) he commented: "the variation of this star is exceedingly remarkable," and published his Light curve of S Arae (Roberts 1901b) in which he derived a period of 0.4519 day and visual range 9.53 to 10.84, compared with the GCVS period of 0.4518587 day and V range 9.92 to 11.24. On this basis S Ara must have been at the very limit of visibility in the 3¼-inch

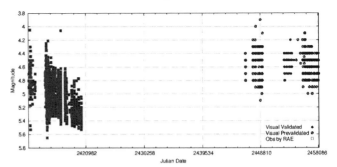

Figure 6. Light curve for V Pup from AAVSO data.

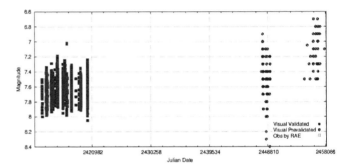

Figure 7. Light curve for RR Cen from AAVSO data.

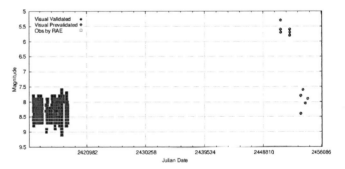

Figure 8. Light curve for X Car from AAVSO data.

when at minimum. Later he published a more complete study in which he concluded:

> The light curve of *S Arae*, and of stars of this definite type of variation, of which the salient characteristics are a distinct stationary period, a very rapid rise to maximum, and a slow fall to minimum phase, exhibits features that, apparently arise from a combination of two other definite types, an eclipse-curve superimposed upon an ordinary short-period curve (Roberts 1911b)

Again the light curve at maximum (Figure 9) shows some variation, which might be explained by the sequence shown in Appendix D. His notes contain one page marked "Comparison Stars," which is a table with changing values for the numbered stars on his rough chart for the years 1901–1914. If the star marked "2" is considered, for example, its magnitude increased from 9.58 in 1901 to 9.22 in 1914, an increase of 0.36 magnitude over the period Roberts observed S Ara, and he presumably used this comparison star when S Ara was near maximum brightness. Whether or not he later adjusted his final estimates to compensate for these fluctuations in comparison star magnitudes is not clear, but it is possible the differences in magnitude at maximum are due to inconsistencies in the comparison magnitudes used over the years. Whatever Roberts concluded, there appear to be no visual observations of S Ara in the AID since he made his last observation.

The second star which Roberts continued to observe long after ceasing others is κ Pav, and like S Ara, he also believed the variability was due to eclipses. This conclusion may have been influenced by the announcement (Wright 1903) that κ Pavonis is a binary star, and Roberts apparently found confirmation of this in explaining the light curve from his observations (Figure 10). He concluded:

> The striking peculiarity of the light curve of κ *Pavonis* is what appears to be a secondary phase, nearly midway between principal minimum and principal maximum…. Now a curve of this form suggests eclipse. (Roberts 1911c)

Inspection of the AAVSO light curve (Figure 11) shows Roberts observed the range slightly wider than others. Plotting light curves in this fashion also highlights erroneous observations in the AID, such as the two points at magnitude 1 and 2, which can be used to improve the reliability of the AAVSO data.

4.5. Two final stars in need of clarification, S Aps and T TrA

The bright R Coronae Borealis star S Aps has been very well observed over the years (Figure 12), but in Roberts' time the mechanism of variability was not understood.

He commented: "Maximum not distinctly marked; increasing and decreasing rate of variation slow and apparently equal. Decreasing phase irregular" (Roberts 1901a). He was only able to monitor S Aps when it was bright, which he observed as brighter than others, but nonetheless, he captured three fades due to the now well-understood mechanism. In an effort to explain the difference in magnitude when S Aps is bright, Roberts' comparison sequence will be investigated to see how it compares with the modern sequence.

As a final example of Roberts' observations, both the historical and current data for T TrA need further scrutiny. The light curve is shown in Figure 13, and shows variations of more than one magnitude.

The GCVS and Hipparcos data, however, give the star as being of constant magnitude, while AAVSO and ASAS-SN data suggest a rapid variable with no discernible period. The spectral type is B9IV, a type not normally associated with large, rapid variations (Anon. 2019). Roberts himself says:

> Gould considered this star to vary between the limits $7^M.2$ to $7^M.4$, in a period of about 1 day (U.A. p. 260). Lovedale observations do not confirm this variation…. It is possible that the apparent variation may be really due to position error; this would be fulfilled in a period of one day. (Roberts 1901a)

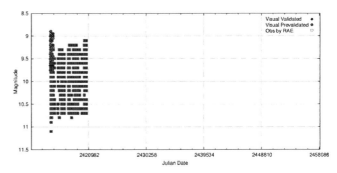

Figure 9. Light curve for S Ara from AAVSO data.

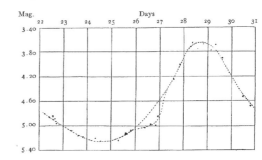

Figure 10. Roberts' light curve for κ Pav.

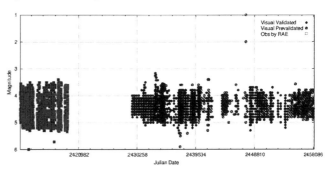

Figure 11. Light curve for κ Pav from AAVSO data.

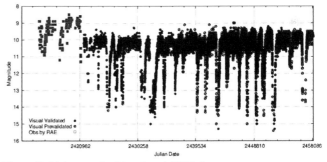

Figure 12. Light curve for S Aps from AAVSO data.

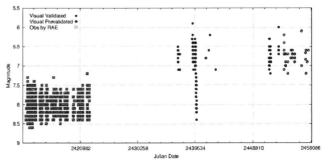

Figure 13. Light curve for T TrA from AAVSO data.

These facts are, however, not consistent with the final observations of Roberts', which show variability over the range 7.2 to 8.6, as well as observations already in the AID covering two periods, around 1966–1967 when the observed range was 5.9 to 8.4, and 1999–2016 when the observed range was 6.0 to 7.2. Roberts' observations and manuscript will be investigated to understand exactly how he arrived at his data, and the star will be observed by the author to clarify the nature of any current variability.

5. Conclusions

The addition of Roberts' observations expanded the AAVSO International Database by around 70,000 observations for 98 separate variable stars. More importantly it extended the database by several decades earlier for some southern variables, and may be useful in studying the longer term changes in variability of some stars. This paper provides a few examples of how Roberts' observations compare to the overall AAVSO observations and highlight some interesting aspects that need further scrutiny, as well as stars that have been neglected that could benefit from increased modern attention. There appear to be some differences in magnitudes at maximum and minimum light of some variables, which may be due to differences in comparison magnitude sequences. Similarly, it should be remembered that in some cases there are large gaps in dates between Roberts' and later AAVSO observations which could cause problems in period determination. A full review of Roberts' magnitude sequences is required before arriving at any conclusions on changes in an individual star's light curve parameters. The construction of light curves over such broad periods is shown to be a potentially useful quality control tool to highlight observations in the AAVSO database which may have been entered in error. Finally, Roberts' manuscripts remain unpublished, and the current digitization project under the auspices of the Centre for Astronomical Heritage will go a long way to opening up Roberts' observations for closer scrutiny.

6. Acknowledgements

The author gratefully acknowledges AAVSO Director, Dr. Stella Kafka, not only for motivation to write this paper, but also for persuading him (over a shared bottle of Chardonnay) to delve deeper into the treasure chest that Alexander Roberts left behind a century ago. The author also wishes to thank Sara Beck for providing the Julian dates of first observations in the AAVSO International Database used in compiling Appendix A, and Auke Slotegraaf for constructive comments and corrections to the original draft of this paper.

References

Anon. 2019, private communications from anonymous referee used in preparation of this paper.

Ashbrook, M. D. 1942, *Ann. Harvard Coll. Obs.*, **109**, 35.

Beck, S. 2019, private communication (April 15), dates of first observations in the AAVSO International Database.

Bedding, T. R., Kiss, L. L., Kjeldsen, H., Brewer, B. J., Dind, Z. E., Kawaler, S. D., and Zijlstra, A. A. 2005, *Mon. Not. Roy. Astron. Soc.*, **361**, 1375.

Cooper, T. P., Fraser, B., Cooper, D., Hoffman M., and van Zyl, B. 2004, *Mon. Not. Astron. Soc. S. Afr.*, **63**, 176.

Eddington, A. S., and Plakidis, S. 1929, *Mon. Not. Roy. Astron. Soc.*, **90**, 65.

Gill, D., and Kapteyn, J. C. 1896–1900, *Ann. Cape Obs.*, **3–5**, 1.

Gould, B. A. 1879, *Result. Obs. Nac. Argentina*, **1**, 1.

McIntyre D. G. 1938, *J. Astron. Soc. S. Afr.*, **4**, 116.

Ramoshebi, P. 2006, M.Sc. dissertation, University of the Free State, South Africa.

Roberts, A. W. 1891, *Trans. S. Afr.. Phil. Soc.*, **8–9**, 24-34.

Roberts, A. W. 1896a, *Mon. Not. Roy. Astron. Soc.*, **56**, 248.

Roberts, A. W. 1896b, *Astrophys. J.*, **4**, 184.

Roberts, A. W. 1896c, *Astron. J.*, **16**, 144.

Roberts, A. W. 1896d, *Astron. J.*, **16**, 205.

Roberts, A. W. 1900. *Mon. Not. Roy. Astron. Soc.*, **60**, 365.

Roberts, A. W. 1901a, "Southern Variable Stars," *Astron. J.*, **21**, 81.

Roberts, A. W. 1901b, *Mon. Not. Roy. Astron. Soc.*, **61**, 163.

Roberts, A. W. 1902, *Mon. Not. Roy. Astron. Soc.*, **62**, 287.

Roberts, A. W. 1905, *Mon. Not. Roy. Astron. Soc.*, **65**, 380.

Roberts, A. W. 1906, *Mon. Not. Roy. Astron. Soc.*, **66**, 215.

Roberts, A. W. 1910, *Mon. Not. Roy. Astron. Soc.*, **70**, 338.

Roberts, A. W. 1911a, *Mon. Not. Roy. Astron. Soc.*, **71**, 315.

Roberts, A. W. 1911b, *Astrophys. J.*, **33**, 197.

Roberts, A. W. 1911c, *Astrophys. J.*, **34**, 164.

Roberts, A. W. 1913, *Mon. Not. Roy. Astron. Soc.*, **73**, 265.

Roberts, A. W. 1923. *Mon. Not. Roy. Astron. Soc.*, **83**, 281.

Samus N. N., Kazarovets E. V., Durlevich O. V., Kireeva N. N., and Pastukhova E. N. 2017, *Astron. Rep.*, **61**, 80.

Snedegar, K. 2015, *Mission, Science, and Race in South Africa: A.W. Roberts of Lovedale, 1883–1938*, Lexington Books, Lanham, MD.

van Zyl, B. 2003, *Mon. Not. Astron. Soc. S. Afr.*, **62**, 186.

Wright, W. H. 1903, *Lick Obs. Bull.*, **3**, 3.

Appendix A: Master list of variable stars observed by A. W. Roberts (RAE).

Star	Type	Period	Roberts First	AAVSO First	Difference
Long Period Variable Stars (55 stars)					
R Cae	M	388.4	2412800.3	2421485.4	8685
R Car	*M*	*305.6*	*2411856.3*	*2407846.9*	*–4009*
S Car	M	150.3	2411856.3	2421417.2	9561
Z Car	M	383.6	2413308.3	2422030.0	8722
AF Car	M	447.6	2417946.4	2425505.5	7559
R Cen	**M**	**501.8**	**2411972.4**	**2421608.8**	**9636**
U Cen	M	220.0	2413311.4	2421405.3	8094
W Cen	M	200.4	2413742.4	2419224.9	5483
X Cen	M	312.0	2413728.4	2421405.3	7677
RS Cen	M	164.1	2414810.3	2419205.0	4395
RT Cen	M	255.0	2415032.6	2419215.0	4182
S Col	M	325.8	2414600.4	2421664.3	7064
T Col	M	225.8	2414570.5	2419770.7	5200
U CrA	**M**	**147.5**	**2414613.3**	**2443257.2**	**28644**
R Gru	M	332.0	2413427.3	2421475.2	8048
S Gru	M	401.5	2414598.4	2419254.0	4656
T Gru	M	136.5	2414593.4	2420039.7	5446
R Hor	M	407.6	2412800.3	2419298.0	6498
S Hor	**M**	**335.8**	**2414612.5**	**2435754.9**	**21142**
T Hor	M	217.6	2414930.3	2419298.0	4368
R Ind	M	216.3	2413803.4	2419254.0	5451
S Ind	M	400.0	2413803.4	2419254.0	5451
R Lup	M	235.6	2413657.3	2419832.0	6175
S Lup	M	339.7	2413657.3	2422170.2	8513
RS Mic	M	228.5	2414930.4	2441900.8	26970
T Nor	M	240.7	2414861.3	2419215.9	4355
R Oct	M	405.4	2412828.4	2421674.6	8846
S Oct	M	259.0	2414016.3	2421699.7	7683
T Oct	M	218.5	2413749.3	2425766.5	12017
R Pav	M	229.5	2413274.4	2419216.0	5942
T Pav	M	243.6	2413829.4	2419216.0	5387
U Pav	M	289.7	2414598.4	2436477.6	21879
R Phe	M	269.3	2413427.3	2419254.0	5827
S Pic	M	428.0	2413652.3	2419298.0	5646
R PsA	M	297.6	2413427.3	2413542.5	115
W Pup	M	119.7	2413586.3	2421480.5	7894
R Ret	*M*	*278.7*	*2412237.2*	*2401302.0*	*–10935*
RT Sgr	M	306.5	2413743.3	2419931.9	6189
RU Sgr	M	240.5	2413743.3	2421763.3	8020
RV Sgr	M	315.8	2414600.3	2421400.3	6800
RR Sco	M	281.4	2413282.4	2419506.0	6224
RS Sco	M	319.9	2412946.4	2419216.0	6270
RT Sco	M	449.0	2414670.6	2420339.7	5669
RU Sco	M	370.8	2413830.3	2421704.3	7874
RW Sco	M	388.4	2413742.5	2421692.3	7950
RZ Sco	M	156.6	2414895.4	2416966.7	2071
S Scl	M	362.6	2413045.3	2420064.7	7019
T Scl	M	202.4	2413787.4	2419330.9	5544
U Scl	M	333.7	2414570.3	2421479.2	6909
V Scl	M	296.1	2413787.4	2421701.3	7914
R Tel	M	467.0	2414727.6	2421782.2	7055
R Tuc	M	286.1	2413801.5	2421477.2	7676
S Tuc	M	240.7	2413747.4	2419298	5551
T Tuc	M	250.3	2416640.5	2421477.2	4837
W Vel	M	394.7	2414961.6	2419204.9	4243
Semiregular Variable Stars (10 stars)					
RR Car	SRB	155.5	2413308.3	2419218.0	5910
T Cen	SR	90.4	2413047.3	2419270.9	6224
R Dor	SRB	338	2411861.3	2421685.2	9824
Z Hya	**SRB**	**75**	**2414966.6**	**2451697.6**	**36731**
S Pav	SRA	380.9	2413274.4	2421705.8	8431
S Phe	SRB	141	2413426.3	2437483.0	24057
R Pic	SR	170.9	2413652.3	2419415.0	5763
L2 Pup	**SRB**	**140.6**	**2411839.4**	**2419816.7**	**7977**
R Scl	SRB	370	2411972.4	2412726.5	754
Y Scl	SRB		2414612.4	2437555.0	22943
Eclipsing Variable Stars (7 stars)					
R Ara	EA	4.42522	2412133.4	2445870.9	33738
X Car	**EB**	**1.0826311**	**2412507.3**	**2451593.5**	**39086**
RR Cen	**EW**	**0.6056845**	**2413029.3**	**2447795.5**	**34766**
V Pup	**EB/SD**	**1.4544859**	**2411921.2**	**2446401.3**	**34480**
RR Pup	EA/SD	6.4296333	2414897.6	2451511.4	36614
RS Sgr	EA/SD	2.4156832	2412324.4	2447681.5	35357
S Vel	EA/SD	5.9336475	2412865.3	2454472.8	41607
Delta Cepheid and Classical Cepheid Variable Stars (20 stars)					
l Car	DCEP	35.55560	2411830.2	2417357.5	5527
U Car	DCEP	38.80942	2411856.3	2429026.2	17170
V Car	DCEP	6.696756	2412130.3	2434869.3	22739
Y Car	DCEP(B)	3.639760	2412633.4	2434517.3	21884
V Cen	DCEP	5.4940	2413029.3	2434513.4	21484
R Cru	DCEP	5.82575	2411942.3	2433776.2	21834
S Cru	DCEP	4.68997	2411942.3	2434587.3	22645
T Cru	DCEP	6.73331	2411942.3	2433776.2	21834
R Mus	DCEP	7.510211	2411857.4	2432029.6	20172
S Mus	DCEP	9.66007	2411882.3	2433776.1	21894
S Nor	DCEP	9.75411	2412264.2	2434561.5	22297
U Nor	DCEP	12.64371	2414896.3	2426423.7	11527
κ Pav	**CEP**	**9.09423**	**2411840.3**	**2429470.5**	**17630**
RS Pup	DCEP	41.3876	2414962.4	2430931.5	15969
RV Sco	DCEP	6.06133	2412974.4	2434564.4	21590
RY Sco	DCEP	20.31322	2414412.6	2435205.7	20793
R TrA	DCEP	3.389287	2411858.3	2427119.6	15261
S TrA	DCEP	6.32344	2412323.2	2434562.4	22239
T Vel	DCEP	4.63974	2412249.3	2433765.1	21516
V Vel	DCEP	4.370991	2412181.3	2434521.2	22340
Other Variable Stars (7 stars)					
S Aps	**RCB**	—	**2414578.3**	**2421643.3**	**7065**
S Ara	*RRAB*	*0.4518587*	*2414962.3*	*2414853.9*	*–108*
RS Car	NA	—	NO OBS	2421566.3	
R CrA	INSA	—	2413743.3	2417117.6	3374
S CrA	INT	—	2413749.3	2419216.0	5467
T CrA	INSB	—	2414818.4	2419216.0	4398
T TrA	**CST**	—	**2412329.2**	**2436805.9**	**24477**

Notes: Stars are given according to the latest GCVS type (Samus et al. 2017). Stars observed within each classification are listed in order of increasing right ascension. Stars discussed further in the text are indicated in bold face. Stars for which Roberts was not the earliest observer are indicated in italics. The period is according to GCVS (Samus et al. 2017). Total stars observed for which estimates submitted to the AAVSO International Database = 98.

Appendix B: Roberts' chart and sequences for U CrA.

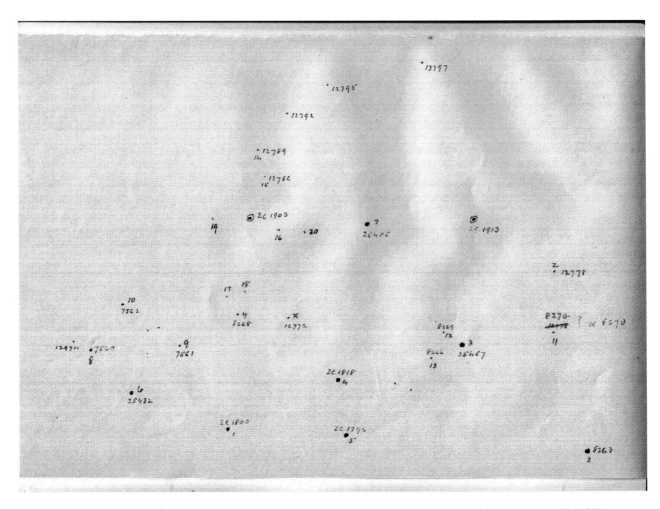

Figure 14. Roberts' hand-drawn chart for U CrA, marked as ZC1903. The star marked ZC1913 is the variable AM CrA (type SR, range 8.6–12.7).

Table 2. Roberts' and AAVSO comparison star sequence for U CrA. AAVSO sequence from chart reference X24407NE.

Star Label	Roberts Sequence Frst	Roberts Sequence Last	AAVSO Sequence (GSC magnitude)
1	6.54	6.60	(7.7)
2	6.95	6.76	—
3	7.00	7.00	8.4 (8.4)
4	7.33	7.30	(8.4)
5	7.41	7.38	(8.8)
6	7.39	7.35	(8.5)
7	7.99	7.94	(9.0)
8	8.83	8.74	9.6
9	9.31	9.24	—
10	9.40	9.31	10.2
11	9.58	9.01	—
12	10.34	9.71	—
13	10.47	9.88	—
14	11.05	10.44	10.9
15	11.50	10.89	—
16	11.62	11.01	—

Appendix C: Roberts' rough chart and sequence for V Pup.

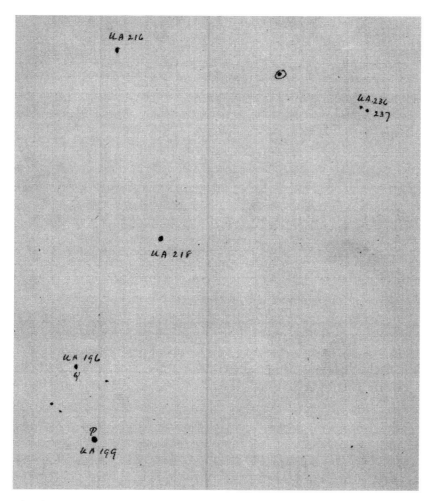

Figure 15. Roberts' hand-drawn chart for V Pup.

Table 3. Roberts' and AAVSO comparison star sequence for V Pup. AAVSO sequence from chart reference X24439AIT.

Star Label	Roberts Sequence	AAVSO Sequence[1]
UA196	5.2	4.7
UA199	4.3	4.1
UA216	5.0	4.6
UA218	4.5	4.2
UA236	6.4	6.0

[1] AAVSO sequence from chart reference X24439AIT

Appendix D: Roberts' charts and sequence for S Ara.

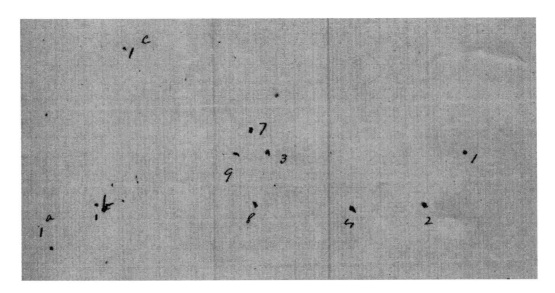

Figure 16. Robert's field for S Ara, as would be noted at the eyepiece.

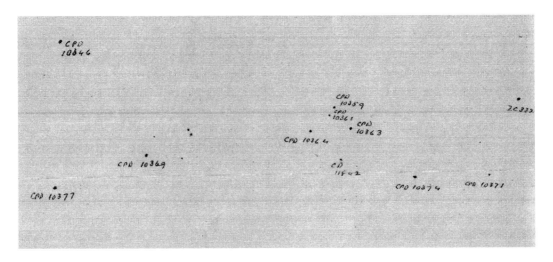

Figure 17. Roberts' hand-drawn chart for S Ara, labelled as CPD10361.

Table 4. Robert's adopted magnitudes for comparison stars for S Ara, which changed over the years as shown.

Star Label	Star Name	1901	1905	Year 1910	1914
1a	CPD10377	—	—	8.93	—
1b	CPD10369	—	—	9.33	—
2	CPD10371	9.65	9.45	9.33	—
3	CPD10363	9.58	9.45	9.42	9.22
4	CPD10374	9.80	9.81	9.77	9.55
7	CPD10359	10.50	10.47	10.57	10.49
8	CD11842	10.53	10.48	10.52	10.40
9	CPD10364	10.58	10.68	10.63	10.49

Note: 1a and 2 are in the same column of the table labelled "Comparison Stars," and have the same magnitude of 9.33. His original rough chart, however, identifies these as two different stars.

Recent Minima of 200 Eclipsing Binary Stars

Gerard Samolyk
P.O. Box 20677, Greenfield, WI 53220; gsamolyk@wi.rr.com

Received August 16, 2019; accepted August 16, 2019

Abstract This paper continues the publication of times of minima for eclipsing binary stars from observations reported to the AAVSO Eclipsing Binaries Section. Times of minima from observations received from February 2019 through July 2019 are presented.

1. Recent observations

The accompanying list contains times of minima calculated from recent CCD observations made by participants in the AAVSO's eclipsing binary program. This list will be web-archived and made available through the AAVSO ftp site at ftp://ftp.aavso.org/public/datasets/gsamj472eb200.txt. This list, along with the eclipsing binary data from earlier AAVSO publications, is also included in the Lichtenknecker database administrated by the Bundesdeutsche Arbeitsgemeinschaft für Veränderliche Sterne e. V. (BAV) at: http://www.bav-astro.de/LkDB/index.php?lang=en. These observations were reduced by the observers or the writer using the method of Kwee and van Worden (1956). The standard error is included when available. Column F in Table 1 indicates the filter used. A "C" indicates a clear filter.

The linear elements in the *General Catalogue of Variable Stars* (GCVS; Kholopov *et al.* 1985) were used to compute the O–C values for most stars. For a few exceptions where the GCVS elements are missing or are in significant error, light elements from another source are used: CD Cam (Baldwin and Samolyk 2007), CW Cas (Samolyk 1992a), DV Cep (Frank and Lichtenknecker 1987), DF Hya (Samolyk 1992b), DK Hya (Samolyk 1990), GU Ori (Samolyk 1985).

The light elements used for V640 Aur, CI CVn, DN Dra, GX Gem, V1053 Her, KM UMa, FQ Vir, and MS Vir are from Kreiner (2004).

The light elements used for CW CMa, V1297 Cas, V1331 Cas, V1342 Cas, V1173 Her, V1282 Her, V470 Hya, IZ Lac, AL Leo, V1848 Ori, KK Psc, and NN Vir are from Paschke (2014).

The light elements used for V459 Aur are from Nelson (2014).

The light elements used for V868 Mon, V1056 Per, and KU Psc are from the AAVSO VSX site (Watson *et al.* 2014). O–C values listed in this paper can be directly compared with values published in the AAVSO *Observed Minima Timings of Eclipsing Binaries* monograph series.

References

Baldwin, M. E., and Samolyk, G. 2007, *Observed Minima Timings of Eclipsing Binaries No. 12*, AAVSO, Cambridge, MA.

Frank, P., and Lichtenknecker, D. 1987, *BAV Mitt.*, No. 47, 1.

Kholopov, P. N., *et al.* 1985, *General Catalogue of Variable Stars*, 4th ed., Moscow.

Kreiner, J. M. 2004, *Acta Astron.*, **54**, 207 (http://www.as.up.krakow.pl/ephem/).

Kwee, K. K., and van Woerden, H. 1956, *Bull. Astron. Inst. Netherlands*, **12**, 327.

Nelson, R. 2014, Eclipsing Binary O–C Files (http://www.aavso.org/bob-nelsons-o-c-files).

Paschke, A. 2014, "O–C Gateway" (http://var.astro.cz/ocgate/).

Samolyk, G. 1985, *J. Amer. Assoc. Var. Star Obs.*, **14**, 12.

Samolyk, G. 1990, *J. Amer. Assoc. Var. Star Obs.*, **19**, 5.

Samolyk, G. 1992a, *J. Amer. Assoc. Var. Star Obs.*, **21**, 34.

Samolyk, G. 1992b, *J. Amer. Assoc. Var. Star Obs.*, **21**, 111.

Watson, C., Henden, A. A., and Price, C. A. 2014, AAVSO International Variable Star Index VSX (Watson+, 2006–2014; http://www.aavso.org/vsx).

Table 1. Recent times of minima of stars in the AAVSO eclipsing binary program.

Star	JD (min) Hel. 2400000+	Cycle	O–C (day)	F	Observer	Standard Error (day)	Star	JD (min) Hel. 2400000+	Cycle	O–C (day)	F	Observer	Standard Error (day)
RT And	58694.6712	27909	–0.0116	V	G. Samolyk	0.0001	SX CMa	58532.4022	18739	0.0244	V	T. Arranz	0.0001
WZ And	58688.8600	25611	0.0832	V	G. Samolyk	0.0003	TZ CMa	58535.4352	16433	–0.2279	V	T. Arranz	0.0002
AB And	58694.7912	68050	–0.0489	V	G. Samolyk	0.0002	UU CMa	58522.3351	6427	–0.0709	V	T. Arranz	0.0001
BD And	58664.8642	51204	0.0148	V	G. Samolyk	0.0002	CW CMa	58561.6669	7777	–0.0088	V	S. Cook	0.0006
DS And	58460.7304	22086	0.0068	V	S. Cook	0.0007	XZ CMi	58528.3652	27788	0.0051	V	L. Corp	0.0002
CX Aqr	58455.6134	39573	0.0165	V	S. Cook	0.0002	XZ CMi	58528.3664	27788	0.0063	V	T. Arranz	0.0001
KO Aql	58676.6704	5862	0.1076	V	G. Samolyk	0.0001	XZ CMi	58530.3897	27791.5	0.0038	V	L. Corp	0.0002
KP Aql	58676.8319	5428.5	–0.0223	V	G. Samolyk	0.0001	XZ CMi	58539.3631	27807	0.0056	V	T. Arranz	0.0001
OO Aql	58655.7676	39548	0.0746	V	G. Samolyk	0.0001	XZ CMi	58559.6206	27842	0.0048	V	G. Samolyk	0.0001
V343 Aql	58672.7224	16388	–0.0486	V	G. Samolyk	0.0001	YY CMi	58549.5993	27903	0.0206	V	G. Samolyk	0.0004
V346 Aql	58690.8325	15160	–0.0146	V	G. Samolyk	0.0001	AK CMi	58531.4099	27266	–0.0233	V	T. Arranz	0.0001
SX Aur	58567.6734	15210	0.0181	V	S. Cook	0.0004	AK CMi	58544.4266	27289	–0.0223	V	T. Arranz	0.0002
SX Aur	58567.6771	15210	0.0218	TG	G. Conrad	0.0002	RZ Cas	58690.7880	12960	0.0806	V	G. Samolyk	0.0002
AP Aur	58527.5648	28248.5	1.7316	V	T. Arranz	0.0002	TW Cas	58462.6949	11520	0.0151	V	S. Cook	0.0011
AP Aur	58536.3874	28264	1.7299	V	T. Arranz	0.0001	CW Cas	58409.5794	52616.5	–0.1185	Ic	G. Lubcke	0.0003
AP Aur	58554.6089	28296	1.7334	V	G. Samolyk	0.0002	CW Cas	58409.5799	52616.5	–0.1180	V	G. Lubcke	0.0004
CL Aur	58520.4725	20535	0.1855	V	T. Arranz	0.0001	CW Cas	58409.5800	52616.5	–0.1179	B	G. Lubcke	0.0003
CL Aur	58530.4282	20543	0.1863	V	T. Arranz	0.0001	DZ Cas	58695.8386	38555	–0.2130	V	G. Samolyk	0.0002
CL Aur	58540.3829	20551	0.1861	V	T. Arranz	0.0001	PV Cas	58687.8263	10546	–0.0332	V	G. Samolyk	0.0001
EM Aur	58559.6884	15131	–1.1249	V	G. Samolyk	0.0004	V1297 Cas	57657.7265	1076.5	0.0075	V	V. Petriew	0.0001
EP Aur	58511.6579	54487	0.0230	V	S. Cook	0.0005	V1297 Cas	57657.8629	1077	0.0076	V	V. Petriew	0.0001
EP Aur	58553.6192	54558	0.0227	V	G. Samolyk	0.0001	V1297 Cas	57658.6803	1080	0.0075	V	V. Petriew	0.0001
HP Aur	58554.6268	11103.5	0.0729	V	G. Samolyk	0.0002	V1297 Cas	57658.8167	1080.5	0.0076	V	V. Petriew	0.0001
V459 Aur	58488.5499	1361	–0.0044	V	G. Samolyk	0.0006	V1297 Cas	57658.9530	1081	0.0076	V	V. Petriew	0.0001
V459 Aur	58592.6850	1459	–0.0077	V	S. Cook	0.0008	V1297 Cas	58005.8483	2354	–0.0022	V	V. Petriew	0.0001
V640 Aur	58121.7586	17139.5	–0.0037	B	K. Alton	0.0004	V1297 Cas	58007.7561	2361	–0.0021	V	V. Petriew	0.0001
V640 Aur	58121.7586	17139.5	–0.0037	V	K. Alton	0.0002	V1297 Cas	58007.8926	2361.5	–0.0018	V	V. Petriew	0.0002
V640 Aur	58121.7598	17139.5	–0.0025	Ic	K. Alton	0.0005	V1331 Cas	57683.7352	5701.5	0.0095	V	V. Petriew	0.0003
V640 Aur	58124.7105	17148.5	–0.0038	B	K. Alton	0.0003	V1331 Cas	57683.8926	5702	0.0086	V	V. Petriew	0.0002
V640 Aur	58124.7105	17148.5	–0.0038	V	K. Alton	0.0003	V1331 Cas	57732.6569	5856	0.0083	V	V. Petriew	0.0002
V640 Aur	58124.7117	17148.5	–0.0026	Ic	K. Alton	0.0004	V1331 Cas	58017.8058	6756.5	0.0104	V	V. Petriew	0.0006
V640 Aur	58124.8739	17149	–0.0044	V	K. Alton	0.0004	V1331 Cas	58025.7221	6781.5	0.0103	V	V. Petriew	0.0003
V640 Aur	58124.8741	17149	–0.0042	B	K. Alton	0.0004	V1331 Cas	58025.8808	6782	0.0107	V	V. Petriew	0.0002
V640 Aur	58124.8747	17149	–0.0036	Ic	K. Alton	0.0002	V1342 Cas	53659.8738	4.5	0.0011	V	V. Petriew	0.0002
V640 Aur	58130.6130	17166.5	–0.0053	Ic	K. Alton	0.0005	V1342 Cas	53668.8243	8.5	0.0013	V	V. Petriew	0.0001
V640 Aur	58130.6141	17166.5	–0.0042	V	K. Alton	0.0001	V1342 Cas	53697.9133	21.5	0.0018	V	V. Petriew	0.0002
V640 Aur	58130.6151	17166.5	–0.0032	B	K. Alton	0.0005	SU Cep	58675.8569	35889	0.0078	V	G. Samolyk	0.0001
V640 Aur	58130.7774	17167	–0.0048	Ic	K. Alton	0.0003	WZ Cep	58687.8561	73416.5	–0.2049	V	G. Samolyk	0.0001
V640 Aur	58130.7785	17167	–0.0037	B	K. Alton	0.0010	XX Cep	58655.7672	5911	0.0275	V	G. Samolyk	0.0001
V640 Aur	58130.7785	17167	–0.0037	V	K. Alton	0.0001	DV Cep	58688.6947	10263	–0.0057	V	G. Samolyk	0.0002
SS Boo	58593.5623	4982	7.3847	V	G. Samolyk	0.0002	EG Cep	58660.7354	29500	0.0089	V	G. Samolyk	0.0002
TU Boo	58533.9460	78534.5	–0.1597	V	G. Samolyk	0.0004	EK Cep	58586.8636	4423	0.0129	V	G. Samolyk	0.0001
TU Boo	58573.8320	78657.5	–0.1610	V	G. Samolyk	0.0001	RW Com	58539.9192	78019	0.0131	V	G. Samolyk	0.0002
TY Boo	58542.6436	75870.5	0.0640	V	T. Arranz	0.0001	RW Com	58543.5979	78034.5	0.0130	V	T. Arranz	0.0001
TY Boo	58568.6509	75952.5	0.0652	V	T. Arranz	0.0001	RW Com	58543.7171	78035	0.0135	V	T. Arranz	0.0001
TZ Boo	58545.6278	63644.5	0.0591	V	T. Arranz	0.0001	RW Com	58566.6207	78131.5	0.0132	TG	G. Conrad	0.0004
TZ Boo	58565.5402	63711.5	0.0616	V	T. Arranz	0.0002	RW Com	58625.6024	78380	0.0145	V	K. Menzies	0.0001
TZ Boo	58565.6887	63712	0.0616	V	T. Arranz	0.0002	RZ Com	58523.9297	69973.5	0.0575	V	G. Samolyk	0.0001
TZ Boo	58570.5896	63728.5	0.0593	V	T. Arranz	0.0002	RZ Com	58573.6900	70120.5	0.0574	V	G. Samolyk	0.0001
TZ Boo	58594.6623	63809.5	0.0619	V	G. Samolyk	0.0002	RZ Com	58626.6665	70277	0.0577	TG	G. Conrad	0.0001
UW Boo	58593.6127	16113	–0.0054	V	G. Samolyk	0.0001	SS Com	58546.9424	81260	0.9626	V	G. Samolyk	0.0002
UW Boo	58617.7294	16137	–0.0018	V	S. Cook	0.0006	SS Com	58606.3887	81404	0.9669	V	T. Arranz	0.0001
VW Boo	58549.9012	79973	–0.2860	V	K. Menzies	0.0001	SS Com	58607.4203	81406.5	0.9665	V	T. Arranz	0.0001
VW Boo	58642.6678	80244	–0.2896	V	G. Samolyk	0.0002	SS Com	58636.7328	81477.5	0.9708	V	S. Cook	0.0006
ZZ Boo	58597.8690	4013	0.0811	V	G. Samolyk	0.0002	CC Com	58530.7796	86082.5	–0.0301	V	G. Samolyk	0.0002
AD Boo	58554.8622	16551	0.0366	V	G. Samolyk	0.0001	CC Com	58530.8905	86083	–0.0295	V	G. Samolyk	0.0002
SV Cam	58582.6553	26958	0.0604	TG	G. Conrad	0.0002	CC Com	58562.8889	86228	–0.0307	V	G. Samolyk	0.0002
CD Cam	58514.7013	7527	–0.0122	TG	G. Conrad	0.0003	CC Com	58571.3861	86266.5	–0.0299	V	G. Coates	0.0001
CD Cam	58527.6904	7544	–0.0143	TG	G. Conrad	0.0005	CC Com	58571.4958	86267	–0.0305	V	G. Coates	0.0001
DN Cam	58577.8160	12196	0.0016	TG	G. Conrad	0.0003	CC Com	58577.4540	86294	–0.0308	V	T. Arranz	0.0001
WW Cnc	58526.6223	2314	0.0365	V	T. Arranz	0.0001	CC Com	58627.6602	86521.5	–0.0308	TG	G. Conrad	0.0002
WY Cnc	58544.3867	38815	–0.0467	V	T. Arranz	0.0001	U CrB	58571.5344	12115	0.1435	V	T. Arranz	0.0001
XZ Cnc	58537.4286	7574	0.0166	V	T. Arranz	0.0001	U CrB	58602.6066	12124	0.1459	V	T. Arranz	0.0001
XZ Cnc	58557.4406	7602	0.0177	V	T. Arranz	0.0001	U CrB	58647.4844	12137	0.1451	V	T. Arranz	0.0001
CI CVn	57853.7022	6561	–0.0029	V	V. Petriew	0.0002	RW CrB	58562.6147	24519	0.0034	V	T. Arranz	0.0001
R CMa	58554.6308	12558	0.1290	V	G. Samolyk	0.0001	TW CrB	58593.7894	35144	0.0591	V	G. Samolyk	0.0001

Table continued on following pages

Table 1. Recent times of minima of stars in the AAVSO eclipsing binary program, cont.

Star	JD (min) Hel. 2400000+	Cycle	O–C (day)	F	Observer	Standard Error (day)	Star	JD (min) Hel. 2400000+	Cycle	O–C (day)	F	Observer	Standard Error (day)
W Crv	58542.8571	48688.5	0.0176	V	G. Samolyk	0.0002	V1053 Her	58248.9022	19975	–0.0009	B	K. Alton	0.0002
W Crv	58573.5154	48767.5	0.0175	V	T. Arranz	0.0001	V1053 Her	58248.9030	19975	–0.0001	Ic	K. Alton	0.0002
W Crv	58603.4010	48844.5	0.0209	V	T. Arranz	0.0001	V1053 Her	58251.7796	19985	–0.0015	V	K. Alton	0.0006
RV Crv	58598.6679	23512	–0.1075	V	G. Samolyk	0.0003	V1053 Her	58251.7804	19985	–0.0007	B	K. Alton	0.0006
RV Crv	58602.4079	23517	–0.1037	V	T. Arranz	0.0001	V1053 Her	58251.7812	19985	0.0001	Ic	K. Alton	0.0003
SX Crv	58598.6488	55527.5	–0.9567	V	G. Samolyk	0.0003	V1053 Her	58251.9231	19985.5	–0.0019	Ic	K. Alton	0.0003
SX Crv	58634.7612	55641.5	–0.9412	V	S. Cook	0.0008	V1053 Her	58251.9240	19985.5	–0.0010	V	K. Alton	0.0001
V Crt	58562.8162	24451	–0.0008	V	G. Samolyk	0.0002	V1053 Her	58251.9248	19985.5	–0.0002	B	K. Alton	0.0002
RV Crt	58581.7743	13707	0.1043	V	S. Cook	0.0006	V1053 Her	58255.8082	19999	–0.0020	Ic	K. Alton	0.0004
Y Cyg	58645.8341	16417.5	0.1212	V	G. Samolyk	0.0001	V1053 Her	58255.8099	19999	–0.0003	B	K. Alton	0.0001
WW Cyg	58688.8042	5519	0.1511	V	G. Samolyk	0.0001	V1053 Her	58255.8103	19999	0.0001	V	K. Alton	0.0002
ZZ Cyg	58597.8831	21631	–0.0761	V	G. Samolyk	0.0001	V1053 Her	58255.9521	19999.5	–0.0020	V	K. Alton	0.0005
AE Cyg	58694.6821	14557	–0.0047	V	G. Samolyk	0.0001	V1053 Her	58255.9525	19999.5	–0.0016	Ic	K. Alton	0.0004
BR Cyg	58685.5699	12867	0.0016	V	T. Arranz	0.0001	V1053 Her	58255.9542	19999.5	0.0001	B	K. Alton	0.0002
BR Cyg	58689.5677	12870	0.0017	V	T. Arranz	0.0001	V1173 Her	57879.8249	24455.5	–0.0021	V	V. Petriew	0.0001
BR Cyg	58693.5656	12873	0.0019	V	T. Arranz	0.0001	V1282 Her	58231.8217	19129	–0.0023	B	K. Alton	0.0005
CG Cyg	58687.6249	30520	0.0795	V	G. Samolyk	0.0001	V1282 Her	58231.8238	19129	–0.0002	V	K. Alton	0.0003
DK Cyg	58676.6790	43929	0.1300	V	G. Samolyk	0.0001	V1282 Her	58231.8242	19129	0.0002	Ic	K. Alton	0.0004
KV Cyg	58664.6580	10284	0.0588	V	G. Samolyk	0.0004	V1282 Her	58231.9624	19129.5	–0.0008	B	K. Alton	0.0004
MY Cyg	58687.7796	6202	0.0010	V	G. Samolyk	0.0001	V1282 Her	58231.9633	19129.5	0.0001	Ic	K. Alton	0.0002
V346 Cyg	58664.6790	8376	0.1990	V	G. Samolyk	0.0002	V1282 Her	58231.9644	19129.5	0.0012	V	K. Alton	0.0005
V387 Cyg	58695.6725	47940	0.0201	V	G. Samolyk	0.0001	V1282 Her	58234.8839	19140	–0.0020	Ic	K. Alton	0.0003
V388 Cyg	58672.6460	19463	–0.1323	V	G. Samolyk	0.0002	V1282 Her	58234.8849	19140	–0.0010	V	K. Alton	0.0001
V388 Cyg	58690.6852	19484	–0.1329	V	G. Samolyk	0.0001	V1282 Her	58234.8858	19140	–0.0001	B	K. Alton	0.0003
V401 Cyg	58676.6960	25469	0.0904	V	G. Samolyk	0.0002	V1282 Her	58243.9321	19172.5	–0.0001	Ic	K. Alton	0.0001
V466 Cyg	58675.6770	21487.5	0.0073	V	G. Samolyk	0.0002	V1282 Her	58243.9331	19172.5	0.0009	V	K. Alton	0.0002
V466 Cyg	58694.4639	21501	0.0081	V	T. Arranz	0.0001	V1282 Her	58243.9339	19172.5	0.0017	B	K. Alton	0.0004
V548 Cyg	58645.6513	7860	0.0241	V	G. Samolyk	0.0002	V1282 Her	58244.9053	19176	–0.0011	Ic	K. Alton	0.0001
V704 Cyg	58690.8158	36361	0.0387	V	G. Samolyk	0.0002	V1282 Her	58244.9057	19176	–0.0007	B	K. Alton	0.0003
V1034 Cyg	58676.8360	16110	0.0186	V	G. Samolyk	0.0002	V1282 Her	58244.9065	19176	0.0001	V	K. Alton	0.0004
V1425 Cyg	57958.7973	14019.5	0.0137	V	V. Petriew	0.0004	TT Hya	58631.7719	2116	0.2099	V	S. Cook	0.0019
W Del	58664.8283	3191	0.0137	V	G. Samolyk	0.0002	WY Hya	58559.6059	25123.5	0.0411	V	G. Samolyk	0.0001
RZ Dra	58586.8318	26157	0.0703	V	G. Samolyk	0.0001	AV Hya	58523.8028	31973	–0.1196	V	G. Samolyk	0.0001
RZ Dra	58686.5406	26338	0.0710	V	T. Arranz	0.0001	AV Hya	58541.5759	31999	–0.1151	V	T. Arranz	0.0001
TW Dra	58650.4491	5171	–0.0517	V	T. Arranz	0.0001	AV Hya	58573.6912	32046	–0.1199	V	G. Samolyk	0.0001
TW Dra	58664.4866	5176	–0.0485	V	T. Arranz	0.0001	AV Hya	58597.6098	32081	–0.1205	V	G. Samolyk	0.0001
UZ Dra	58600.8073	5222	0.0032	V	G. Samolyk	0.0001	DF Hya	58542.6135	47654.5	0.0105	V	G. Samolyk	0.0001
UZ Dra	58672.5559	5244	0.0031	V	T. Arranz	0.0001	DF Hya	58562.6149	47715	0.0103	V	G. Samolyk	0.0002
AI Dra	58637.6930	12801	0.0403	V	G. Samolyk	0.0002	DF Hya	58562.7808	47715.5	0.0109	V	G. Samolyk	0.0001
BH Dra	58653.7597	10254	–0.0028	V	S. Cook	0.0015	DF Hya	58568.4008	47732.5	0.0106	V	T. Arranz	0.0001
BH Dra	58664.6626	10260	–0.0033	V	G. Samolyk	0.0002	DF Hya	58597.6598	47821	0.0110	V	G. Samolyk	0.0001
RW Gem	58528.5074	14038	0.0027	V	T. Arranz	0.0001	DI Hya	58542.6848	44490	–0.0365	V	G. Samolyk	0.0002
RW Gem	58531.3727	14039	0.0025	V	T. Arranz	0.0001	DI Hya	58554.3634	44509	–0.0374	V	T. Arranz	0.0001
WW Gem	58521.3781	26286	0.0212	V	T. Arranz	0.0001	DK Hya	58523.7824	30043	0.0008	V	G. Samolyk	0.0002
AF Gem	58533.3473	25228	–0.0680	V	T. Arranz	0.0001	V470 Hya	58559.5905	14815.5	0.0151	V	G. Samolyk	0.0004
AF Gem	58559.4608	25249	–0.0681	V	T. Arranz	0.0001	CO Lac	58655.8302	20180	0.0100	V	G. Samolyk	0.0001
AF Gem	58569.4094	25257	–0.0675	V	T. Arranz	0.0001	CO Lac	58672.7940	20191	0.0096	V	G. Samolyk	0.0001
GX Gem	58605.6730	1512	0.0171	V	S. Cook	0.0012	IZ Per	58045.5954	31639	0.0071	V	V. Petriew	0.0007
SZ Her	58637.7403	20503	–0.0339	V	G. Samolyk	0.0001	Y Leo	58530.6513	7766	–0.0678	V	G. Samolyk	0.0001
SZ Her	58656.5567	20526	–0.0338	V	T. Arranz	0.0001	Y Leo	58542.4537	7773	–0.0681	V	T. Arranz	0.0001
SZ Her	58661.4656	20532	–0.0335	V	T. Arranz	0.0001	Y Leo	58552.5703	7779	–0.0682	V	T. Arranz	0.0001
SZ Her	58687.6442	20564	–0.0340	V	G. Samolyk	0.0001	Y Leo	58569.4319	7789	–0.0676	V	T. Arranz	0.0001
TT Her	58660.6650	20464	0.0448	V	G. Samolyk	0.0001	UU Leo	58531.5716	7819	0.2215	V	T. Arranz	0.0001
TU Her	58608.5418	6417	–0.2624	V	T. Arranz	0.0001	UV Leo	58563.4148	33533	0.0455	V	T. Arranz	0.0001
TU Her	58676.5494	6447	–0.2648	V	T. Arranz	0.0001	UV Leo	58566.4161	33538	0.0464	V	T. Arranz	0.0001
UX Her	58642.8192	12248	0.1516	V	G. Samolyk	0.0001	UV Leo	58594.6197	33585	0.0460	V	G. Samolyk	0.0001
UX Her	58664.5029	12262	0.1514	V	T. Arranz	0.0001	UZ Leo	58564.4619	30360.5	0.0005	V	T. Arranz	0.0001
UX Her	58695.4815	12282	0.1531	V	T. Arranz	0.0001	UZ Leo	58586.4026	30396	0.0006	V	L. Corp	0.0001
CC Her	58600.5449	10918	0.3276	V	T. Arranz	0.0001	VZ Leo	58528.5405	25107	–0.0454	V	T. Arranz	0.0001
CC Her	58671.6416	10959	0.3300	V	G. Samolyk	0.0002	VZ Leo	58540.5288	25118	–0.0461	V	T. Arranz	0.0001
CT Her	58623.5380	9013	0.0099	V	T. Arranz	0.0001	XY Leo	58537.4565	47388	0.1820	V	L. Corp	0.0001
CT Her	58637.8305	9021	0.0114	V	G. Samolyk	0.0002	XY Leo	58551.8021	47438.5	0.1807	V	G. Samolyk	0.0003
V1053 Her	58247.8951	19971.5	–0.0007	V	K. Alton	0.0001	XY Leo	58576.6650	47526	0.1851	V	G. Samolyk	0.0002
V1053 Her	58247.8954	19971.5	–0.0004	Ic	K. Alton	0.0002	XY Leo	58593.7107	47586	0.1850	TG	G. Conrad	0.0002
V1053 Her	58247.8959	19971.5	0.0001	B	K. Alton	0.0003	XZ Leo	58199.6461	27011	0.0753	B	G. Lubcke	0.0002
V1053 Her	58248.9013	19975	–0.0018	V	K. Alton	0.0003	XZ Leo	58199.6464	27011	0.0756	Ic	G. Lubcke	0.0001

Table continued on following pages

Table 1. Recent times of minima of stars in the AAVSO eclipsing binary program, cont.

Star	JD (min) Hel. 2400000+	Cycle	O–C (day)	F	Observer	Standard Error (day)	Star	JD (min) Hel. 2400000+	Cycle	O–C (day)	F	Observer	Standard Error (day)
XZ Leo	58199.6468	27011	0.0760	V	G. Lubcke	0.0001	V1848 Ori	57433.4989	12641.5	–0.0011	Ic	K. Alton	0.0001
XZ Leo	58551.7946	27733	0.0791	V	G. Samolyk	0.0006	V1848 Ori	57433.4993	12641.5	–0.0007	B	K. Alton	0.0004
XZ Leo	58576.6710	27784	0.0810	V	G. Samolyk	0.0002	V1848 Ori	57433.4997	12641.5	–0.0003	V	K. Alton	0.0003
XZ Leo	58593.7393	27819	0.0785	TG	G. Conrad	0.0002	BX Peg	58687.6682	51681	–0.1363	V	G. Samolyk	0.0001
AL Leo	58537.4245	6672.5	–0.0070	V	L. Corp	0.0002	KW Peg	58687.7009	12897	0.2289	V	G. Samolyk	0.0002
AM Leo	58570.5632	43951	0.0127	V	L. Corp	0.0002	RV Per	58523.6094	8349	0.0070	V	G. Samolyk	0.0001
AP Leo	58567.4011	44221	0.0157	V	L. Corp	0.0002	XZ Per	58531.6227	13046	–0.0734	V	K. Menzies	0.0001
T LMi	58533.7328	4350	–0.1350	V	G. Samolyk	0.0002	KW Per	58694.8633	17493	0.0204	V	G. Samolyk	0.0001
Z Lep	58526.3957	31299	–0.2011	V	T. Arranz	0.0001	V1056 Per	58050.7120	5039.5	0.0720	V	V. Petriew	0.0001
Z Lep	58529.3783	31302	–0.1996	V	T. Arranz	0.0001	V1056 Per	58050.8967	5040	0.0711	V	V. Petriew	0.0002
Z Lep	58530.3728	31303	–0.1988	V	T. Arranz	0.0001	KK Psc	58055.6474	15460.5	–0.0095	V	K. Alton	0.0002
SS Lib	58608.8295	12137	0.1841	V	G. Samolyk	0.0001	KK Psc	58055.6483	15460.5	–0.0086	B	K. Alton	0.0002
SS Lib	58676.4147	12184	0.1834	V	T. Arranz	0.0002	KK Psc	58055.6493	15460.5	–0.0076	Ic	K. Alton	0.0005
ES Lib	58658.7493	20757	0.1124	V	S. Cook	0.0005	KK Psc	58055.7917	15461	–0.0092	B	K. Alton	0.0001
RY Lyn	58551.7531	10916	–0.0189	V	G. Samolyk	0.0003	KK Psc	58055.7924	15461	–0.0085	V	K. Alton	0.0003
RY Lyn	58597.6739	10948	–0.0177	V	G. Samolyk	0.0002	KK Psc	58055.7928	15461	–0.0081	Ic	K. Alton	0.0004
UZ Lyr	58674.4405	7923	–0.0546	V	T. Arranz	0.0001	KK Psc	58068.7484	15506	–0.0103	B	K. Alton	0.0003
EW Lyr	58671.4639	16509	0.2989	V	T. Arranz	0.0001	KK Psc	58068.7494	15506	–0.0093	Ic	K. Alton	0.0002
FL Lyr	58593.8289	9353	–0.0017	V	G. Samolyk	0.0001	KK Psc	58068.7504	15506	–0.0083	V	K. Alton	0.0004
V582 Lyr	58041.6230	21654	–0.008	Ic	K. Alton	0.0008	KK Psc	58070.7650	15513	–0.0094	Ic	K. Alton	0.0002
V582 Lyr	58041.6241	21654	–0.007	V	K. Alton	0.0001	KK Psc	58070.7657	15513	–0.0087	B	K. Alton	0.0004
V582 Lyr	58041.6251	21654	–0.006	B	K. Alton	0.0005	KK Psc	58070.7661	15513	–0.0083	V	K. Alton	0.0003
V582 Lyr	58042.6468	21658	–0.008	V	K. Alton	0.0004	KK Psc	58072.7796	15520	–0.0104	Ic	K. Alton	0.0003
V582 Lyr	58042.6479	21658	–0.007	B	K. Alton	0.0002	KK Psc	58072.7806	15520	–0.0094	B	K. Alton	0.0002
V582 Lyr	58042.6489	21658	–0.006	Ic	K. Alton	0.0009	KK Psc	58072.7817	15520	–0.0083	Ic	K. Alton	0.0004
V582 Lyr	58066.5719	21751.5	–0.010	B	K. Alton	0.0009	KK Psc	58102.5831	15623.5	–0.0100	Ic	K. Alton	0.0002
V582 Lyr	58066.5735	21751.5	–0.008	V	K. Alton	0.0002	KK Psc	58102.5841	15623.5	–0.0090	V	K. Alton	0.0001
V582 Lyr	58066.5760	21751.5	–0.006	Ic	K. Alton	0.0008	KK Psc	58102.5852	15623.5	–0.0079	B	K. Alton	0.0004
V582 Lyr	58068.6210	21759.5	–0.008	V	K. Alton	0.0004	KK Psc	58102.7258	15624	–0.0112	Ic	K. Alton	0.0004
V582 Lyr	58068.6221	21759.5	–0.007	B	K. Alton	0.0005	KK Psc	58102.7268	15624	–0.0102	V	K. Alton	0.0002
V582 Lyr	58068.6230	21759.5	–0.006	Ic	K. Alton	0.0004	KK Psc	58102.7279	15624	–0.0091	B	K. Alton	0.0003
V582 Lyr	58072.5871	21775	–0.008	V	K. Alton	0.0006	KU Psc	58439.6342	19243.5	–0.0445	V	K. Alton	0.0004
V582 Lyr	58072.5872	21775	–0.008	Ic	K. Alton	0.0008	KU Psc	58439.6351	19243.5	–0.0436	B	K. Alton	0.0001
V582 Lyr	58072.5881	21775	–0.007	B	K. Alton	0.0003	KU Psc	58439.6360	19243.5	–0.0427	Ic	K. Alton	0.0004
RU Mon	58542.4005	4686.5	–0.7204	V	T. Arranz	0.0001	KU Psc	58439.7790	19244	–0.0443	B	K. Alton	0.0002
RW Mon	58520.5785	13032	–0.0882	V	T. Arranz	0.0002	KU Psc	58439.7799	19244	–0.0434	V	K. Alton	0.0001
RW Mon	58539.6384	13042	–0.0892	V	G. Samolyk	0.0001	KU Psc	58439.7809	19244	–0.0424	V	K. Alton	0.0006
RW Mon	58543.4506	13044	–0.0892	V	T. Arranz	0.0001	KU Psc	58441.6588	19250.5	–0.0448	B	K. Alton	0.0005
RW Mon	58545.3566	13045	–0.0893	V	T. Arranz	0.0001	KU Psc	58441.6597	19250.5	–0.0439	Ic	K. Alton	0.0001
BB Mon	58522.4503	43457	–0.0043	V	T. Arranz	0.0001	KU Psc	58441.6607	19250.5	–0.0429	V	K. Alton	0.0003
BB Mon	58536.3730	43476	–0.0028	V	T. Arranz	0.0001	KU Psc	58441.8034	19251	–0.0449	Ic	K. Alton	0.0003
BO Mon	58523.3647	6748	–0.0121	V	T. Arranz	0.0001	KU Psc	58441.8044	19251	–0.0439	V	K. Alton	0.0001
BO Mon	58543.3913	6757	–0.0125	V	T. Arranz	0.0001	KU Psc	58441.8054	19251	–0.0429	B	K. Alton	0.0002
EP Mon	58533.6942	22337	0.0241	V	G. Samolyk	0.0002	UZ Pup	58530.7364	17509	–0.0116	V	G. Samolyk	0.0001
V868 Mon	58540.3815	7518.5	0.0449	V	L. Corp	0.0004	UZ Pup	58532.3268	17511	–0.0109	V	T. Arranz	0.0001
SX Oph	58656.4585	12241	–0.0003	V	T. Arranz	0.0001	UZ Pup	58559.3516	17545	–0.0110	V	T. Arranz	0.0001
SX Oph	58664.7106	12245	–0.0014	V	G. Samolyk	0.0002	AV Pup	58554.6984	49208	0.2428	V	G. Samolyk	0.0001
V501 Oph	58688.6578	28697	–0.0098	V	G. Samolyk	0.0001	U Sge	58671.4819	12288	0.0202	V	T. Arranz	0.0001
V508 Oph	58683.5314	39447	–0.0267	V	T. Arranz	0.0001	V1968 Sgr	58694.6949	37060	–0.0154	V	G. Samolyk	0.0003
V839 Oph	58661.5117	44530.5	0.3327	V	T. Arranz	0.0001	AO Ser	58597.7775	27821	–0.0119	V	G. Samolyk	0.0001
V839 Oph	58683.3929	44584	0.3327	V	T. Arranz	0.0001	AO Ser	58606.5713	27831	–0.0116	V	T. Arranz	0.0001
V1010 Oph	58642.7705	29792	–0.2058	V	G. Samolyk	0.0001	AO Ser	58671.6429	27905	–0.0117	V	G. Samolyk	0.0002
V1010 Oph	58695.6825	29872	–0.2079	V	G. Samolyk	0.0001	CC Ser	58563.6297	40854	1.1424	V	T. Arranz	0.0001
EQ Ori	58518.3004	15509	–0.0406	V	T. Arranz	0.0001	CC Ser	58586.8523	40899	1.1447	V	G. Samolyk	0.0002
ER Ori	58549.5935	39969.5	0.1442	V	G. Samolyk	0.0001	CC Ser	58602.5913	40929.5	1.1455	V	T. Arranz	0.0001
ET Ori	58545.3764	33505	–0.0039	V	T. Arranz	0.0001	CC Ser	58662.7088	41046	1.1483	V	S. Cook	0.0006
ET Ori	58559.6402	33520	–0.0041	V	S. Cook	0.0006	Y Sex	58598.6411	40094	–0.0222	V	G. Samolyk	0.0003
FL Ori	58518.3519	8492	0.0422	V	T. Arranz	0.0001	AC Tau	58542.6101	6316	0.1836	V	G. Samolyk	0.0002
FT Ori	58518.4679	5450	0.0231	V	T. Arranz	0.0001	GR Tau	58487.3786	32370	–0.0539	V	L. Corp	0.0002
GU Ori	58521.3510	32828	–0.0679	V	T. Arranz	0.0001	HU Tau	58523.6039	8388	0.0401	V	G. Samolyk	0.0002
V1848 Ori	57406.5971	12540.5	–0.0015	B	K. Alton	0.0001	V781 Tau	58542.3742	42525.5	–0.0500	V	L. Corp	0.0002
V1848 Ori	57406.5971	12540.5	–0.0015	V	K. Alton	0.0001	X Tri	58520.3224	16487	–0.0994	V	T. Arranz	0.0001
V1848 Ori	57406.5978	12540.5	–0.0008	Ic	K. Alton	0.0002	RV Tri	58694.8623	16800	–0.0426	V	G. Samolyk	0.0001
V1848 Ori	57417.6505	12582	–0.0017	Ic	K. Alton	0.0002	W UMa	58200.6330	37271	–0.1084	Ic	G. Lubcke	0.0002
V1848 Ori	57417.6515	12582	–0.0007	V	K. Alton	0.0002	W UMa	58200.6332	37271	–0.1082	B	G. Lubcke	0.0002
V1848 Ori	57417.6525	12582	0.0003	B	K. Alton	0.0005	W UMa	58200.6335	37271	–0.1079	V	G. Lubcke	0.0002

Table continued on next page

Table 1. Recent times of minima of stars in the AAVSO eclipsing binary program, cont.

Star	JD (min) Hel. 2400000+	Cycle	O–C (day)	F	Observer	Standard Error (day)	Star	JD (min) Hel. 2400000+	Cycle	O–C (day)	F	Observer	Standard Error (day)
W UMa	58516.7505	38218.5	–0.1125	TG	G. Conrad	0.0002	VV Vir	58643.7922	61467	–0.0532	V	S. Cook	0.0009
TY UMa	58546.6490	53629.5	0.4247	V	G. Samolyk	0.0002	AG Vir	58608.6709	20503	–0.0120	V	G. Samolyk	0.0002
TY UMa	58546.8274	53630	0.4258	V	G. Samolyk	0.0001	AH Vir	58600.6576	31375	0.3012	V	G. Samolyk	0.0001
TY UMa	58564.3771	53679.5	0.4258	V	T. Arranz	0.0001	AK Vir	58553.8713	13386	–0.0409	V	G. Samolyk	0.0002
TY UMa	58576.6097	53714	0.4268	V	G. Samolyk	0.0001	AW Vir	58569.6103	38268.5	0.0330	V	T. Arranz	0.0002
UX UMa	58554.7243	107397	–0.0016	V	G. Samolyk	0.0001	AW Vir	58571.5567	38274	0.0324	V	T. Arranz	0.0001
UX UMa	58554.9214	107398	–0.0011	V	G. Samolyk	0.0001	AW Vir	58600.4076	38355.5	0.0326	V	T. Arranz	0.0001
UX UMa	58559.6406	107422	–0.0020	V	T. Arranz	0.0002	AW Vir	58606.4253	38372.5	0.0323	V	T. Arranz	0.0001
UX UMa	58627.6887	107768	–0.0022	V	G. Samolyk	0.0001	AX Vir	58582.7885	44144	0.0279	V	G. Samolyk	0.0001
VV UMa	58516.6614	18478	–0.0827	TG	G. Conrad	0.0002	AZ Vir	58573.8141	41746.5	–0.0204	V	G. Samolyk	0.0001
VV UMa	58545.5296	18520	–0.0845	V	T. Arranz	0.0001	AZ Vir	58608.4304	41845.5	–0.0210	V	T. Arranz	0.0001
XZ UMa	58539.3756	10121	–0.1511	V	T. Arranz	0.0001	BH Vir	58533.8683	18734	–0.0134	V	G. Samolyk	0.0002
XZ UMa	58545.4881	10126	–0.1502	V	T. Arranz	0.0002	BH Vir	58629.4422	18851	–0.0135	V	T. Arranz	0.0001
XZ UMa	58562.5995	10140	–0.1513	V	G. Samolyk	0.0001	FQ Vir	58648.8372	8202	–0.0022	V	S. Cook	0.0006
XZ UMa	58567.4878	10144	–0.1523	V	T. Arranz	0.0001	MS Vir	58647.7092	19676.5	0.0029	V	S. Cook	0.0009
XZ UMa	58572.3772	10148	–0.1522	V	T. Arranz	0.0001	NN Vir	58661.7566	21139	0.0063	V	S. Cook	0.0004
ZZ UMa	58562.4054	9834	–0.0014	V	T. Arranz	0.0001	Z Vul	58693.4085	6414	–0.0159	V	T. Arranz	0.0001
AF UMa	58573.7017	6044	0.6393	V	G. Samolyk	0.0003	AW Vul	58637.8466	15317	–0.0346	V	G. Samolyk	0.0001
KM UMa	57793.7580	15044.5	–0.0077	V	V. Petriew	0.0006	BE Vul	58695.6630	11974	0.1071	V	G. Samolyk	0.0001
KM UMa	57793.9355	15045	–0.0062	V	V. Petriew	0.0001	BO Vul	58660.7382	11651	–0.0125	V	G. Samolyk	0.0001
RU UMi	58576.6342	32348	–0.0144	V	G. Samolyk	0.0001	BS Vul	58676.8353	32366	–0.0353	V	G. Samolyk	0.0001
UW Vir	58657.7266	7904	–0.0559	V	S. Cook	0.0004	BT Vul	58690.6550	20407	0.0066	V	G. Samolyk	0.0002
VV Vir	58570.6283	61303	–0.0508	V	T. Arranz	0.0001	CD Vul	58675.6560	18102	–0.0010	V	G. Samolyk	0.0002

Recent Minima of 267 Recently Discovered Eclipsing Variable Stars

Gerard Samolyk
P.O. Box 20677, Greenfield, WI 53220; gsamolyk@wi.rr.com

Vance Petriew
P.O. Box 9 Station Main, White City, SK S4L 5B1, Canada; vance.petriew@sasktel.net

Received October 4, 2019; revised October 30, 2019, accepted October 30, 2019

Abstract The AAVSO Eclipsing Variable Section has received observations of newly discovered variable stars that have not yet been listed in the GCVS. This paper contains times of minima for these eclipsing variables.

1. Recent observations

The accompanying list contains times of minima calculated from recent CCD observations made by participants in the AAVSO's eclipsing binary program.

Many of these eclipsing stars are recent discoveries by all-sky surveys. Many of them have precise periods derived from the survey data but not all of them are accurate. Most of the stars in this paper have little or no published times of minima in literature, so these times of minima were observed to help determine accurate orbital periods for these stars. Further observations are encouraged to check the stability and further refine the accuracy of these periods so these stars can eventually be added to the GCVS. In a few cases, these stars happened to be in the field of another variable during a time series observing run.

This list will be web-archived and made available through the AAVSO ftp site at ftp://ftp.aavso.org/public/datasets/gsamj472nongcvs.txt. These observations were reduced by the observers or the writer using the method of Kwee and van Woerden (1956) or the Fourier Fit in Minima 2.5 (Nelson 2007). The standard error is included. Column F indicates the filter used. A "C" indicates a clear filter.

Many of these stars have multiple identifications from various catalogs. The AAVSO AUID (AAVSO Unique IDentifier) is also included in this table to provide a unique identification for each variable.

References

Kwee, K. K., and van Woerden, H. 1956, *Bull. Astron. Inst. Netherlands*, **12**, 327.

Nelson, B. 2007, Software by Bob Nelson (https://www.variablestarssouth.org/software-by-bob-nelson).

Watson, C., Henden, A. A., and Price, C. A. 2014, AAVSO International Variable Star Index VSX (Watson+, 2006–2014; http://www.aavso.org/vsx).

Table 1. Recent times of minima of stars in the AAVSO eclipsing binary program.

Star	AUID	JD (min) Hel. 2400000+	F	Observer	Standard Error (day)
1SWASP J022311.92+425529.9	000-BJX-274	57306.7373	V	V. Petriew	0.0005
1SWASP J022311.92+425529.9	000-BJX-274	57306.8808	V	V. Petriew	0.0005
1SWASP J022311.92+425529.9	000-BJX-274	57309.7345	V	V. Petriew	0.0005
1SWASP J022311.92+425529.9	000-BJX-274	57309.8770	V	V. Petriew	0.0004
1SWASP J182358.52+351518.2	000-BLY-330	57569.7731	V	V. Petriew	0.0005
1SWASP J182358.52+351518.2	000-BLY-330	57573.7458	V	V. Petriew	0.0007
1SWASP J182358.52+351518.2	000-BLY-330	57575.7310	V	V. Petriew	0.0016
1SWASP J182358.52+351518.2	000-BLY-330	57575.8840	V	V. Petriew	0.0009
1SWASP J211703.96+404916.8	000-BLY-378	57930.8061	V	V. Petriew	0.0011
1SWASP J211703.96+404916.8	000-BLY-378	57932.7629	V	V. Petriew	0.0010
1SWASP J211703.96+404916.8	000-BLY-378	57954.8292	V	V. Petriew	0.0010
1SWASP J211703.96+404916.8	000-BLY-378	57960.8356	V	V. Petriew	0.0007
1SWASP J211703.96+404916.8	000-BLY-378	57961.8121	V	V. Petriew	0.0010
1SWASP J211703.96+404916.8	000-BLY-378	57968.7958	V	V. Petriew	0.0010
1SWASP J211703.96+404916.8	000-BLY-378	57975.7800	V	V. Petriew	0.0010
1SWASP J211703.96+404916.8	000-BLY-378	57975.9195	V	V. Petriew	0.0008
1SWASP J211703.96+404916.8	000-BLY-378	57979.8299	V	V. Petriew	0.0007
1SWASP J211703.96+404916.8	000-BLY-378	58323.8247	V	V. Petriew	0.0012
1SWASP J211703.96+404916.8	000-BLY-378	58325.7803	V	V. Petriew	0.0012
1SWASP_J211703.96+404916.8	000-BLY-378	57577.7330	V	V. Petriew	0.0021
1SWASP_J211703.96+404916.8	000-BLY-378	57577.8748	V	V. Petriew	0.0007
1SWASP_J211703.96+404916.8	000-BLY-378	57584.8581	V	V. Petriew	0.0007
1SWASP_J211703.96+404916.8	000-BLY-378	57586.8126	V	V. Petriew	0.0009

Table continued on following pages

Table 1. Recent times of minima of stars in the AAVSO eclipsing binary program, cont.

Star	AUID	JD (min) Hel. 2400000+	F	Observer	Standard Error (day)
1SWASP_J211703.96+404916.8	000-BLY-378	57590.8613	V	V. Petriew	0.0009
1SWASP_J211703.96+404916.8	000-BLY-378	57594.7730	V	V. Petriew	0.0010
1SWASP_J211703.96+404916.8	000-BLY-378	57594.9129	V	V. Petriew	0.0010
1SWASP_J211703.96+404916.8	000-BLY-378	57597.7071	V	V. Petriew	0.0009
1SWASP_J211703.96+404916.8	000-BLY-378	57597.8456	V	V. Petriew	0.0010
1SWASP_J211703.96+404916.8	000-BLY-378	57599.8003	V	V. Petriew	0.0012
1SWASP_J211703.96+404916.8	000-BLY-378	57601.8964	V	V. Petriew	0.0007
ASAS J003953+2626.6	000-BLX-779	57357.5991	V	V. Petriew	0.0005
ASAS J003953+2626.6	000-BLX-779	57358.6314	V	V. Petriew	0.0004
ASAS J003953+2626.6	000-BLX-779	57361.5530	V	V. Petriew	0.0003
ASAS J003953+2626.6	000-BLX-779	57361.7249	V	V. Petriew	0.0010
ASAS J003953+2626.6	000-BLX-779	57362.5850	V	V. Petriew	0.0004
ASAS J041529+3129.9	000-BMC-816	57696.8157	V	V. Petriew	0.0006
ASAS J041529+3129.9	000-BMC-816	57696.9906	V	V. Petriew	0.0016
ASAS J041529+3129.9	000-BMC-816	57711.6855	V	V. Petriew	0.0010
ASAS J041529+3129.9	000-BMC-816	57711.8719	V	V. Petriew	0.0004
ASAS J092544+2542.6	000-BLX-470	57349.0144	V	V. Petriew	0.0004
ASAS J092544+2542.6	000-BLX-470	57362.0315	V	V. Petriew	0.0008
ASAS J092544+2542.6	000-BLX-470	57390.7810	V	V. Petriew	0.0006
ASAS J092544+2542.6	000-BLX-470	57390.9145	V	V. Petriew	0.0007
ASAS J092544+2542.6	000-BLX-470	57391.0522	V	V. Petriew	0.0007
ASAS J092544+2542.6	000-BLX-470	57757.0535	V	V. Petriew	0.0009
ASAS J092544+2542.6	000-BLX-470	57785.8022	V	V. Petriew	0.0004
ASAS J092544+2542.6	000-BLX-470	57785.9383	V	V. Petriew	0.0003
ASAS J092544+2542.6	000-BLX-470	57792.7171	V	V. Petriew	0.0005
ASAS J092544+2542.6	000-BLX-470	57792.8542	V	V. Petriew	0.0005
ASAS J092933+0929.0	000-BLT-858	57102.6665	V	V. Petriew	0.0004
ASAS J132229+1543.0	000-BLX-452	57492.6908	V	V. Petriew	0.0007
ASAS J150239+1546.5	000-BLP-887	57109.7895	V	V. Petriew	0.0021
ASAS J150239+1546.5	000-BLP-887	57109.9521	V	V. Petriew	0.0045
ASAS J184728+2621.4	000-BLX-425	57190.7575	V	V. Petriew	0.0003
ASAS J184728+2621.4	000-BLX-425	57199.8084	V	V. Petriew	0.0004
ASAS J184728+2621.4	000-BLX-425	57200.7873	V	V. Petriew	0.0005
ASAS J192826+3939.6	000-BLX-710	57237.8089	V	V. Petriew	0.0002
ASAS J192826+3939.6	000-BLX-710	57238.7711	V	V. Petriew	0.0002
ASAS J192826+3939.6	000-BLX-710	57238.9040	V	V. Petriew	0.0003
ASAS J193512+4146.4	000-BJH-676	56546.3060	C	Y. Ogmen	0.0001
ASAS J195751+4512.7	000-BLX-709	57226.8718	V	V. Petriew	0.0009
ASAS J195751+4512.7	000-BLX-709	57234.6946	V	V. Petriew	0.0009
ASAS J195751+4512.7	000-BLX-709	57235.7728	V	V. Petriew	0.0003
ASAS J195751+4512.7	000-BLX-709	57236.8519	V	V. Petriew	0.0003
ASAS_J053803+2734.8	000-BML-463	57719.7628	V	V. Petriew	0.0001
CSS_J022048.7+332751	000-BLX-773	57304.7650	V	V. Petriew	0.0001
CSS_J024657.7+343112	000-BLX-772	57275.8134	V	V. Petriew	0.0006
CSS_J024657.7+343112	000-BLX-772	57303.8807	V	V. Petriew	0.0005
CSS_J152658.4+403531	000-BLX-619	57498.7689	V	V. Petriew	0.0001
CSS_J152658.4+403531	000-BLX-619	57498.9002	V	V. Petriew	0.0001
CSS_J160353.4+353058	000-BLX-913	57525.7393	V	V. Petriew	0.0004
CSS_J160353.4+353058	000-BLX-913	57525.8664	V	V. Petriew	0.0003
GSC 02707-01173	000-BCN-774	58012.3927	V	T. Arranz	0.0003
GSC 03374-00372	000-BMF-212	57717.7642	V	V. Petriew	0.0005
GSC 03374-00372	000-BMF-212	57718.9751	V	V. Petriew	0.0008
GSC 03704-01240	000-BLX-929	57316.7591	V	V. Petriew	0.0016
GSC 03704-01240	000-BLX-929	57321.7139	V	V. Petriew	0.0034
GSC 03704-01240	000-BLX-929	57342.7716	V	V. Petriew	0.0011
GSC 03704-01240	000-BLX-929	57349.7063	V	V. Petriew	0.0011
GSC 03704-01240	000-BLX-929	57349.9559	V	V. Petriew	0.0015
GSC 3028 0134	000-BMV-507	58205.8272	Ic	K. Alton	0.0001
GSC 3028 0134	000-BMV-507	58205.8282	V	K. Alton	0.0004
GSC 3028 0134	000-BMV-507	58208.9002	Ic	K. Alton	0.0006
GSC 3028 0134	000-BMV-507	58208.9013	V	K. Alton	0.0001
GSC 3028 0134	000-BMV-507	58208.9032	B	K. Alton	0.0009
GSC 3028 0134	000-BMV-507	58211.9746	V	K. Alton	0.0006
GSC 3028 0134	000-BMV-507	58211.9752	Ic	K. Alton	0.0003
GSC 3028 0134	000-BMV-507	58211.9772	B	K. Alton	0.0003
GSC 3028 0134	000-BMV-507	58217.9612	Ic	K. Alton	0.0006

Table continued on following pages

Table 1. Recent times of minima of stars in the AAVSO eclipsing binary program, cont.

Star	AUID	JD (min) Hel. 2400000+	F	Observer	Standard Error (day)
GSC 3028 0134	000-BMV-507	58217.9621	V	K. Alton	0.0007
GSC 3028 0134	000-BMV-507	58217.9633	B	K. Alton	0.0004
GSC 3028-0134	000-BMV-507	58205.8261	B	K. Alton	0.0007
GSC 3253 0322	000-BMZ-962	58396.8892	Ic	K. Alton	0.0004
GSC 3253 0322	000-BMZ-962	58396.8898	B	K. Alton	0.0003
GSC 3253 0322	000-BMZ-962	58396.8918	V	K. Alton	0.0004
GSC 3253 0322	000-BMZ-962	58398.8322	V	K. Alton	0.0001
GSC 3253 0322	000-BMZ-962	58398.8326	Ic	K. Alton	0.0002
GSC 3253 0322	000-BMZ-962	58398.8332	B	K. Alton	0.0004
GSC 3253 0322	000-BMZ-962	58400.7740	Ic	K. Alton	0.0003
GSC 3253 0322	000-BMZ-962	58400.7751	V	K. Alton	0.0001
GSC 3253 0322	000-BMZ-962	58400.7760	B	K. Alton	0.0001
GSC 3253 0322	000-BMZ-962	58400.9228	B	K. Alton	0.0005
GSC 3253 0322	000-BMZ-962	58400.9237	Ic	K. Alton	0.0001
GSC 3253 0322	000-BMZ-962	58400.9248	V	K. Alton	0.0003
GSC 3253 0322	000-BMZ-962	58401.8203	Ic	K. Alton	0.0001
GSC 3253 0322	000-BMZ-962	58401.8208	B	K. Alton	0.0001
GSC 3253 0322	000-BMZ-962	58401.8213	V	K. Alton	0.0003
GSC 4394 1100	000-BMZ-960	58223.7848	V	K. Alton	0.0001
GSC 4394 1100	000-BMZ-960	58223.7857	B	K. Alton	0.0003
GSC 4394 1100	000-BMZ-960	58226.7244	V	K. Alton	0.0002
GSC 4394 1100	000-BMZ-960	58226.7247	Ic	K. Alton	0.0002
GSC 4394 1100	000-BMZ-960	58226.7255	B	K. Alton	0.0007
GSC 4394 1100	000-BMZ-960	58226.8625	B	K. Alton	0.0005
GSC 4394 1100	000-BMZ-960	58226.8632	V	K. Alton	0.0002
GSC 4394 1100	000-BMZ-960	58226.8651	Ic	K. Alton	0.0004
GSC 4394 1100	000-BMZ-960	58229.6617	V	K. Alton	0.0006
GSC 4394 1100	000-BMZ-960	58229.6627	B	K. Alton	0.0001
GSC 4394 1100	000-BMZ-960	58229.6637	Ic	K. Alton	0.0003
GSC 4394 1100	000-BMZ-960	58229.8022	Ic	K. Alton	0.0004
GSC 4394 1100	000-BMZ-960	58229.8032	V	K. Alton	0.0001
GSC 4394 1100	000-BMZ-960	58229.8042	B	K. Alton	0.0002
GSC 4394-1100	000-BMZ-960	58223.7837	Ic	K. Alton	0.0003
KIC 4656194	000-BLX-575	57237.7561	V	V. Petriew	0.0030
KIC 4656194	000-BLX-575	57238.7134	V	V. Petriew	0.0027
KIC 4656194	000-BLX-575	57238.8075	V	V. Petriew	0.0018
KIC 4656194	000-BLX-575	57238.9050	V	V. Petriew	0.0037
NSV 19494	000-BLV-211	57123.7293	V	V. Petriew	0.0009
NSV 19494	000-BLV-211	57130.6720	V	V. Petriew	0.0008
NSVS 10464245	000-BGZ-513	57499.7251	V	V. Petriew	0.0003
NSVS 10464245	000-BGZ-513	57499.8617	V	V. Petriew	0.0003
NSVS 1825826	000-BMB-887	58025.8958	V	V. Petriew	0.0003
NSVS 1932777	000-BLX-708	57316.7817	V	V. Petriew	0.0002
NSVS 1932777	000-BLX-708	57316.9345	V	V. Petriew	0.0002
NSVS 1932777	000-BLX-708	57342.7028	V	V. Petriew	0.0002
NSVS 1932777	000-BLX-708	57342.8562	V	V. Petriew	0.0002
NSVS 1932777	000-BLX-708	57342.8562	V	V. Petriew	0.0002
NSVS 1932777	000-BLX-708	57349.5653	V	V. Petriew	0.0002
NSVS 1932777	000-BLX-708	57349.7164	V	V. Petriew	0.0002
NSVS 1932777	000-BLX-708	57349.8704	V	V. Petriew	0.0002
NSVS 2423495	000-BLX-456	57443.5834	V	V. Petriew	0.0004
NSVS 2423495	000-BLX-456	57460.7467	V	V. Petriew	0.0004
NSVS 2443858	000-BLP-236	57083.7530	V	V. Petriew	0.0002
NSVS 2569022	000-BBR-218	57449.6605	V	V. Petriew	0.0003
NSVS 2569022	000-BBR-218	57461.7487	V	V. Petriew	0.0005
NSVS 2569022	000-BBR-218	57473.6929	V	V. Petriew	0.0007
NSVS 2729390	000-BLX-354	57480.7407	V	V. Petriew	0.0002
NSVS 2729390	000-BLX-354	57480.9144	V	V. Petriew	0.0002
NSVS 3752379	000-BML-464	58036.7809	V	V. Petriew	0.0007
NSVS 3752379	000-BML-464	58043.6757	V	V. Petriew	0.0001
NSVS 3752379	000-BML-464	58051.6607	V	V. Petriew	0.0002
NSVS 4333194	000-BLS-738	57314.8380	V	V. Petriew	0.0003
NSVS 4333194	000-BLS-738	57314.9816	V	V. Petriew	0.0002
NSVS 4333194	000-BLS-738	57354.8333	V	V. Petriew	0.0002
NSVS 4333194	000-BLS-738	57354.9771	V	V. Petriew	0.0002
NSVS 4333194	000-BLS-738	57355.8400	V	V. Petriew	0.0002

Table continued on following pages

Table 1. Recent times of minima of stars in the AAVSO eclipsing binary program, cont.

Star	AUID	JD (min) Hel. 2400000+	F	Observer	Standard Error (day)
NSVS 4333194	000-BLS-738	57355.9848	V	V. Petriew	0.0002
NSVS 4333194	000-BLS-738	57356.7033	V	V. Petriew	0.0002
NSVS 4333194	000-BLS-738	57356.8479	V	V. Petriew	0.0002
NSVS 4333194	000-BLS-738	57356.9918	V	V. Petriew	0.0002
NSVS 4666563	000-BMY-869	57739.7689	V	V. Petriew	0.0002
NSVS 4666563	000-BMY-869	57740.7984	V	V. Petriew	0.0002
NSVS 4666563	000-BMY-869	57740.9698	V	V. Petriew	0.0002
NSVS 4666563	000-BMY-869	57744.7442	V	V. Petriew	0.0002
NSVS 4922327	000-BNB-318	57835.6793	V	V. Petriew	0.0005
NSVS 4922327	000-BNB-318	57835.8825	V	V. Petriew	0.0003
NSVS 4922327	000-BNB-318	57846.6484	V	V. Petriew	0.0006
NSVS 4993806	000-BLX-472	57444.8553	V	V. Petriew	0.0004
NSVS 4993806	000-BLX-472	57444.9966	V	V. Petriew	0.0003
NSVS 5079550	000-BLX-454	57474.8433	V	V. Petriew	0.0005
NSVS 5079550	000-BLX-454	57474.9836	V	V. Petriew	0.0005
NSVS 5138234	000-BNB-313	57844.8766	V	V. Petriew	0.0002
NSVS 5138234	000-BNB-313	57846.8216	V	V. Petriew	0.0002
NSVS 5138234	000-BNB-313	57847.7059	V	V. Petriew	0.0002
NSVS 5138234	000-BNB-313	57847.8827	V	V. Petriew	0.0002
NSVS 5138234	000-BNB-313	57848.7670	V	V. Petriew	0.0002
NSVS 5138234	000-BNB-313	57848.9444	V	V. Petriew	0.0002
NSVS 5138234	000-BNB-313	57855.8411	V	V. Petriew	0.0002
NSVS 5886937	000-BLM-950	57591.8193	V	V. Petriew	0.0004
NSVS 5943354	000-BLX-696	57242.7600	V	V. Petriew	0.0003
NSVS 5943354	000-BLX-696	57242.9152	V	V. Petriew	0.0006
NSVS 5943354	000-BLX-696	57253.8569	V	V. Petriew	0.0003
NSVS 5943354	000-BLX-696	57257.9058	V	V. Petriew	0.0011
NSVS 5943354	000-BLX-696	57976.7369	V	V. Petriew	0.0004
NSVS 5943354	000-BLX-696	57976.8790	V	V. Petriew	0.0006
NSVS 5943354	000-BLX-696	57978.8368	V	V. Petriew	0.0006
NSVS 6143186	000-BLX-778	57355.6889	V	V. Petriew	0.0003
NSVS 6143186	000-BLX-778	57356.6172	V	V. Petriew	0.0003
NSVS 6571865	000-BLX-774	57303.7601	V	V. Petriew	0.0002
NSVS 6571865	000-BLX-774	57303.9270	V	V. Petriew	0.0002
NSVS 6601998	000-BLX-928	57319.8448	V	V. Petriew	0.0002
NSVS 6731062	000-BLX-776	57312.7873	V	V. Petriew	0.0001
NSVS 6731062	000-BLX-776	57312.9362	V	V. Petriew	0.0001
NSVS 6731062	000-BLX-776	57313.8237	V	V. Petriew	0.0001
NSVS 6731062	000-BLX-776	57313.9705	V	V. Petriew	0.0001
NSVS 6731062	000-BLX-776	57317.8165	V	V. Petriew	0.0002
NSVS 6731062	000-BLX-776	57317.9652	V	V. Petriew	0.0001
NSVS 6731062	000-BLX-776	57333.7914	V	V. Petriew	0.0002
NSVS 6731062	000-BLX-776	57333.9393	V	V. Petriew	0.0001
NSVS 6731062	000-BLX-776	57334.8271	V	V. Petriew	0.0001
NSVS 6731062	000-BLX-776	57334.9746	V	V. Petriew	0.0002
NSVS 6731062	000-BLX-776	57339.7076	V	V. Petriew	0.0002
NSVS 7377756	000-BMZ-610	57760.8591	V	V. Petriew	0.0002
NSVS 7377756	000-BMZ-610	57761.0035	V	V. Petriew	0.0001
NSVS 7377756	000-BMZ-610	57771.8037	V	V. Petriew	0.0002
NSVS 7377756	000-BMZ-610	57771.9476	V	V. Petriew	0.0002
NSVS 7440793	000-BLX-457	57441.7305	V	V. Petriew	0.0004
NSVS 7440793	000-BLX-457	57441.8578	V	V. Petriew	0.0004
NSVS 7440793	000-BLX-457	57459.6536	V	V. Petriew	0.0006
NSVS 7456642	000-BLX-453	57417.8420	V	V. Petriew	0.0002
NSVS 7456642	000-BLX-453	57418.0161	V	V. Petriew	0.0002
NSVS 7456642	000-BLX-453	57418.8840	V	V. Petriew	0.0002
NSVS 7456642	000-BLX-453	57474.6271	V	V. Petriew	0.0019
NSVS 7456642	000-BLX-453	57475.6703	V	V. Petriew	0.0002
NSVS 7874563	000-BLV-212	57122.7765	V	V. Petriew	0.0004
NSVS 7874563	000-BLV-212	57123.8564	V	V. Petriew	0.0005
NSVS 7874563	000-BLV-212	57132.9269	V	V. Petriew	0.0010
NSVS 7874563	000-BLV-212	57166.8372	V	V. Petriew	0.0004
NSVS 8110779	000-BLY-331	57569.7561	V	V. Petriew	0.0002
NSVS 8110779	000-BLY-331	57573.7509	V	V. Petriew	0.0003
NSVS 8110779	000-BLY-331	57575.8129	V	V. Petriew	0.0003
NSVS 8183892	000-BLX-663	57210.8288	V	V. Petriew	0.0001

Table continued on next page

Table 1. Recent times of minima of stars in the AAVSO eclipsing binary program, cont.

Star	AUID	JD (min) Hel. 2400000+	F	Observer	Standard Error (day)
NSVS 8183892	000-BLX-663	57211.7856	V	V. Petriew	0.0001
NSVS 8183892	000-BLX-663	57212.7424	V	V. Petriew	0.0001
NSVS 8183892	000-BLX-663	57212.9020	V	V. Petriew	0.0002
NSVS 8183892	000-BLX-663	57213.8587	V	V. Petriew	0.0001
ROTSE1 J140627.09+372840.5	000-BLX-352	57486.6681	V	V. Petriew	0.0006
ROTSE1 J140627.09+372840.5	000-BLX-352	57486.8146	V	V. Petriew	0.0004
ROTSE1 J143554.74+284045.7	000-BLX-336	57489.7758	V	V. Petriew	0.0002
ROTSE1 J153727.94+254022.9	000-BMY-927	57883.8291	V	V. Petriew	0.0009
ROTSE1 J153727.94+254022.9	000-BMY-927	57888.8099	V	V. Petriew	0.0008
ROTSE1 J153727.94+254022.9	000-BMY-927	57891.7539	V	V. Petriew	0.0009
ROTSE1 J153727.94+254022.9	000-BMY-927	57896.7408	V	V. Petriew	0.0010
ROTSE1 J153727.94+254022.9	000-BMY-927	57903.7588	V	V. Petriew	0.0010
ROTSE1 J153727.94+254022.9	000-BMY-927	57905.7969	V	V. Petriew	0.0009
ROTSE1 J155551.87+331100.5	000-BLX-846	57513.8032	V	V. Petriew	0.0006
ROTSE1 J161118.72+353818.9	000-BLX-162	57508.7259	V	V. Petriew	0.0014
ROTSE1 J161118.72+353818.9	000-BLX-162	57508.8754	V	V. Petriew	0.0025
ROTSE1 J161118.72+353818.9	000-BLX-162	57509.8295	V	V. Petriew	0.0003
ROTSE1 J161118.72+353818.9	000-BLX-162	57140.7979	V	V. Petriew	0.0014
ROTSE1 J161118.72+353818.9	000-BLX-162	57141.7508	V	V. Petriew	0.0014
ROTSE1 J161118.72+353818.9	000-BLX-162	57141.9115	V	V. Petriew	0.0032
ROTSE1 J161118.72+353818.9	000-BLX-162	57144.9241	V	V. Petriew	0.0031
ROTSE1 J161118.72+353818.9	000-BLX-162	57145.7245	V	V. Petriew	0.0012
ROTSE1 J161118.72+353818.9	000-BLX-162	57145.8801	V	V. Petriew	0.0012
ROTSE1 J161118.72+353818.9	000-BLX-162	57146.8343	V	V. Petriew	0.0013
ROTSE1 J161118.72+353818.9	000-BLX-162	57152.7124	V	V. Petriew	0.0024
ROTSE1 J161118.72+353818.9	000-BLX-162	57152.8725	V	V. Petriew	0.0017
ROTSE1 J161118.72+353818.9	000-BLX-162	57162.7207	V	V. Petriew	0.0025
ROTSE1 J161118.72+353818.9	000-BLX-162	57162.8797	V	V. Petriew	0.0014
ROTSE1 J162406.50+344241.7	000-BLX-901	57523.8117	V	V. Petriew	0.0004
ROTSE1 J162406.50+344241.7	000-BLX-901	57533.7436	V	V. Petriew	0.0004
ROTSE1 J162406.50+344241.7	000-BLX-901	57538.8155	V	V. Petriew	0.0004
ROTSE1 J173334.66+391012.9	000-BLX-355	57169.7587	V	V. Petriew	0.0018
ROTSE1 J173334.66+391012.9	000-BLX-355	57177.7963	V	V. Petriew	0.0011
ROTSE1 J173334.66+391012.9	000-BLX-355	57184.8216	V	V. Petriew	0.0037
ROTSE1 J173334.66+391012.9	000-BLX-355	57540.7788	V	V. Petriew	0.0016
S_J182320.7+351132	000-BLY-332	CS57569.7928	V	V. Petriew	0.0093
S_J182320.7+351132	000-BLY-332	CS57573.7523	V	V. Petriew	0.0002
S_J182320.7+351132	000-BLY-332	CS57575.8019	V	V. Petriew	0.0001
T-CrB0-01199	000-BLX-907	57524.7563	V	V. Petriew	0.0007
T-CrB0-01199	000-BLX-907	57524.8823	V	V. Petriew	0.0009
VSX_J005225.3+503636	000-BKN-470	58743.8782	V	K. Menzies	0.0003
WHAT236 50	000-BLX-501	57184.8221	V	V. Petriew	0.0010
WHAT236 50	000-BLX-501	57497.7449	V	V. Petriew	0.0012
WHAT236 50	000-BLX-501	57497.8675	V	V. Petriew	0.0012
WISE J211738.0+405313	000-BND-257	57594.7551	V	V. Petriew	0.0019
WISE J211738.0+405313	000-BND-257	57599.7593	V	V. Petriew	0.0007

Abstracts of Papers and Posters Presented at the Joint Meeting of the Royal Astronomical Society of Canada (RASC) and the American Association of Variable Star Observers (AAVSO 108th Spring Meeting), Held in Toronto, Ontario, Canada, June 13-16, 2019, and Other Abstract Received

AAVSO Paper Sessions

A Robotic Observatory System that Anyone Can Use for Education, Astro-Imaging, Research, or Fun!

David Lane
FRASC Halifax Centre, Director, Burke-Gaffney Observatory, Saint Mary's University

Abstract A Robotic Observatory System developed for the Burke-Gaffney and Abbey Ridge Observatories will be described. Since its announcement at the 2015 RASC General Assembly, over 8,000 observation requests have been completed! Its goal is to make the acquisition and processing of astronomical images easy and fun, whether used for research, education, astro-imaging, or just for fun. Communication is via a "chat bot" style interface (with a companion website) that responds to a comprehensive set of commands—simple enough for beginners or children but flexible enough to handle the needs of research projects (e.g. variable stars, exoplanets, etc.). Observers communicate by Twitter, email, an Android/MacOS/Windows app, or even by SMS text from any cell phone! Observation requests are queued and observations completed using a responsive queue engine. Some recent usage examples and future plans will also be presented.

DASCH Project Update (poster)

Edward Los
Harvard College Observatory, 60 Garden Street, Cambridge, MA 02138

Abstract The "Digital Acesss to a Sky Century @ Harvard" (DASCH) project is an effort to scan the glass plate negatives in the archive of the Harvard College Observatory for the purpose of generating light curves which can span 100+ years. The logbook digitization project is complete. The scanning project is 77% complete. With the recent DR6 release, all of the photometry North of the galactic plane is now available. We have recalibrated nearly all of the photometry with the ATLAS refcat2 catalog. Parallel to the scanning, there are efforts to improve the yield with new algorithms for multiple exposure processing and astrometry.

The Eye is Mightier than the Algorithm: Lessons Learned from the Erroneous Classification of ASAS Variables as "Miscellaneous" (poster)

Kristine Larsen
Copernican Observatory and Planetarium, Central Connecticut State University, 1615 Stanley St. New Britain, CT 06050; larsen@ccsu.edu

Abstract The superior ability of the human eye and brain to detect patterns in data is central to many crowd sourcing and citizen science projects such as Galaxy Zoo and Planet Hunters TESS. This poster summarizes lessons learned in a much smaller scale project, the largely student-driven identification and proper reclassification of selected variable stars that have been erroneously flagged as "miscellaneous" by the algorithms of the ASAS (All Sky Automated Survey). Examples of stars whose light curves allow for trivial classification by the human eye as well as those that require a deeper investigation are discussed.

Contributions of the AAVSO to Year 1 of TESS (poster)

Dennis Conti
Stella Kafka
AAVSO, 49 Bay State Road, Cambridge, MA 02138

Abstract This poster depicts contributions that the American Association of Variable Star Observers (AAVSO) and its members have made during the first year of TESS' operation. With its well-known legacy and contributions in the area of variable star observing, the AAVSO added in early 1996 an additional focus, that of exoplanet observing. This was done in recognition of the valuable data such amateur astronomers were providing in helping to candidate exoplanets, as well as in the refinement of the ephemerides of known exoplanets. An AAVSO Exoplanet Section was thus established at that time to provide a "home" for such observers where they could be trained in "best practices," share ideas, and foster communication with the professional community. Over 125 AAVSO members have since completed an Exoplanet Observing Course that covers such best practices, as well as an introduction to the ASTROIMAGEJ (AIJ) software. When TESS began its science operation this past year, a cadre of AAVSO members were then well-trained to participate in TESS' Follow-up Program (TFOP) Subgroup-1 (SG1), the "Seeing Limited Subgroup." With SG1's primary objective being to help in the confirmation of candidate exoplanets, especially by ruling out false positives,

AAVSO members have and continue to contribute to this objective. This poster depicts where in the TESS pipeline these observations are made, as well as examples of some of the methods used to identify true exoplanet transits from false positives. Because systems such as near-by eclipsing binaries (NEBs) can contaminate TESS' relatively large aperture, a means was needed to reliably eliminate stars near a TESS target that, when blended with the target, could mimic an exoplanet transit. For example, it was found that for some targets, there could potentially be hundreds of near-by stars that needed to be cleared as potential false positives. In order to automate what was a manual process that could take an hour or more and to provide a more comprehensive analysis using Gaia's Data Release 2 data, the AAVSO liaison to SG1, Dennis Conti, developed an add-on to AIJ that completed this evaluation in minutes. This poster then also displays this automated "NEB clearing" process.

The Project PHaEDRA Collection: An Anchor to Connect Modern Science with Historical Data (poster)

Peggy Wargelin
Wolbach Library, Harvard-Smithsonian Center for Astrophysics, 60 Garden Street, Cambridge, MA 02138 (presented by Edward Los)

Abstract Project PHaEDRA (Preserving Harvard's Early Data and Research in Astronomy) is an initiative by the Wolbach Library, in collaboration with many partners, to catalog, digitize, transcribe, and enrich the metadata of over 2,500 logbooks and notebooks produced by the Harvard Computers and early Harvard astronomers. Our goal is to ensure that this remarkable set of items, created by a remarkable group of people, is as accessible and useful as possible. We are working with the Smithsonian Transcription Center to transcribe the PHaEDRA collection. The transcription markup will allow the collection to be full-text searchable on the NASA Astrophysics Data System, and for the notebooks to eventually be linked to their original source material: 500,000 glass plate photographs representing the first ever picture of the visible universe. These glass plate photos are currently being digitized through a project called DASCH: Digital Access to a Sky Century at Harvard. DASCH will enable scientists to produce full photometry results for the entire sky from over 100 years of observations. This will allow researchers in Time Domain Astrophysics to better understand how the Universe is evolving. Physical conservation of the Project PHaEDRA collection is being generously supported by the Smithsonian Women's Committee.

The Eclipsing Triple Star System b Persei (HD 26961)

Donald F. Collins
Emeritus Professor of Physics, Warren Williams College, 701 Warren Wilson Road, Swannanoa, NC 28778

Abstract The triple star system, b Persei (HD 26961, HR 1324, HIP 20070, SAO 24531), magnitude 4.6 V, spectral class A1 III, consists of two non-eclipsing stars in a close orbit (1.52-day period) and a third star that orbits the inner pair with a period of approximately 700 days. Many AAVSO observers have been observing high time-resolved eclipses or transits of the third star transiting both behind the inner pair (secondary eclipse) and in front of the inner pair (primary eclipse) since the first observations in February 2013. The AAVSO eclipse light curve observations have refined the measured period for the third star's orbit to be 704.9 ± 0.1 d, a significant departure from the earlier spectroscopic radial velocity observations of 701.8 ± 0.4 d (Hill, *et al.* 1976, *AJ* **208**, 152). A simple simulation of the light curves was developed to fit the timings of the dips in the light curves by adjusting a limited number of the orbital parameters for the star system. It is hoped to continue observing these transits to examine a possible evolution of the orbits and to obtain better knowledge of the astrophysical parameters for the three separate stars in the system.

Photometric Accuracy of CMOS and CCD Cameras (Preliminary Results)

Kenneth T. Menzies
318A Potter Road, Framingham, MA 01701; kenmenstar@gmail.com

Abstract Photometrists regularly ask the question as to how accurate their magnitudes are. Methods are described that will help you evaluate your accuracy by imaging standard fields in multiple filters, using VPHOT to conduct photometry, and using EXCEL/PIVOT tables to reduce the data and compare the accuracy of measured magnitudes compared across different cameras and compared to Landolt Standards (SA101). Specific examples of data from BSM cameras, including a CMOS camera and three CCD cameras, are compared. Based on preliminary results, it has been determined that: (1) Landolt standard fields provide effective analysis of accuracy of photometric systems;, (2) VPHOT and EXCEL/PIVOT table tools facilitate analysis; (3) Differences between CMOS and CCD camera magnitudes are negligible/within experimental error (if SNR>50); (4) Differences between known and measured magnitudes are negligible/within experimental error (if SNR>50); and (5) As expected, selection of proper comp SNR and photometric aperture, as well as image quality, impacts experimental error and bias.

Measuring and Correcting Non-linearities in ABG Cameras

Richard Wagner
126 Powell Bay Road, Elgin, ON K0G 1E0, Canada

Abstract Most of the commercially available astronomical CCD cameras for the amateur astronomer incorporate an anti-blooming gate (ABG). It is widely accepted that the ABG introduces significant non-linearities and as a result these cameras are not recommended for astronomical photometry. Having two of these cameras, the SBIG ST2000XM and the QSI583ws, I investigated their non-linearities with a view to determining just how large the problem is and how it could be corrected. In response tests to a constant light source both cameras show a linearly decreasing flux (ADU/s) as a function of increasing signal. A quadratic function is fitted which corrects the non-linearities to a remaining non-linearity of <8ADU RMS from 0 to 55000ADU signal.

A Simple Objective Prism Spectrograph for Educational Purposes

John Thompson
Ottawa Centre RASC

Abstract This paper presents the construction and use of a simple low resolution spectrograph that can be made from a few dollars' worth of parts. The device is placed in front of a standard DSLR on an adjustable tripod, no telescope, slit, or drive being required, and spectrum width is produced by trailing from the Earth's motion. The spectrograph can reach to at least third magnitude, and has shown hydrogen Balmer lines, nebular emission lines, molecular absorption bands, and atmospheric absorption lines. It is well suited for educational purposes for schools and science camps, to introduce stellar classification and the fundamentals of astrophysics.

Studies of Pulsating Red Giants using AAVSO and ASAS-SN Data

John R. Percy
Lucas Fenaux
Department of Astronomy and Astrophysics, and Dunlap Institute of Astronomy and Astrophysics, University of Toronto, 50 St. George Street, Toronto, ON M5S 3H4, Canada; john.percy@utoronto.ca

Abstract We review a decade of analysis and interpretation of AAVSO data on pulsating red giant stars (PRGs), with emphasis on several unexplained long-term phenomena which have been discovered using many decades of AAVSO visual observations. More recently, we have explored the use of data from the All-Sky Automated Survey for Supernovae (ASAS-SN), which provides over 2,000 days of sustained observations of hundreds of thousands of variables. ASAS-SN has used automated methods to analyze and classify variable stars. By examining two small samples of PRGs, we show that many of the ASAS-SN analyses and classifications of these types of variables are problematic for various reasons; we will show specific examples. We are now using ASAS-SN data to investigate the nature of semi-regular and irregular PRGs on the AAVSO visual program which are sparsely-observed and therefore poorly-understood. Both the AAVSO and ASAS-SN data, along with the AAVSO time-series software package VSTAR, are ideally suited for student projects at the senior secondary and post-secondary levels.

Variable Stars in the College Classroom

Kristine Larsen
Copernican Observatory and Planetarium, Central Connecticut State University, 1615 Stanley St. New Britain, CT 06050; larsen@ccsu.edu

Abstract Variable stars play a starring role in our understanding of the lives of stars. From the T Tauri protostar stage to starspot cycles of main sequence stars like our Sun, instability strip variations of Cepheids to the dying gasps of red giants, observations of the myriad species found in the variable star zoo have allowed us to piece together the details of stellar evolution. It is therefore important not only to introduce students to the properties of such stars but also demonstrate how our sustained monitoring of these objects continues to add to our understanding of stars, galaxies, and even the universe at large. This talk describes how variable stars have been integrated into a lab-based college level course in Stellar and Galactic Astronomy that targets non-science majors.

A Study of RR Lyrae Variable Stars in a Professor-Student Collaboration using the American Public University System (APUS) Observatory

David Syndergaard
Melanie Crowson
American Public University System (APUS), 111 W. Congress Street, Charles Town, WV 25414; davesyndergaard@me.com

Abstract American Public University System (APUS), in support of astronomical research, has recently acquired an observatory that includes a 24-inch CDK telescope. This research project was the first instructor-student research collaboration using the APUS Observatory. The objective was to add to the body of knowledge of RR Lyrae variable stars. Collaboration using a remotely-controlled observatory presents both opportunities and challenges. This project proved that a professor and student, living in different states (Utah and Florida, respectively), using a telescope located in a third state, could successfully work together to analyze images and produce important results. The collaboration was a valuable learning experience for the instructor, the student, and the APUS STEM department. RR Lyrae variable stars provide insight into how stars form and evolve, including the general progression of their lives as compared to other stars more like our Sun. For this reason, it is important to study their variation in apparent magnitude (brightness). This project focused on obtaining photometric data and magnitude variations over the course of 10 months for 10 selected RR Lyrae stars. Light curves were constructed for four of the stars (NSV 13602, NSV 5394, NSV 7465, and NSV 7647). The results of this project include both scientific and educational lessons learned which are being applied in subsequent APUS scientific research.

The Rotating Frame

Christa Van Laerhoven
Department of Physics and Astronomy, University of British Columbia, 325–6224 Agricultural Road. Vancouver, BC V6T 1Z1, Canada; cvl@phas.ubc.ca

Abstract We generally think of planets (and the smaller stuff, too) as going about their orbits, minding their own business. In actuality, our Solar System is a giant interact-a-thon, as fellow planetary bodies perturb each other via gravity. I will talk about a kind of interaction called orbital resonance and how we can diagnose that this particular kind of interaction is happening via something called the "rotating frame." I will show how to translate from actual motion to motion in the rotating frame and will give some examples from our Solar System where it is helpful to look at orbits in this way.

Five Years of CCD Photometry

Damien Lemay
195 4E Rang O , Saint-Anaclet, QC GOK 1HO, Canada; damien.lemay@globetrotter.net

Abstract During my life as an amateur astronomer I have built three permanent observatories (the first in 1976), each improving on the previous one. When my present one got in service at the beginning of 2014, I had long ago decided that it would be used for science. And this is what I have done, specifically photometry almost every clear night from dusk to dawn. During the five years since first light, I have accumulated more than 75,000 photos, each one providing photometric data for the AAVSO and/or CBA. There is room for two instruments, each one resting on a 20-foot concrete pillar. Recently an unexpected donation of equipment allowed me to add spectroscopy to my favorite hobby. While my talk will be essentially about my work in photometry, I will show some preliminary results with spectroscopy.

How-to Hour—Planning an Observing Program

Kenneth T. Menzies (Moderator/Coordinator)
318A Potter Road, Framingham, MA 01701; kenmenstar@gmail.com

Michael J. Cook
Newcastle Observatory, 9 Laking Drive, Newcastle, ON L1B 1M5, Canada; michael.cook@newcastleobservatory.ca

Franz-Josef (Josch) Hambsch
Oude Bleken 12, Mol 2400, Belgium; hambsch@telenet.be

Abstract This "How-to Hour" presentation is designed to provide a discussion of principles, procedures, and skills that can be understood and utilized by amateurs to advance their practice of variable star photometry. It is presented by experienced observers who will describe the techniques needed to answer questions that all of us may ask and struggle with when advancing through our avocation.

The first presentation, "Planning a Visual Observing Program" by Michael Cook, identifies/explains the fundamental principles of visual observing including: 1. How to select targets, 2. How to find them, 3. How to pick comps, 4. How to compare targets and comps, 5. How to measure with the eye, binoculars, scope, 6. How to report, and 7. How to deal with issues with color detection.

The second presentation, "Planning a CCD Observing Program" by Josch Hambsch, identifies: 1. Types of variables of interest, 2. Different observing techniques needed to measure them, 3. How to identify organizations that request support, 4. How to identify professionals who need collaboration, 5. How to set up an observing plan, and 6. What software tools are available to observe efficiently.

RASC Paper Sessions (invited talks)

The Zooniverse

L. Clifton Johnson
Center for Interdisciplinary Exploration and Research in Astrophysics (CIERA), Northwestern University, 1800 Sherman Avenue, Evanston, IL 60201; lcj@northwestern.edu

Abstract Zooniverse has become the world's largest and most popular platform for people-powered research since its modest inception in 2007 as Galaxy Zoo. Zooniverse research is made possible by volunteers—hundreds of thousands of people around the world who come together to assist professional researchers. The goal of Zooniverse is to enable research that would not be possible, or practical, otherwise. Zooniverse research results in new discoveries, and datasets useful to the wider research community. This presentation will highlight the history and highlights of the Zooniverse, as well as presenting some personal perspectives of a researcher who has benefited from citizen science work on star clusters in the Andromeda Galaxy and Local Group. It will also highlight future efforts regarding variable stars and upcoming all-sky surveys.

The BRITE Constellation Mission

Gregg Wade
Department of Physics and Space Science, Royal Military College, P.O. Box 17000, Station Forces, Kingston, ON K7K 7B4, Canada, and Department of Physics, Engineering Physics and Astronomy, Queen's University, Stirling Hall, 64 Bader Lane, Kingston, ON K7L 3N6, Canada; Gregg.Wade@rmc.ca

Abstract The BRITE (BRIght Target Explorer) Constellation mission is described, as are some of the results to date. A collaboration between scientists in Canada, Austria, and Poland, BRITE is a network of five nanosatellites investigating the stellar structure and evolution of the brightest stars in the sky, and their interaction with the local environment. Forms of stellar variability—micropulsation and wind phenomena among others—are recorded via high-precision photometry in two colours (red and blue). The massive BRITE target stars are extremely important in producing the chemical elements in our universe, and recycling them in winds and supernovae. They are "laboratories" that help us to understand a wide range of the astrophysical processes in our universe.

The Importance of Religion and Art in Sowing the Seeds of Success of the U.S. Space Program in the 1940s and 1950s

Catherine L. Newell
Department of Religious Studies, University of Miami, P.O. Box 248264, Coral Gables, FL 33124; clnewell@miami.edu

Abstract An exploration of the importance of religion and art in sowing the seeds of success of the U.S. space program in the 1940s–1950s, and the roles of Chesley Bonestell, Willy Ley, and Werner von Braun in inspiring a technological and scientific faith that awoke a deep-seated belief in a sense of divine destiny to conquer the "final frontier."

Public Outreach Through the Cronyn Observatory

Jan Cami
Department of Physics and Astronomy, Western University, and Director of the Cronyn Observatory, PAB Room 138, 1151 Richmond Street, London, ON N6A 3K7, Canada; cronyn@uwo.ca

Abstract Historic facilities and instrumentation can be more than static artifacts for display. They can be brought to life as powerful engines for education and public outreach (EPO). Jan Cami talks about the challenges and rewards of repurposing the historic Cronyn Observatory (Western University) as an instrument of effective EPO, and the cooperative synergy of the groups involved—the university astrophysics community, and the RASC.

Canada's Role in the James Webb Space Telescope

Nathalie Ouellette
Institut de recherche sur les exoplanètes (iREx), Département de physique, Université de Montréal, Pavillon Roger-Gaudry, P.O. Box 6128, Centre-Ville STN, Montreal, QC H3C 3J7, Canada; nathalie@astro.umontreal.ca

Abstract Dr. Ouellette will speak on the Canadian contribution to the James Web Space Telescope (JWST), and possible citizen science opportunities.

Helen Sawyer Hogg: One Woman's Journey with the Stars

Maria J. Cahill
College of Science and Humanities, Husson University, 117 Beardsley Meeting House, 1 College Circle, Bangor, Maine 04401; cahillm@husson.edu

Abstract Helen Hogg (1905–1993), Canada's premier female astronomer and science writer, appeared to live a blessed life with a successful career. Not only did she rise through the academic ranks at the University of Toronto, but a list of her awards and honors fills more than one page. As we look back at Helen's life, we see success, but we do not immediately notice that her journey as an astronomer and science writer was complex; the road she travelled was full of twists and turns which she strategically navigated. This paper will illuminate some of the personal and professional forces that influenced Helen's work and supported her along her journey, demonstrating how she remains a model for future scientists.

Some Memories of the David Dunlap Observatory

Donald C. Morton
National Research Council, Herzberg Institute of Astrophysics; dcm123@dcmorton.com

Abstract Dr. Morton will talk on the history of the David Dunlap Observatory from the standpoint of someone who worked there.

Canada's Role in the Maunakea Spectroscopic Explorer

Alan McConnachie
National Research Council Herzberg Institute of Astrophysics (Dominion Astrophysical Observatory), and Department of Physics and Astronomy, University of Victoria, Elliott Building, room 101, 3800 Finnerty Road, Victoria, BC V8P 5C2, Canada; alan.mcconnachie@nrc-cnrc.gc.ca

Abstract The Canada-France-Hawaii Telescope (CFHT, 1979–), a 3.5-metre class instrument, has been a consistently productive and competitive instrument during its decades-long life. Science evolves, and the instruments for science must evolve too. The CFHT will be decommissioned in 2021, and the plan is to replace it with the Maunakea Spectroscopic Explorer (MSE), an ambitious 11.25-metre dedicated multi-object spectroscopic facility, which will be the leading instrument of its kind at first light in 2026. Canada is the originator of, and one of the leaders in, the project, which represents an exciting opportunity for the country and its international partners to explore cutting edge science.

RASC Paper Sessions (other presentations)

Shadow Chasers 2017

David Shuman
Paul Simard

Abstract   This presentation features a screening of Shuman and Simard's documentary about their involvement in the great American eclipse of 2017, with a question and answer session with the filmmakers.

Toronto's Astronomical Heritage

John R. Percy
Department of Astronomy and Astrophysics, and Dunlap Institute of Astronomy and Astrophysics, University of Toronto, 50 St. George Street, Toronto, ON M5S 3H4, Canada; john.percy@utoronto.ca

Abstract   As a project for the International Year of Astronomy 2009, I developed an astronomical walking tour for Heritage Toronto, which highlighted the remarkable individuals, organizations, facilities, and structures which have enabled Toronto to become a "centre of the universe" for astronomical research, education, and public outreach. This profusely-illustrated "presentation" version of the walk has been given over a dozen times, in various venues, especially during Canada's sesquicentennial year in 2007. The nature and growth of astronomy in Toronto closely parallels the growth and maturation of the city, and the country from the 1830s to today. It illustrates the many ways that astronomy impacts society and culture. For those who are interested, I will provide links to further information, and to a brochure with a map for a self-guided version of the walking tour.

On the Construction of the Heav'ns: A Multi-media, Interdisciplinary Concert and Outreach Event

James E. Hesser
David Lee
W. John McDonald
RASC Victoria Centre

Abstract   Polymath William Herschel (1733–1822) began life as a professional musician who, after immigrating to England at age 19, became in midlife, a renowned astronomer, eventually working in close partnership with his sister, Caroline (1750–1848). Sadly, his delightful music is very rarely heard today, a situation which a Victoria team aims to correct. Over the past two years, multi-media concerts of chamber and symphonic music based upon original research in the British Library by Michael Jarvis (harpsichord and continuo) and Paul Luchkow (violin) (The Luchkow-Jarvis Duo) have provided exceptional opportunities for engaging the public with images and sketches by Victoria Centre members accompanied by dialogue comprised of enlightening quotes from the Herschels' writings. Pre-concert explanations put the science and the music in context. This presentation will include recordings and images from the 2018 concert. Inspired by the excellent public reception of the Victoria concerts, the team is preparing to take the program on tour starting in fall 2020, with the intent of involving local astrophotographers and astronomy clubs in each presentation. Expressions of interest in partnering with us are welcome!

Who doth not see the measures of the Moon/ Which thirteen times she danceth every year?/ And ends her pavan thirteen times as soon

David H. Levy
P.O. Box 895, Vail, AZ 85641; jarnacqc@outlook.com

R. A. Rosenfeld

Abstract   The later Elizabethan and early Jacobean periods (1570s–1620s) were marked by an unusual richness of dramatic naked-eye astronomical events—supernovae, great comets, and eclipses. It was also the period of Tycho Brahe, Johannes Kepler, and Galileo, and is seen as one of the most dynamic periods for the development of astronomical theory and instrumentation. It was also a stellar era for English as a literary language. Astronomy and literature were not worlds apart at the time. Curious about the Elizabethan's curiosity about the night sky? David Levy reads some of the most memorable expressions of early-modern astronomically inspired poetry and prose, accompanied by R. A. Rosenfeld on a reproduction of a 16th-century flute.

Dorner Telescope Museum

Rudolph Dorner
Randall Rosenfeld

Abstract   Part I: Rudolph Dorner presents the rationale for a major new gift to the RASC, which will enable it to tell the story of the telescope in Canada through a world-class institution. Part II: Randall Rosenfeld, Director of the Dorner Telescope Museum, discusses the potential of the museum, its purpose and contemplated programs, some of the challenges to be overcome in its realization, acquisitions achieved and acquisitions to come, what the museum could look like, telescope museum precedents and the unique nature of the Droner vision, and the potential place of the Dorner Telescope Museum in the cultural landscape of the country.

Science of the Stars: The National Heritage of the Dominion Observatory

Sharon Odell

Abstract After the Dominion 15-inch telescope was removed from the Dominion Observatory in 1974, there was an enquiry as to whether this architecture should still stand if it no longer served its original purpose: keeping time, seismology, and astronomy for the country. Astronomers joined fellow Canadian citizens to challenge this declaration, by claiming that the architecture alone was worthy of national heritage designation. This presentation examines the documents, actions, and historical scientific merit put forth to the Canadian Government, in a determined effort to safeguard Canada's National Observatory. Additionally, it compares this past challenge to a similar one facing it today, with the construction of a new hospital super-campus that is about to be placed on its grounds.

RASC Robotic Telescope Project

Paul Mortfield
34 Portree Crescent, Thornhill, ON L3T 3G2, Canada; paul@industrialstars.com

Abstract An up-to-the-minute account is given of the instrumentation, operation, and results of this exciting national initiative, the RASC Robotic Telescope (https:www.rasc.ca/robotic-telescope-nav).

The Past, Present and Future of Youth Outreach at the RASC

Jenna Hinds
RASC Youth Committee

Abstract With about 8% youth membership nationally at the RASC, youth members can be few and far between. While many volunteers focus their effort on engaging young people by reaching out to cubs, scouts, schools, and camp groups, we often don't see an increase in membership because of these outreach efforts. What have we offered to our youth members in the past, and what are we doing now to engage more young people in the future?

Bringing Astronomy Alive—Student Engagement at John Abbott College

Karim Jaffer
Physics Department, John Abbott College, 21 275 Lakeshore Road, Sainte-Anne-de-Bellevue, QC H9X 3L9, Canada

Abstract The approach to Introductory Astronomy at John Abbott College puts the depth and direction of content in the hands of the students, utilizing the partnership with RASC Montreal Centre to mentor students in observational techniques and public outreach. Recognizing that students have access to recent discoveries and current data that best represent the direction of space sciences, each student is given an opportunity to put their stamp on the learning outcomes of the course. Combining this with regular (weather pending), non-compulsory observing opportunities and public outreach events with RASC Montreal Centre allows students to embrace the field in a manner that best suits their interests and motivation, resulting in incredibly diverse term projects, and in several initiatives that have moved beyond the scope of the Introductory Astronomy Course. Other departments and courses are now taking advantage of the Astronomy activities to engage their students beyond the classroom.

A Story of Spectroscopy: How Stellar Composition is Made Clear through Deceptively Simple Tools

Julien da Silva

Abstract Astronomy's unique position as both an empirical science and an accessible one, casting a wide net for individuals to engage with the material. In this talk, I would like to propose spectroscopy to be added to the list of ways astronomy can be communicated to the public. I will speak about my experience working with spectroscopy on the analysis of stellar composition as a student at John Abbott College in Astronomy. Furthermore, I will talk about the benefits of integrating stellar spectroscopy into public outreach events, the deceptive simplicity of the spectroscopy process through a program called RSPEC, and potential methods of incorporating stellar spectroscopy into public outreach.

Things You May Not Know about the Apollo 11 Mission

Randy Attwood
Executive Director, RASC

Abstract The 50th anniversary of the first Moon landing happens this summer. There will be much coverage, but perhaps little or no coverage of some of the aspects of the mission that you may not have heard before. For example, what changes were made to the lunar module while it stood on the launch pad as a result of issues found during the Apollo 10 mission in May 1969? Why was there nearly no television from the Moon during the moonwalk? Why was Armstrong chosen to take the first steps on the Moon? Why are there no good pictures of him standing on the lunar surface? Finally, Neil Armstrong and Buzz Aldrin faced five problems during the landing, each of which, if not resolved by the crew and Mission Control, could have meant no lunar landing. What were they?

Teaching Neil to Land—Simulators Used in the Apollo Program

Ron MacNaughton

Abstract The lunar module never landed until Neil Armstrong brought Eagle down on the Sea of Tranquility. It did not even have enough thrust to lift against Earth's gravity. To practice landings, the astronaut crews spent many hours in simulators whose interiors and controls were identical to the spacecraft. As astronauts made adjustments to the controls, NASA computers calculated what the new orientation would be and changed the dials and views through the windows. This was before computers could generate images. The author has inspected actual components of the simulator from Cape Kennedy and talked to people who built them. The talk will explain some history of simulators and how the Apollo system worked. Light from models and film was projected into the spacecraft through eyepieces roughly the size of the windows of the lunar module. Those might have been the largest eyepieces ever made.

RASC Public lecture

First Man—The Life of Neil Armstrong

James R. Hansen
Director, The University Honors College, Auburn University, Ralph B. Draughon Library, Auburn University, AL 36849; hansejr@auburn.edu

Abstract Neil Armstrong—the first human to set foot on the Moon—made history and inspired an entire planet. Meet James Hansen, author of the acclaimed biography, *First Man: The Life of Neil A. Armstrong*, upon which the recent film was based. Author and professor of history at Auburn University, Hansen will speak about the amazing life of Neil Armstrong as we reflect on the 50 years since Apollo 11. Following his talk, Hansen will be interviewed by RASC member and CBC science journalist Nicole Mortillaro. The Ruth Northcott Memorial Lecture is sponsored by the RASC. It is held biennially in odd-numbered years. The public is invited to attend.

Other abstract received

Flare Stars: A Short Review

Krstinja Dzombeta
Department of Astronomy and Astrophysics, University of Toronto, Toronto, ON M5S 3H4, Canada; krstinja.dzombeta@mail.utoronto.ca

John R. Percy
Department of Astronomy and Astrophysics, and Dunlap Institute of Astronomy and Astrophysics, University of Toronto, Toronto ON M5S 3H4, Canada; percy@astro.utoronto.ca

Abstract We present a short review of flare stars. Flare stars, or UV Ceti stars, are a type of eruptive variable star, defined by their flaring behavior—a rapid (minutes) increase in brightness, followed by a slower (hours) decrease. This short review outlines current knowledge about flare stars, their importance, recent research developments, future research directions, and some practical activities for skilled amateur astronomers and students. Over the past decade, flare stars have been the subject of intensive research, as a result of an abundance of new data, especially from the Kepler and TESS space telescopes. The large statistical samples of data have clarified the relation between flaring and stellar spectral type, luminosity, and rotation. They have allowed for the expansion of the range of spectral types of flare stars, from K and M type dwarfs, to F and G, and possibly even A. They have confirmed the greater frequency of flares on M dwarfs, compared to K, and that flare stars' energies follow a decreasing power law fit for the number of high-energy flares, although a break in the relationship has also been demonstrated. Current problems in flare-star research include improved modelling of the new observational results, using the dynamo theory which produces the stars' magnetic field. What is the difference, if any, between the dynamo in completely-convective stars such as M dwarfs, and in stars such as the sun with only partial convective zones?

Authors' Note: This short review was prepared by author Krstinja Dzombeta, as a senior thesis for the Astronomy Major program at the University of Toronto, under the supervision of author John Percy, who has also revised and edited it. The full review is permanently archived at the University of Toronto Library, at: http://hdl.handle.net/1807/97060

Erratum: KAO-EGYPT J064512.06+341749.2 is a Low Amplitude and Multi-Periodic δ Scuti Variable Star

Ahmed Essam
Mohamed Abdel-Sabour
Gamal Bakr Ali
National Research Institute of Astronomy and Geophysics, 11421 Helwan, Cairo, Egypt; essam60@yahoo.com, sabour2000@hotmail.com, mrezk9@yahoo.com

In the article "KAO-EGYPT J064512.06+341749.2 is a Low Amplitude and Multi-Periodic δ Scuti Variable Star" (*JAAVSO*, 2019, **47**, 66–69), the name of the second author was given incorrectly. "Mouhamed Abdel-Sabour" should be "Mohamed Abdel-Sabour."

Index to Volume 47

Author

Abdel-Sabour, Mohamed, and Ahmed Essam, Gamal Bakr Ali
  Erratum: KAO-EGYPT J064512.06+341749.2 is a Low Amplitude and Multi-Periodic δ Scuti Variable Star 283
  KAO-EGYPT J064512.06+341749.2 is a Low Amplitude and Multi-Periodic δ Scuti Variable Star 66
Adkins, J. Kevin, and Jennifer J. Birriel
  Sky Brightness at Zenith During the January 2019 Total Lunar Eclipse 94
Akiba, Tatsuya, and Andrew Neugarten, Charlyn Ortmann, Vayujeet Gokhale
  Multi-filter Photometric Analysis of W Ursae Majoris (W UMa) Type Eclipsing Binary Stars KID 11405559 and V342 Boo 186
Akins, Hollis B., and Donald A. Smith
  Automated Data Reduction at a Small College Observatory 248
Ali, Gamal Bakr, and Ahmed Essam, Mohamed Abdel-Sabour
  Erratum: KAO-EGYPT J064512.06+341749.2 is a Low Amplitude and Multi-Periodic δ Scuti Variable Star 283
  KAO-EGYPT J064512.06+341749.2 is a Low Amplitude and Multi-Periodic δ Scuti Variable Star 66
Alton, Kevin B.
  CCD Photometry, Light Curve Modeling, and Period Study of the Overcontact Binary Systems V647 Virginis and V948 Monocerotis 7
  CCD Photometry, Period Analysis, and Evolutionary Status of the Pulsating Variable V544 Andromedae 231
Alton, Kevin B., and Kazimierz Stępień
  CCD Photometry, Light Curve Deconvolution, Period Analysis, Kinematics, and Evolutionary Status of the HADS Variable V460 Andromedae 53
Anon.
  Index to Volume 47 284
Attwood, Randy
  Things You May Not Know about the Apollo 11 Mission (Abstract) 281
Axelsen, Roy A.
  A New Candidate δ Scuti Star in Centaurus: HD 121191 173
Barnes, Thomas G., III, in Elizabeth J. Jeffery et al.
  Radial Velocities for Four δ Sct Variable Stars 111
Berrington, Robert C., and Mathew F. Knote, Ronald H. Kaitchuck
  Observations and Preliminary Modeling of the Light Curves of Eclipsing Binary Systems NSVS 7322420 and NSVS 5726288 194
Berry, Richard L.
  Conducting the Einstein Gravitational Deflection Experiment (Abstract) 133
Billings, Gary
  Apsidal Motion Analysis of the Eccentric Eclipsing Binary V1103 Cassiopeiae (Abstract) 130
Birriel, Jennifer J., and J. Kevin Adkins
  Sky Brightness at Zenith During the January 2019 Total Lunar Eclipse 94

Cahill, Maria J.
  Helen Sawyer Hogg: One Woman's Journey with the Stars (Abstract) 279
Cami, Jan
  Public Outreach Through the Cronyn Observatory (Abstract) 279
Caples, David, in Vayujeet Gokhale et al.
  Sky Brightness Measurements and Ways to Mitigate Light Pollution in Kirksville, Missouri 241
Caton, Daniel B., and Ronald G. Samec, Danny R. Faulkner
  BVR_cI_c Photometric Observations and Analyses of the Totally Eclipsing, Solar Type Binary, OR Leonis 147
  Observations and Analysis of a Detached Eclipsing Binary, V385 Camelopardalis 176
Caton, Daniel B., in Ronald G. Samec et al.
  VR_cI_c Photometric Study of the Totally Eclipsing Pre-W UMa Binary, V616 Camelopardalis: Is it Detached? 157
Chamberlain, Heather A., and Ronald G. Samec, Walter Van Hamme
  The Southern Solar-type, Totally Eclipsing Binary PY Aquarii 29
Chaplin, Geoff B.
  Medium-term Variation in Times of Minimum of Algol-type Binaries: XZ And, RZ Cas, U Cep, TW Dra, U Sge 222
  SSA Analysis and Significance Tests for Periodicity in S, RS, SU, AD, BU, KK, and PR Persei 17
Clem, James L., and Arlo U. Landolt
  Multi-color Photometry of the Hot R Coronae Borealis Star and Proto-planetary Nebula V348 Sagittarii 83
Collins, Donald F.
  The Eclipsing Triple Star System b Persei (HD 26961) (Abstract) 276
Conti, Dennis, and Stella Kafka
  Contributions of the AAVSO to Year 1 of TESS (Abstract) 275
Cook, Lewis M., and Enrique de Miguel, Geoffrey Stone, Gary E. Walker
  The Faint Cataclysmic Variable Star V677 Andromedae (Abstract) 131
Cook, Michael J., and Kenneth T. Menzies, Franz-Josef (Josch) Hambsch
  How-to Hour—Planning an Observing Program (Abstract) 278
Cooper, Tim
  The Contribution of A. W. Roberts' Observations to the AAVSO International Database 254
Crawford, Tim R.
  The History of AAVSO Charts, Part III: The Henden Era 122
Crowson, Melanie, and David Syndergaard
  A Study of RR Lyrae Variable Stars in a Professor-Student Collaboration using the American Public University System (APUS) Observatory (Abstract) 277
da Silva, Julien
  A Story of Spectroscopy: How Stellar Composition is Made Clear through Deceptively Simple Tools (Abstract) 281
De Lee, Nathan, and Pradip Karmakar, Horace A. Smith
  The Long-term Period Changes and Evolution of V1, a W Virginis Star in the Globular Cluster M12 167

de Miguel, Enrique, in Lewis M. Cook *et al.*
    The Faint Cataclysmic Variable Star V677 Andromedae (Abstract) 131
Dorner, Rudolph, and Randall Rosenfeld
    Dorner Telescope Museum (Abstract) 280
Dzombeta, Krstinja, and John R. Percy
    Flare Stars: A Short Review (Abstract) 282
Essam, Ahmed, and Mohamed Abdel-Sabour, Gamal Bakr Ali
    Erratum: KAO-EGYPT J064512.06+341749.2 is a Low Amplitude and Multi-Periodic δ Scuti Variable Star 283
    KAO-EGYPT J064512.06+341749.2 is a Low Amplitude and Multi-Periodic δ Scuti Variable Star 66
Faulkner, Danny R., and Ronald G. Samec, Daniel B. Caton
    BVR_cI_c Photometric Observations and Analyses of the Totally Eclipsing, Solar Type Binary, OR Leonis 147
    Observations and Analysis of a Detached Eclipsing Binary, V385 Camelopardalis 176
Faulkner, Danny R., in Ronald G. Samec *et al.*
    VR_cI_c Photometric Study of the Totally Eclipsing Pre-W UMa Binary, V616 Camelopardalis: Is it Detached? 157
Fenaux, Lucas, and John R. Percy
    Period Analysis of All-Sky Automated Survey for Supernovae (ASAS-SN) Data on Pulsating Red Giants 202
    Studies of Pulsating Red Giants using AAVSO and ASAS-SN Data (Abstract) 277
Garcia, Jorge, and Joyce A. Guzik, Jason Jackiewicz
    Discoveries for δ Scuti Variable Stars in the NASA Kepler 2 Mission (Abstract) 129
Gentry, Davis R., in Ronald G. Samec *et al.*
    VR_cI_c Photometric Study of the Totally Eclipsing Pre-W UMa Binary, V616 Camelopardalis: Is it Detached? 157
Goins, Jordan, in Vayujeet Gokhale *et al.*
    Sky Brightness Measurements and Ways to Mitigate Light Pollution in Kirksville, Missouri 241
Gokhale, Vayujeet, and Jordan Goins, Ashley Herdmann, Eric Hilker, Emily Wren, David Caples, James Tompkins
    Sky Brightness Measurements and Ways to Mitigate Light Pollution in Kirksville, Missouri 241
Gokhale, Vayujeet, in Tatsuya Akiba *et al.*
    Multi-filter Photometric Analysis of W Ursae Majoris (W UMa) Type Eclipsing Binary Stars KID 11405559 and V342 Boo 186
Gollapudy, Sriram, and Wayne Osborn
    Observations of the Suspected Variable Star Ross 114 (NSV 13523) 3
Gorodenski, Stanley A.
    The Fun of Processing a Stellar Spectrum—the Hard Way (Abstract) 131
Guzik, Joyce A., and Jorge Garcia, Jason Jackiewicz
    Discoveries for δ Scuti Variable Stars in the NASA Kepler 2 Mission (Abstract) 129
Hambsch, Franz-Josef (Josch), and Kenneth T. Menzies, Michael J. Cook
    How-to Hour—Planning an Observing Program (Abstract) 278
Hambsch, Franz-Josef, and Christopher S. Jeffery
    New Intense Multiband Photometric Observations of the Hot Carbon Star V348 Sagittarii (Abstract) 132

Hansen, James R.
    First Man—The Life of Neil Armstrong (Abstract) 282
Harris, Jesse D'Shawn, and Lucian Undreiu
    β Cepheid and Mira Variable Stars: A Spectral Analysis (Abstract) 131
Henden, Arne A.
    APASS DR10 Has Arrived! (Abstract) 130
Henden, Arne, in Stephen Levine *et al.*
    Solar System Objects and the AAVSO Photometric All-Sky Survey (APASS) (Abstract) 132
Herdmann, Ashley, in Vayujeet Gokhale *et al.*
    Sky Brightness Measurements and Ways to Mitigate Light Pollution in Kirksville, Missouri 241
Hesser, James E., and David Lee, W. John McDonald
    On the Construction of the Heav'ns: A Multi-media, Interdisciplinary Concert and Outreach Event (Abstract) 280
Hilker, Eric, in Vayujeet Gokhale *et al.*
    Sky Brightness Measurements and Ways to Mitigate Light Pollution in Kirksville, Missouri 241
Hill, Brian, and Ariana Hofelmann
    Stepping Stones to TFOP: Experience of the Saint Mary's College Geissberger Observatory (Abstract) 129
Hinds, Jenna
    The Past, Present and Future of Youth Outreach at the RASC (Abstract) 281
Hofelmann, Ariana, and Brian Hill
    Stepping Stones to TFOP: Experience of the Saint Mary's College Geissberger Observatory (Abstract) 129
Howe, Rodney
    Comparison of North-South Hemisphere Data from AAVSO Visual Observers and the SDO Satellite Computer-Generated Wolf Numbers (Abstract) 129
Jackiewicz, Jason, and Joyce A. Guzik, Jorge Garcia
    Discoveries for δ Scuti Variable Stars in the NASA Kepler 2 Mission (Abstract) 129
Jaffer, Karim
    Bringing Astronomy Alive—Student Engagement at John Abbott College (Abstract) 281
Jeffery, Christopher S., and Franz-Josef Hambsch
    New Intense Multiband Photometric Observations of the Hot Carbon Star V348 Sagittarii (Abstract) 132
Jeffery, Elizabeth J., and Thomas G. Barnes III, Ian Skillen, Thomas J. Montemayor
    Radial Velocities for Four δ Sct Variable Stars 111
Johnson, L. Clifton
    The Zooniverse (Abstract) 278
Joner, Michael
    Small Observatory Operations: 2018 Highlights from the West Mountain Observatory (Abstract) 129
Kafka, Stella, and Dennis Conti
    Contributions of the AAVSO to Year 1 of TESS (Abstract) 275
Kaitchuck, Ronald H., and Mathew F. Knote, Robert C. Berrington
    Observations and Preliminary Modeling of the Light Curves of Eclipsing Binary Systems NSVS 7322420 and NSVS 5726288 194

Karmakar, Pradip, and Horace A. Smith, Nathan De Lee
 The Long-term Period Changes and Evolution of V1, a
 W Virginis Star in the Globular Cluster M12   167
Kline, Adam, and Isobel Snellenberger
 Camera Characterization and First Observation after
 Upgrade of Feder Observatory (Abstract)   132
Kloppenborg, Brian, in Stephen Levine et al.
 Solar System Objects and the AAVSO Photometric All-Sky
 Survey (APASS) (Abstract)   132
Knote, Mathew F., and Ronald H. Kaitchuck,
  Robert C. Berrington
 Observations and Preliminary Modeling of the Light
 Curves of Eclipsing Binary Systems NSVS 7322420 and
 NSVS 5726288   194
Landolt, Arlo U., and James L. Clem
 Multi-color Photometry of the Hot R Coronae Borealis Star
 and Proto-planetary Nebula V348 Sagittarii   83
Lane, David
 A Robotic Observatory System that Anyone Can Use for
 Education, Astro-Imaging, Research, or Fun! (Abstract) 275
Lanning, Chlöe M., and Edward J. Michaels, Skyler N. Self
 A Photometric Study of the Contact Binary V384 Serpentis 43
Larsen, Kristine
 The Eye is Mightier than the Algorithm: Lessons Learned
 from the Erroneous Classification of ASAS Variables as
 "Miscellaneous" (Abstract)   275
 Variable Stars and Cultural Astronomy (Abstract)   130
 Variable Stars in the College Classroom (Abstract)   277
Lee, David, and James E. Hesser, W. John McDonald
 On the Construction of the Heav'ns: A Multi-media,
 Interdisciplinary Concert and Outreach Event (Abstract) 280
Lemay, Damien
 Five Years of CCD photometry (Abstract)   278
Levine, Stephen, and Arne Henden, Dirk Terrell, Doug Welch,
  Brian Kloppenborg
 Solar System Objects and the AAVSO Photometric All-Sky
 Survey (APASS) (Abstract)   132
Levy, David H., and R. A. Rosenfeld
 Who doth not see the measures of the Moon/ Which thirteen
 times she danceth every year?/ And ends her pavan thirteen
 times as soon (Abstract)   280
Los, Edward
 DASCH Project Update (Abstract)   275
MacNaughton, Ron
 Teaching Neil to Land—Simulators Used in the Apollo
 Program (Abstract)   282
McConnachie, Alan
 Canada's Role in the Maunakea Spectroscopic Explorer
 (Abstract)   279
McDonald, W. John, and James E. Hesser, David Lee
 On the Construction of the Heav'ns: A Multi-media,
 Interdisciplinary Concert and Outreach Event (Abstract) 280
Menzies, Kenneth T., and Michael J. Cook, Franz-Josef
  (Josch) Hambsch
 How-to Hour—Planning an Observing Program (Abstract) 278
Menzies, Kenneth T.
 Photometric Accuracy of CMOS and CCD Cameras
 (Preliminary Results) (Abstract)   276

Michaels, Edward J.
 A Photometric Study of Five Low Mass Contact Binaries 209
Michaels, Edward J., and Chlöe M. Lanning, Skyler N. Self
 A Photometric Study of the Contact Binary V384 Serpentis 43
Montemayor, Thomas J., in Elizabeth J. Jeffery et al.
 Radial Velocities for Four δ Sct Variable Stars   111
Morrison, Nancy D.
 Life, Variable Stars, and Everything   137
Mortfield, Paul
 RASC Robotic Telescope Project (Abstract)   281
Morton, Donald C.
 Some Memories of the David Dunlap Observatory
 (Abstract)   279
Neugarten, Andrew, in Tatsuya Akiba et al.
 Multi-filter Photometric Analysis of W Ursae Majoris
 (W UMa) Type Eclipsing Binary Stars KID 11405559 and
 V342 Boo   186
Newell, Catherine L.
 The Importance of Religion and Art in Sowing the Seeds
 of Success of the U.S. Space Program in the 1940s and
 1950s (Abstract)   279
Nicholas, Michael
 Bright Star Monitor Network (Abstract)   132
Odell, Sharon
 Science of the Stars: The National Heritage of the Dominion
 Observatory (Abstract)   281
Ortmann, Charlyn, in Tatsuya Akiba et al.
 Multi-filter Photometric Analysis of W Ursae Majoris
 (W UMa) Type Eclipsing Binary Stars KID 11405559 and
 V342 Boo   186
Osborn, Wayne, and Sriram Gollapudy
 Observations of the Suspected Variable Star Ross 114
 (NSV 13523)   3
Ouellette, Nathalie
 Canada's Role in the James Webb Space Telescope
 (Abstract)   279
Percy, John R.
 Citizen Science   1
 So Long, and Thanks for All the Manuscripts   135
 Toronto's Astronomical Heritage (Abstract)   280
Percy, John R., and Arthur Lei Qiu
 Long-Term Changes in the Variability of Pulsating Red
 Giants (and One RCB Star)   76
Percy, John R., and Krstinja Dzombeta
 Flare Stars: A Short Review (Abstract)   282
Percy, John R., and Lucas Fenaux
 Period Analysis of All-Sky Automated Survey for Supernovae
 (ASAS-SN) Data on Pulsating Red Giants   202
 Studies of Pulsating Red Giants using AAVSO and ASAS-
 SN Data (Abstract)   277
Petriew, Vance, and Gerard Samolyk
 Recent Minima of 267 Recently Discovered Eclipsing
 Variable Stars   270
Polakis, Tom
 Discovery and Period Analysis of Seven Variable Stars 117
Qiu, Arthur Lei, and John R. Percy
 Long-Term Changes in the Variability of Pulsating Red
 Giants (and One RCB Star)   76

Rea, Bill
   Low Resolution Spectroscopy of Miras—X Octantis   70
Rosenfeld, R. A., and David H. Levy
   Who doth not see the measures of the Moon/ Which thirteen times she danceth every year?/ And ends her pavan thirteen times as soon (Abstract)   280
Rosenfeld, Randall, and Rudolph Dorner
   Dorner Telescope Museum (Abstract)   280
Samec, Ronald G., and Daniel B. Caton, Danny R. Faulkner
   BVR_cI_c Photometric Observations and Analyses of the Totally Eclipsing, Solar Type Binary, OR Leonis   147
   Observations and Analysis of a Detached Eclipsing Binary, V385 Camelopardalis   176
Samec, Ronald G., and Daniel B. Caton, Davis R. Gentry, Danny R. Faulkner
   VR_cI_c Photometric Study of the Totally Eclipsing Pre-W UMa Binary, V616 Camelopardalis: Is it Detached?   157
Samec, Ronald G., and Heather A. Chamberlain, Walter Van Hamme
   The Southern Solar-type, Totally Eclipsing Binary PY Aquarii   29
Samolyk, Gerard
   Erratum: Recent Minima of 266 Eclipsing Binary Stars   134
   Recent Maxima of 85 Short Period Pulsating Stars   103
   Recent Minima of 200 Eclipsing Binary Stars   265
   Recent Minima of 242 Eclipsing Binary Stars   106
   Visual Times of Maxima for Short Period Pulsating Stars V   98
Samolyk, Gerard, and Vance Petriew
   Recent Minima of 267 Recently Discovered Eclipsing Variable Stars   270
Self, Skyler N., and Edward J. Michaels, Chlöe M. Lanning
   A Photometric Study of the Contact Binary V384 Serpentis   43
Shuman, David, and Paul Simard
   Shadow Chasers 2017 (Abstract)   280
Simard, Paul, and David Shuman
   Shadow Chasers 2017 (Abstract)   280
Skillen, Ian, in Elizabeth J. Jeffery et al.
   Radial Velocities for Four δ Sct Variable Stars   111
Smith, Donald A., and Hollis B. Akins
   Automated Data Reduction at a Small College Observatory   248
Smith, Horace A., and Pradip Karmakar, Nathan De Lee
   The Long-term Period Changes and Evolution of V1, a W Virginis Star in the Globular Cluster M12   167
Snellenberger, Isobel, and Adam Kline
   Camera Characterization and First Observation after Upgrade of Feder Observatory (Abstract)   132
Steiner, Ken
   Cold War Spy in the Sky now Provides an Eye on the Cosmos (Abstract)   130
Stępień, Kazimierz, and Kevin B. Alton
   CCD Photometry, Light Curve Deconvolution, Period Analysis, Kinematics, and Evolutionary Status of the HADS Variable V460 Andromedae   53

Stone, Geoffrey, in Lewis M. Cook et al.
   The Faint Cataclysmic Variable Star V677 Andromedae (Abstract)   131
Syndergaard, David, and Melanie Crowson
   A Study of RR Lyrae Variable Stars in a Professor-Student Collaboration using the American Public University System (APUS) Observatory (Abstract)   277
Terrell, Dirk, in Stephen Levine et al.
   Solar System Objects and the AAVSO Photometric All-Sky Survey (APASS) (Abstract)   132
Thompson, John
   A Simple Objective Prism Spectrograph for Educational Purposes (Abstract)   277
Tompkins, James, in Vayujeet Gokhale et al.
   Sky Brightness Measurements and Ways to Mitigate Light Pollution in Kirksville, Missouri   241
Undreiu, Lucian, and Jesse D'Shawn Harris
   β Cepheid and Mira Variable Stars: A Spectral Analysis (Abstract)   131
Van Hamme, Walter, and Ronald G. Samec, Heather A. Chamberlain
   The Southern Solar-type, Totally Eclipsing Binary PY Aquarii   29
Van Laerhoven, Christa
   The Rotating Frame (Abstract)   278
Vander Haagen, Gary
   High Occurrence Optical Spikes and Quasi-Periodic Pulses (QPPs) on X-ray Star 47 Cassiopeiae   141
Wade, Gregg
   The BRITE Constellation Mission (Abstract)   279
Wadhwa, Surjit S.
   Photometric Analysis of Two Contact Binary Systems: USNO-A2.0 1200-16843637 and V1094 Cassiopeiae   40
   Photometric Analysis of V633 Virginis   138
Wagner, Richard
   Measuring and Correcting Non-linearities in ABG Cameras (Abstract)   277
Walker, Gary
   Is sCMOS Really sCMAS? (Abstract)   131
Walker, Gary E., in Lewis M. Cook et al.
   The Faint Cataclysmic Variable Star V677 Andromedae (Abstract)   131
Wargelin, Peggy
   The Project PHaEDRA Collection: An Anchor to Connect Modern Science with Historical Data (Abstract)   276
Welch, Doug, in Stephen Levine et al.
   Solar System Objects and the AAVSO Photometric All-Sky Survey (APASS) (Abstract)   132
Wren, Emily, in Vayujeet Gokhale et al.
   Sky Brightness Measurements and Ways to Mitigate Light Pollution in Kirksville, Missouri   241

Subject

AAVSO

APASS DR10 Has Arrived! (Abstract)
  Arne A. Henden   130
β Cepheid and Mira Variable Stars: A Spectral Analysis
  (Abstract)
  Jesse D'Shawn Harris and Lucian Undreiu   131
Bright Star Monitor Network (Abstract)
  Michael Nicholas   132
Citizen Science
  John R. Percy   1
Contributions of the AAVSO to Year 1 of TESS (Abstract)
  Dennis Conti and Stella Kafka   275
Helen Sawyer Hogg: One Woman's Journey with the Stars
  (Abstract)
  Maria J. Cahill   279
The History of AAVSO Charts, Part III: The Henden Era
  Tim R. Crawford   122
Life, Variable Stars, and Everything
  Nancy D. Morrison   137
Long-Term Changes in the Variability of Pulsating Red
  Giants (and One RCB Star)
  John R. Percy and Arthur Lei Qiu   76
Low Resolution Spectroscopy of Miras—X Octantis
  Bill Rea   70
Recent Minima of 267 Recently Discovered Eclipsing
  Variable Stars
  Gerard Samolyk and Vance Petriew   270
So Long, and Thanks for All the Manuscripts
  John R. Percy   135
Solar System Objects and the AAVSO Photometric All-Sky
  Survey (APASS) (Abstract)
  Stephen Levine et al.   132
Stepping Stones to TFOP: Experience of the Saint Mary's
  College Geissberger Observatory (Abstract)
  Ariana Hofelmann and Brian Hill   129

AAVSO INTERNATIONAL DATABASE

APASS DR10 Has Arrived! (Abstract)
  Arne A. Henden   130
CCD Photometry, Period Analysis, and Evolutionary Status
  of the Pulsating Variable V544 Andromedae
  Kevin B. Alton   231
The Contribution of A. W. Roberts' Observations
  to the AAVSO International Database
  Tim Cooper   254
Discovery and Period Analysis of Seven Variable Stars
  Tom Polakis   117
The Eclipsing Triple Star System b Persei (HD 26961)
  (Abstract)
  Donald F. Collins   276
Erratum: Recent Minima of 266 Eclipsing Binary Stars
  Gerard Samolyk   134
Five Years of CCD photometry (Abstract)
  Damien Lemay   278

Long-Term Changes in the Variability of Pulsating Red
  Giants (and One RCB Star)
  John R. Percy and Arthur Lei Qiu   76
Multi-color Photometry of the Hot R Coronae Borealis Star
  and Proto-planetary Nebula V348 Sagittarii
  Arlo U. Landolt and James L. Clem   83
Period Analysis of All-Sky Automated Survey for
  Supernovae (ASAS-SN) Data on Pulsating Red Giants
  John R. Percy and Lucas Fenaux   202
Recent Maxima of 85 Short Period Pulsating Stars
  Gerard Samolyk   103
Recent Minima of 200 Eclipsing Binary Stars
  Gerard Samolyk   265
Recent Minima of 242 Eclipsing Binary Stars
  Gerard Samolyk   106
Recent Minima of 267 Recently Discovered Eclipsing
  Variable Stars
  Gerard Samolyk and Vance Petriew   270
SSA Analysis and Significance Tests for Periodicity in S, RS,
  SU, AD, BU, KK, and PR Persei
  Geoff B. Chaplin   17
Studies of Pulsating Red Giants using AAVSO and ASAS-SN
  Data (Abstract)
  John R. Percy and Lucas Fenaux   277
Visual Times of Maxima for Short Period Pulsating Stars V
  Gerard Samolyk   98

AAVSO, JOURNAL OF

Index to Volume 47
  Anon.   284
Life, Variable Stars, and Everything
  Nancy D. Morrison   137
So Long, and Thanks for All the Manuscripts
  John R. Percy   135

ACTIVE GALACTIC NUCLEI (AGNs)
[See also EXTRAGALACTIC]

Small Observatory Operations: 2018 Highlights from the
  West Mountain Observatory (Abstract)
  Michael Joner   129

AMPLITUDE ANALYSIS

Long-Term Changes in the Variability of Pulsating Red
  Giants (and One RCB Star)
  John R. Percy and Arthur Lei Qiu   76

ARCHAEOASTRONOMY
[See also ASTRONOMY, HISTORY OF]

Variable Stars and Cultural Astronomy (Abstract)
  Kristine Larsen   130

ASTEROIDS

Discovery and Period Analysis of Seven Variable Stars
  Tom Polakis   117
Small Observatory Operations: 2018 Highlights from the
  West Mountain Observatory (Abstract)
  Michael Joner   129

Solar System Objects and the AAVSO Photometric All-Sky Survey (APASS) (Abstract)
   Stephen Levine et al.   132

ASTRONOMERS, AMATEUR; PROFESSIONAL-AMATEUR COLLABORATION
Citizen Science
   John R. Percy   1
Contributions of the AAVSO to Year 1 of TESS (Abstract)
   Dennis Conti and Stella Kafka   275
The Eclipsing Triple Star System b Persei (HD 26961) (Abstract)
   Donald F. Collins   276
How-to Hour—Planning an Observing Program (Abstract)
   Kenneth J. Menzies, Michael J. Cook, and Franz-Josef (Josch) Hambsch   278
The Zooniverse (Abstract)
   L. Clifton Johnson   278

ASTRONOMY, HISTORY OF
[See also ARCHAEOASTRONOMY; OBITUARIES]
Canada's Role in the James Webb Space Telescope (Abstract)
   Nathalie Ouellette   279
Canada's Role in the Maunakea Spectroscopic Explorer (Abstract)
   Alan McConnachie   279
Citizen Science
   John R. Percy   1
Cold War Spy in the Sky now Provides an Eye on the Cosmos (Abstract)
   Ken Steiner   130
Conducting the Einstein Gravitational Deflection Experiment (Abstract)
   Richard L. Berry   133
The Contribution of A. W. Roberts' Observations to the AAVSO International Database
   Tim Cooper   254
DASCH Project Update (Abstract)
   Edward Los   275
First Man—The Life of Neil Armstrong (Abstract)
   James R. Hansen   282
Helen Sawyer Hogg: One Woman's Journey with the Stars (Abstract)
   Maria J. Cahill   279
The History of AAVSO Charts, Part III: The Henden Era
   Tim R. Crawford   122
The Importance of Religion and Art in Sowing the Seeds of Success of the U.S. Space Program in the 1940s and 1950s (Abstract)
   Catherine L. Newell   279
On the Construction of the Heav'ns: A Multi-media, Interdisciplinary Concert and Outreach Event (Abstract)
   James E. Hesser, David Lee, and W. John McDonald   280
The Project PHaEDRA Collection: An Anchor to Connect Modern Science with Historical Data (Abstract)
   Peggy Wargelin   276
Science of the Stars: The National Heritage of the Dominion Observatory (Abstract)
   Sharon Odell   281
Some Memories of the David Dunlap Observatory (Abstract)
   Donald C. Morton   279
Teaching Neil to Land—Simulators Used in the Apollo Program (Abstract)
   Ron MacNaughton   282
Things You May Not Know about the Apollo 11 Mission (Abstract)
   Randy Attwood   281
Toronto's Astronomical Heritage (Abstract)
   John R. Percy   280
Variable Stars and Cultural Astronomy (Abstract)
   Kristine Larsen   130
Who doth not see the measures of the Moon/ Which thirteen times she danceth every year?/ And ends her pavan thirteen times as soon (Abstract)
   David H. Levy and R. A. Rosenfeld   280

ASTRONOMY, WOMEN IN
Helen Sawyer Hogg: One Woman's Journey with the Stars (Abstract)
   Maria J. Cahill   279

BIOGRAPHY [See also ASTRONOMY, HISTORY OF]
The Contribution of A. W. Roberts' Observations to the AAVSO International Database
   Tim Cooper   254
First Man—The Life of Neil Armstrong (Abstract)
   James R. Hansen   282
Helen Sawyer Hogg: One Woman's Journey with the Stars (Abstract)
   Maria J. Cahill   279
Life, Variable Stars, and Everything
   Nancy D. Morrison   137
On the Construction of the Heav'ns: A Multi-media, Interdisciplinary Concert and Outreach Event (Abstract)
   James E. Hesser, David Lee, and W. John McDonald   280
So Long, and Thanks for All the Manuscripts
   John R. Percy   135
Things You May Not Know about the Apollo 11 Mission (Abstract)
   Randy Attwood   281

CATACLYSMIC VARIABLES
[See also VARIABLE STARS (GENERAL)]
The Contribution of A. W. Roberts' Observations to the AAVSO International Database
   Tim Cooper   254
The Faint Cataclysmic Variable Star V677 Andromedae (Abstract)
   Lewis M. Cook et al.   131

CATALOGUES, DATABASES, SURVEYS
APASS DR10 Has Arrived! (Abstract)
   Arne A. Henden   130

β Cepheid and Mira Variable Stars: A Spectral Analysis (Abstract)
 Jesse D'Shawn Harris and Lucian Undreiu 131
The BRITE Constellation Mission (Abstract)
 Gregg Wade 279
CCD Photometry, Light Curve Deconvolution, Period Analysis, Kinematics, and Evolutionary Status of the HADS Variable V460 Andromedae
 Kevin B. Alton and Kazimierz Stępień 53
CCD Photometry, Light Curve Modeling, and Period Study of the Overcontact Binary Systems V647 Virginis and V948 Monocerotis
 Kevin B. Alton 7
CCD Photometry, Period Analysis, and Evolutionary Status of the Pulsating Variable V544 Andromedae
 Kevin B. Alton 231
Contributions of the AAVSO to Year 1 of TESS (Abstract)
 Dennis Conti and Stella Kafka 275
The Contribution of A. W. Roberts' Observations to the AAVSO International Database
 Tim Cooper 254
DASCH Project Update (Abstract)
 Edward Los 275
Discoveries for δ Scuti Variable Stars in the NASA Kepler 2 Mission (Abstract)
 Joyce A. Guzik, Jorge Garcia, and Jason Jackiewicz 129
Discovery and Period Analysis of Seven Variable Stars
 Tom Polakis 117
Erratum: Recent Minima of 266 Eclipsing Binary Stars
 Gerard Samolyk 134
The Eye is Mightier than the Algorithm: Lessons Learned from the Erroneous Classification of ASAS Variables as "Miscellaneous" (Abstract)
 Kristine Larsen 275
Five Years of CCD photometry (Abstract)
 Damien Lemay 278
The History of AAVSO Charts, Part III: The Henden Era
 Tim R. Crawford 122
The Long-term Period Changes and Evolution of V1, a W Virginis Star in the Globular Cluster M12
 Pradip Karmakar, Horace A. Smith, and Nathan De Lee 167
Observations and Preliminary Modeling of the Light Curves of Eclipsing Binary Systems NSVS 7322420 and NSVS 5726288
 Mathew F. Knote, Ronald H. Kaitchuck, and Robert C. Berrington 194
Observations of the Suspected Variable Star Ross 114 (NSV 13523)
 Sriram Gollapudy and Wayne Osborn 3
Period Analysis of All-Sky Automated Survey for Supernovae (ASAS-SN) Data on Pulsating Red Giants
 John R. Percy and Lucas Fenaux 202
Photometric Analysis of Two Contact Binary Systems: USNO-A2.0 1200-16843637 and V1094 Cassiopeiae
 Surjit S. Wadhwa 40
Photometric Analysis of V633 Virginis
 Surjit S. Wadhwa 138
A Photometric Study of the Contact Binary V384 Serpentis
 Edward J. Michaels, Chlöe M. Lanning, and Skyler N. Self 43
The Project PHaEDRA Collection: An Anchor to Connect Modern Science with Historical Data (Abstract)
 Peggy Wargelin 276
Recent Maxima of 85 Short Period Pulsating Stars
 Gerard Samolyk 103
Recent Minima of 200 Eclipsing Binary Stars
 Gerard Samolyk 265
Recent Minima of 242 Eclipsing Binary Stars
 Gerard Samolyk 106
Recent Minima of 267 Recently Discovered Eclipsing Variable Stars
 Gerard Samolyk and Vance Petriew 270
SSA Analysis and Significance Tests for Periodicity in S, RS, SU, AD, BU, KK, and PR Persei
 Geoff B. Chaplin 17
Solar System Objects and the AAVSO Photometric All-Sky Survey (APASS) (Abstract)
 Stephen Levine et al. 132
Studies of Pulsating Red Giants using AAVSO and ASAS-SN Data (Abstract)
 John R. Percy and Lucas Fenaux 277
Visual Times of Maxima for Short Period Pulsating Stars V
 Gerard Samolyk 98
The Zooniverse (Abstract)
 L. Clifton Johnson 278

CEPHEID VARIABLES
[See also VARIABLE STARS (GENERAL)]
β Cepheid and Mira Variable Stars: A Spectral Analysis (Abstract)
 Jesse D'Shawn Harris and Lucian Undreiu 131
The Contribution of A. W. Roberts' Observations to the AAVSO International Database
 Tim Cooper 254
The Long-term Period Changes and Evolution of V1, a W Virginis Star in the Globular Cluster M12
 Pradip Karmakar, Horace A. Smith, and Nathan De Lee 167

CHARTS, VARIABLE STAR
APASS DR10 Has Arrived! (Abstract)
 Arne A. Henden 130
The Contribution of A. W. Roberts' Observations to the AAVSO International Database
 Tim Cooper 254
The History of AAVSO Charts, Part III: The Henden Era
 Tim R. Crawford 122
How-to Hour—Planning an Observing Program (Abstract)
 Kenneth J. Menzies, Michael J. Cook, and Franz-Josef (Josch) Hambsch 278

CHARTS; COMPARISON STAR SEQUENCES
How-to Hour—Planning an Observing Program (Abstract)
 Kenneth J. Menzies, Michael J. Cook, and Franz-Josef (Josch) Hambsch 278

Photometric Accuracy of CMOS and CCD Cameras
(Preliminary Results) (Abstract)
Kenneth T. Menzies 276

CLUSTERS, GLOBULAR

The Long-term Period Changes and Evolution of V1,
a W Virginis Star in the Globular Cluster M12
Pradip Karmakar, Horace A. Smith, and
Nathan De Lee 167

COMETS

Solar System Objects and the AAVSO Photometric All-Sky
Survey (APASS) (Abstract)
Stephen Levine et al. 132

COMPUTERS; SOFTWARE; INTERNET, WORLD WIDE WEB

Automated Data Reduction at a Small College
Observatory
Donald A. Smith and Hollis B. Akins 248
Citizen Science
John R. Percy 1
The Fun of Processing a Stellar Spectrum—the Hard Way
(Abstract)
Stanley A. Gorodenski 131
The History of AAVSO Charts, Part III: The Henden Era
Tim R. Crawford 122
How-to Hour—Planning an Observing Program (Abstract)
Kenneth J. Menzies, Michael J. Cook, and Franz-Josef
(Josch) Hambsch 278
Long-Term Changes in the Variability of Pulsating Red
Giants (and One RCB Star)
John R. Percy and Arthur Lei Qiu 76
Medium-term Variation in Times of Minimum of Algol-type
Binaries: XZ And, RZ Cas, U Cep, TW Dra, U Sge
Geoff B. Chaplin 222
Photometric Accuracy of CMOS and CCD Cameras
(Preliminary Results) (Abstract)
Kenneth T. Menzies 276
The Project PHaEDRA Collection: An Anchor to Connect
Modern Science with Historical Data (Abstract)
Peggy Wargelin 276
SSA Analysis and Significance Tests for Periodicity in S, RS,
SU, AD, BU, KK, and PR Persei
Geoff B. Chaplin 17
Stepping Stones to TFOP: Experience of the Saint Mary's
College Geissberger Observatory (Abstract)
Ariana Hofelmann and Brian Hill 129
A Story of Spectroscopy: How Stellar Composition is Made
Clear through Deceptively Simple Tools (Abstract)
Julien da Silva 281

DATA MANAGEMENT [See also AAVSO; COMPUTERS]

Automated Data Reduction at a Small College
Observatory
Donald A. Smith and Hollis B. Akins 248

The Contribution of A. W. Roberts' Observations
to the AAVSO International Database
Tim Cooper 254

DATA MINING

DASCH Project Update (Abstract)
Edward Los 275
Discoveries for δ Scuti Variable Stars in the NASA
Kepler 2 Mission (Abstract)
Joyce A. Guzik, Jorge Garcia, and Jason Jackiewicz 129
Discovery and Period Analysis of Seven Variable Stars
Tom Polakis 117
The Eye is Mightier than the Algorithm: Lessons Learned
from the Erroneous Classification of ASAS Variables as
"Miscellaneous" (Abstract)
Kristine Larsen 275
Flare Stars: A Short Review (Abstract)
Krstinja Dzombeta and John R. Percy 282
Multi-filter Photometric Analysis of W Ursae Majoris
(W UMa) Type Eclipsing Binary Stars KID 11405559
and V342 Boo
Tatsuya Akiba et al. 186
The Project PHaEDRA Collection: An Anchor to Connect
Modern Science with Historical Data (Abstract)
Peggy Wargelin 276
Solar System Objects and the AAVSO Photometric All-Sky
Survey (APASS) (Abstract)
Stephen Levine et al. 132
Stepping Stones to TFOP: Experience of the Saint Mary's
College Geissberger Observatory (Abstract)
Ariana Hofelmann and Brian Hill 129

DATA REDUCTION

Contributions of the AAVSO to Year 1 of TESS (Abstract)
Dennis Conti and Stella Kafka 275
The Long-term Period Changes and Evolution of V1,
a W Virginis Star in the Globular Cluster M12
Pradip Karmakar, Horace A. Smith, and
Nathan De Lee 167
Period Analysis of All-Sky Automated Survey for
Supernovae (ASAS-SN) Data on Pulsating Red Giants
John R. Percy and Lucas Fenaux 202
Photometric Accuracy of CMOS and CCD Cameras
(Preliminary Results) (Abstract)
Kenneth T. Menzies 276
A Story of Spectroscopy: How Stellar Composition is Made
Clear through Deceptively Simple Tools (Abstract)
Julien da Silva 281
Studies of Pulsating Red Giants using AAVSO and ASAS-SN
Data (Abstract)
John R. Percy and Lucas Fenaux 277

DATABASES [See CATALOGUES]

δ SCUTI STARS
[See also VARIABLE STARS (GENERAL)]

CCD Photometry, Light Curve Deconvolution, Period
Analysis, Kinematics, and Evolutionary Status of the HADS
Variable V460 Andromedae
 Kevin B. Alton and Kazimierz Stępień 53

CCD Photometry, Period Analysis, and Evolutionary Status
of the Pulsating Variable V544 Andromedae
 Kevin B. Alton 231

Discoveries for δ Scuti Variable Stars in the NASA
Kepler 2 Mission (Abstract)
 Joyce A. Guzik, Jorge Garcia, and Jason Jackiewicz 129

Erratum: KAO-EGYPT J064512.06+341749.2 is a
Low Amplitude and Multi-Periodic δ Scuti Variable Star
 Ahmed Essam, Mohamed Abdel-Sabour, and
 Gamal Bakr Ali 283

KAO-EGYPT J064512.06+341749.2 is a Low Amplitude
and Multi-Periodic δ Scuti Variable Star
 Ahmed Essam, Mohamed Abdel-Sabour, and
 Gamal Bakr Ali 66

A New Candidate delta Scuti Star in Centaurus:
HD 121191
 Roy A. Axelsen 173

Radial Velocities for Four δ Sct Variable Stars
 Elizabeth J. Jeffery et al. 111

Recent Maxima of 85 Short Period Pulsating Stars
 Gerard Samolyk 103

ECLIPSING BINARIES
[See also VARIABLE STARS (GENERAL)]

Apsidal Motion Analysis of the Eccentric Eclipsing Binary
V1103 Cassiopeiae (Abstract)
 Gary Billings 130

Automated Data Reduction at a Small College
Observatory
 Donald A. Smith and Hollis B. Akins 248

CCD Photometry, Light Curve Modeling, and Period Study
of the Overcontact Binary Systems V647 Virginis and
V948 Monocerotis
 Kevin B. Alton 7

The Contribution of A. W. Roberts' Observations
to the AAVSO International Database
 Tim Cooper 254

Discovery and Period Analysis of Seven Variable Stars
 Tom Polakis 117

Erratum: Recent Minima of 266 Eclipsing Binary Stars
 Gerard Samolyk 134

Medium-term Variation in Times of Minimum of Algol-type
Binaries: XZ And, RZ Cas, U Cep, TW Dra, U Sge
 Geoff B. Chaplin 222

Multi-filter Photometric Analysis of W Ursae Majoris
(W UMa) Type Eclipsing Binary Stars KID 11405559 and
V342 Boo
 Tatsuya Akiba et al. 186

Observations and Analysis of a Detached Eclipsing
Binary, V385 Camelopardalis
 Ronald G. Samec, Daniel B. Caton, and
 Danny R. Faulkner 176

Observations and Preliminary Modeling of the Light
Curves of Eclipsing Binary Systems NSVS 7322420
and NSVS 5726288
 Mathew F. Knote, Ronald H. Kaitchuck, and
 Robert C. Berrington 194

Photometric Analysis of Two Contact Binary Systems:
USNO-A2.0 1200-16843637 and V1094 Cassiopeiae
 Surjit S. Wadhwa 40

Photometric Analysis of V633 Virginis
 Surjit S. Wadhwa 138

A Photometric Study of the Contact Binary V384 Serpentis
 Edward J. Michaels, Chlöe M. Lanning, and
 Skyler N. Self 43

A Photometric Study of Five Low Mass Contact Binaries
 Edward J. Michaels 209

Recent Minima of 200 Eclipsing Binary Stars
 Gerard Samolyk 265

Recent Minima of 242 Eclipsing Binary Stars
 Gerard Samolyk 106

Recent Minima of 267 Recently Discovered Eclipsing
Variable Stars
 Gerard Samolyk and Vance Petriew 270

The Southern Solar-type, Totally Eclipsing Binary
PY Aquarii
 Ronald G. Samec, Heather A. Chamberlain, and
 Walter Van Hamme 29

VR_cI_c Photometric Study of the Totally Eclipsing Pre-
W UMa Binary, V616 Camelopardalis: Is it Detached?
 Ronald G. Samec et al. 157

EDITORIAL
Citizen Science
 John R. Percy 1

Life, Variable Stars, and Everything
 Nancy D. Morrison 137

So Long, and Thanks for All the Manuscripts
 John R. Percy 135

EDUCATION
Citizen Science
 John R. Percy 1

EDUCATION, VARIABLE STARS IN
Automated Data Reduction at a Small College Observatory
 Donald A. Smith and Hollis B. Akins 248

Bringing Astronomy Alive—Student Engagement at John
Abbott College (Abstract)
 Karim Jaffer 281

Camera Characterization and First Observation after Upgrade
of Feder Observatory (Abstract)
 Isobel Snellenberger and Adam Kline 132

Citizen Science
 John R. Percy 1

The Eye is Mightier than the Algorithm: Lessons Learned
from the Erroneous Classification of ASAS Variables as
"Miscellaneous" (Abstract)
 Kristine Larsen 275

Flare Stars: A Short Review (Abstract)

    Krstinja Dzombeta and John R. Percy   282
Multi-filter Photometric Analysis of W Ursae Majoris
  (W UMa) Type Eclipsing Binary Stars KID 11405559 and
  V342 Boo
    Tatsuya Akiba et al.   186
Observations of the Suspected Variable Star Ross 114
  (NSV 13523)
    Sriram Gollapudy and Wayne Osborn   3
The Past, Present and Future of Youth Outreach at the RASC
  (Abstract)
    Jenna Hinds   281
Period Analysis of All-Sky Automated Survey for
  Supernovae (ASAS-SN) Data on Pulsating Red Giants
    John R. Percy and Lucas Fenaux   202
The Project PHaEDRA Collection: An Anchor to Connect
  Modern Science with Historical Data (Abstract)
    Peggy Wargelin   276
Public Outreach Through the Cronyn Observatory
  (Abstract)
    Jan Cami   279
A Robotic Observatory System that Anyone Can Use for
  Education, Astro-Imaging, Research, or Fun! (Abstract)
    David Lane   275
A Simple Objective Prism Spectrograph for Educational
  Purposes (Abstract)
    John Thompson   277
Sky Brightness Measurements and Ways to Mitigate
  Light Pollution in Kirksville, Missouri
    Vayujeet Gokhale et al.   241
Small Observatory Operations: 2018 Highlights from the
  West Mountain Observatory (Abstract)
    Michael Joner   129
Stepping Stones to TFOP: Experience of the Saint Mary's
  College Geissberger Observatory (Abstract)
    Ariana Hofelmann and Brian Hill   129
A Story of Spectroscopy: How Stellar Composition is Made
  Clear through Deceptively Simple Tools (Abstract)
    Julien da Silva   281
Studies of Pulsating Red Giants using AAVSO and ASAS-SN
  Data (Abstract)
    John R. Percy and Lucas Fenaux   277
A Study of RR Lyrae Variable Stars in a Professor-Student
  Collaboration using the American Public University System
  (APUS) Observatory (Abstract)
    David Syndergaard and Melanie Crowson   277
Variable Stars and Cultural Astronomy (Abstract)
    Kristine Larsen   130
Variable Stars in the College Classroom (Abstract)
    Kristine Larsen   277

EQUIPMENT [See INSTRUMENTATION]

ERRATA
Erratum: KAO-EGYPT J064512.06+341749.2 is a
  Low Amplitude and Multi-Periodic δ Scuti Variable Star
    Ahmed Essam, Mohamed Abdel-Sabour, and
    Gamal Bakr Ali   283
Erratum: Recent Minima of 266 Eclipsing Binary Stars
    Gerard Samolyk   134

EVOLUTION, STELLAR
CCD Photometry, Period Analysis, and Evolutionary Status
  of the Pulsating Variable V544 Andromedae
    Kevin B. Alton   231
Variable Stars in the College Classroom (Abstract)
    Kristine Larsen   277

EXTRASOLAR PLANETS
[See PLANETS, EXTRASOLAR]

FLARE STARS
[See also VARIABLE STARS (GENERAL)]
Flare Stars: A Short Review (Abstract)
    Krstinja Dzombeta and John R. Percy   282
High Occurrence Optical Spikes and Quasi-Periodic
  Pulses (QPPs) on X-ray Star 47 Cassiopeiae
    Gary Vander Haagen   141

FLARES, EXTRASOLAR
High Occurrence Optical Spikes and Quasi-Periodic
  Pulses (QPPs) on X-ray Star 47 Cassiopeiae
    Gary Vander Haagen   141

GIANTS, RED
Long-Term Changes in the Variability of Pulsating Red
  Giants (and One RCB Star)
    John R. Percy and Arthur Lei Qiu   76
Period Analysis of All-Sky Automated Survey for
  Supernovae (ASAS-SN) Data on Pulsating Red Giants
    John R. Percy and Lucas Fenaux   202
Studies of Pulsating Red Giants using AAVSO and ASAS-SN
  Data (Abstract)
    John R. Percy and Lucas Fenaux   277

GRAVITY
Conducting the Einstein Gravitational Deflection Experiment
  (Abstract)
    Richard L. Berry   133

INDEX, INDICES
Index to Volume 47
    Anon.   284

INSTRUMENTATION
[See also CCD; VARIABLE STAR OBSERVING]
Automated Data Reduction at a Small College Observatory
    Donald A. Smith and Hollis B. Akins   248
Bright Star Monitor Network (Abstract)
    Michael Nicholas   132
CCD Photometry, Light Curve Modeling, and Period Study
  of the Overcontact Binary Systems V647 Virginis and
  V948 Monocerotis
    Kevin B. Alton   7

Camera Characterization and First Observation after Upgrade
of Feder Observatory (Abstract)
   Isobel Snellenberger and Adam Kline   132
Canada's Role in the Maunakea Spectroscopic Explorer
(Abstract)
   Alan McConnachie   279
Cold War Spy in the Sky now Provides an Eye on the Cosmos
(Abstract)
   Ken Steiner   130
Conducting the Einstein Gravitational Deflection Experiment
(Abstract)
   Richard L. Berry   133
DASCH Project Update (Abstract)
   Edward Los   275
Five Years of CCD photometry (Abstract)
   Damien Lemay   278
High Occurrence Optical Spikes and Quasi-Periodic
Pulses (QPPs) on X-ray Star 47 Cassiopeiae
   Gary Vander Haagen   141
Is sCMOS Really sCMAS? (Abstract)
   Gary Walker   131
Low Resolution Spectroscopy of Miras—X Octantis
   Bill Rea   70
Measuring and Correcting Non-linearities in ABG Cameras
(Abstract)
   Richard Wagner   277
Period Analysis of All-Sky Automated Survey for
Supernovae (ASAS-SN) Data on Pulsating Red Giants
   John R. Percy and Lucas Fenaux   202
Photometric Accuracy of CMOS and CCD Cameras
(Preliminary Results) (Abstract)
   Kenneth T. Menzies   276
RASC Robotic Telescope Project (Abstract)
   Paul Mortfield   281
A Robotic Observatory System that Anyone Can Use for
Education, Astro-Imaging, Research, or Fun! (Abstract)
   David Lane   275
A Simple Objective Prism Spectrograph for Educational
Purposes (Abstract)
   John Thompson   277
Sky Brightness Measurements and Ways to Mitigate
Light Pollution in Kirksville, Missouri
   Vayujeet Gokhale et al.   241
Sky Brightness at Zenith During the January 2019 Total
Lunar Eclipse
   Jennifer J. Birriel and J. Kevin Adkins   94
Small Observatory Operations: 2018 Highlights from the
West Mountain Observatory (Abstract)
   Michael Joner   129
Stepping Stones to TFOP: Experience of the Saint Mary's
College Geissberger Observatory (Abstract)
   Ariana Hofelmann and Brian Hill   129
Studies of Pulsating Red Giants using AAVSO and ASAS-SN
Data (Abstract)
   John R. Percy and Lucas Fenaux   277
Teaching Neil to Land—Simulators Used in the Apollo
Program (Abstract)
   Ron MacNaughton   282

IRREGULAR VARIABLES
[See also VARIABLE STARS (GENERAL)]
The Contribution of A. W. Roberts' Observations
to the AAVSO International Database
   Tim Cooper   254
Period Analysis of All-Sky Automated Survey for
Supernovae (ASAS-SN) Data on Pulsating Red Giants
   John R. Percy and Lucas Fenaux   202
Studies of Pulsating Red Giants using AAVSO and ASAS-SN
Data (Abstract)
   John R. Percy and Lucas Fenaux   277

LIGHT POLLUTION
Sky Brightness Measurements and Ways to Mitigate
Light Pollution in Kirksville, Missouri
   Vayujeet Gokhale et al.   241

LONG-PERIOD VARIABLES
[See MIRA VARIABLES; SEMIREGULAR VARIABLES]

LUNAR
First Man—The Life of Neil Armstrong (Abstract)
   James R. Hansen   282
Sky Brightness at Zenith During the January 2019 Total
Lunar Eclipse
   Jennifer J. Birriel and J. Kevin Adkins   94
Teaching Neil to Land—Simulators Used in the Apollo
Program (Abstract)
   Ron MacNaughton   282
Things You May Not Know about the Apollo 11 Mission
(Abstract)
   Randy Attwood   281

MINOR PLANETS [See ASTEROIDS]

MIRA VARIABLES
[See also VARIABLE STARS (GENERAL)]
β Cepheid and Mira Variable Stars: A Spectral Analysis
(Abstract)
   Jesse D'Shawn Harris and Lucian Undreiu   131
The Contribution of A. W. Roberts' Observations
to the AAVSO International Database
   Tim Cooper   254
Low Resolution Spectroscopy of Miras—X Octantis
   Bill Rea   70
Observations of the Suspected Variable Star Ross 114
(NSV 13523)
   Sriram Gollapudy and Wayne Osborn   3

MODELS, STELLAR
BVR_cI_c Photometric Observations and Analyses of the
Totally Eclipsing, Solar Type Binary, OR Leonis
   Ronald G. Samec, Daniel B. Caton, and
   Danny R. Faulkner   147
CCD Photometry, Light Curve Deconvolution, Period
Analysis, Kinematics, and Evolutionary Status of the HADS
Variable V460 Andromedae
   Kevin B. Alton and Kazimierz Stępień   53

CCD Photometry, Light Curve Modeling, and Period Study
  of the Overcontact Binary Systems V647 Virginis and
  V948 Monocerotis
    Kevin B. Alton   7
Observations and Analysis of a Detached Eclipsing
  Binary, V385 Camelopardalis
    Ronald G. Samec, Daniel B. Caton, and
      Danny R. Faulkner   176
Observations and Preliminary Modeling of the Light
  Curves of Eclipsing Binary Systems NSVS 7322420
  and NSVS 5726288
    Mathew F. Knote, Ronald H. Kaitchuck, and
      Robert C. Berrington   194
Photometric Analysis of V633 Virginis
    Surjit S. Wadhwa   138
A Photometric Study of the Contact Binary V384 Serpentis
    Edward J. Michaels, Chlöe M. Lanning, and
      Skyler N. Self   43
A Photometric Study of Five Low Mass Contact Binaries
    Edward J. Michaels   209
The Southern Solar-type, Totally Eclipsing Binary
  PY Aquarii
    Ronald G. Samec, Heather A. Chamberlain, and
      Walter Van Hamme   29
VR_cI_c Photometric Study of the Totally Eclipsing Pre-
  W UMa Binary, V616 Camelopardalis: Is it Detached?
    Ronald G. Samec et al.   157

MULTI-SITE OBSERVATIONS
[See COORDINATED OBSERVATIONS]

MULTIPLE STAR SYSTEMS
The Eclipsing Triple Star System b Persei (HD 26961)
  (Abstract)
    Donald F. Collins   276

NETWORKS, COMMUNICATION
Bright Star Monitor Network (Abstract)
    Michael Nicholas   132
Citizen Science
    John R. Percy   1

OBSERVATORIES
Automated Data Reduction at a Small College Observatory
    Donald A. Smith and Hollis B. Akins   248
Bringing Astronomy Alive—Student Engagement at John
  Abbott College (Abstract)
    Karim Jaffer   281
Camera Characterization and First Observation after Upgrade
  of Feder Observatory (Abstract)
    Isobel Snellenberger and Adam Kline   132
Public Outreach Through the Cronyn Observatory
  (Abstract)
    Jan Cami   279
A Robotic Observatory System that Anyone Can Use for
  Education, Astro-Imaging, Research, or Fun! (Abstract)
    David Lane   275
Science of the Stars: The National Heritage of the Dominion
  Observatory (Abstract)
    Sharon Odell   281
Small Observatory Operations: 2018 Highlights from the
  West Mountain Observatory (Abstract)
    Michael Joner   129
Some Memories of the David Dunlap Observatory
  (Abstract)
    Donald C. Morton   279
A Story of Spectroscopy: How Stellar Composition is Made
  Clear through Deceptively Simple Tools (Abstract)
    Julien da Silva   281
A Study of RR Lyrae Variable Stars in a Professor-Student
  Collaboration using the American Public University System
  (APUS) Observatory (Abstract)
    David Syndergaard and Melanie Crowson   277

PERIOD ANALYSIS; PERIOD CHANGES
Apsidal Motion Analysis of the Eccentric Eclipsing Binary
  V1103 Cassiopeiae (Abstract)
    Gary Billings   130
BVR_cI_c Photometric Observations and Analyses of the
  Totally Eclipsing, Solar Type Binary, OR Leonis
    Ronald G. Samec, Daniel B. Caton, and
      Danny R. Faulkner   147
CCD Photometry, Light Curve Deconvolution, Period
  Analysis, Kinematics, and Evolutionary Status of the HADS
  Variable V460 Andromedae
    Kevin B. Alton and Kazimierz Stępień   53
CCD Photometry, Light Curve Modeling, and Period Study
  of the Overcontact Binary Systems V647 Virginis and V948
  Monocerotis
    Kevin B. Alton   7
CCD Photometry, Period Analysis, and Evolutionary Status
  of the Pulsating Variable V544 Andromedae
    Kevin B. Alton   231
Discovery and Period Analysis of Seven Variable Stars
    Tom Polakis   117
The Eclipsing Triple Star System b Persei (HD 26961)
  (Abstract)
    Donald F. Collins   276
Erratum: KAO-EGYPT J064512.06+341749.2 is a
  Low Amplitude and Multi-Periodic δ Scuti Variable Star
    Ahmed Essam, Mohamed Abdel-Sabour, and
      Gamal Bakr Ali   283
Erratum: Recent Minima of 266 Eclipsing Binary Stars
    Gerard Samolyk   134
The Faint Cataclysmic Variable Star V677 Andromedae
  (Abstract)
    Lewis M. Cook et al.   131
High Occurrence Optical Spikes and Quasi-Periodic
  Pulses (QPPs) on X-ray Star 47 Cassiopeiae
    Gary Vander Haagen   141
KAO-EGYPT J064512.06+341749.2 is a Low Amplitude
  and Multi-Periodic δ Scuti Variable Star
    Ahmed Essam, Mohamed Abdel-Sabour, and
      Gamal Bakr Ali   66

Long-Term Changes in the Variability of Pulsating Red
  Giants (and One RCB Star)
    John R. Percy and Arthur Lei Qiu   76
The Long-term Period Changes and Evolution of V1,
  a W Virginis Star in the Globular Cluster M12
    Pradip Karmakar, Horace A. Smith, and
      Nathan De Lee   167
Medium-term Variation in Times of Minimum of Algol-type
  Binaries: XZ And, RZ Cas, U Cep, TW Dra, U Sge
    Geoff B. Chaplin   222
Multi-color Photometry of the Hot R Coronae Borealis Star
  and Proto-planetary Nebula V348 Sagittarii
    Arlo U. Landolt and James L. Clem   83
A New Candidate δ Scuti Star in Centaurus: HD 121191
    Roy A. Axelsen   173
New Intense Multiband Photometric Observations of the Hot
  Carbon Star V348 Sagittarii (Abstract)
    Franz-Josef Hambsch and Christopher S. Jeffery 132
Observations and Analysis of a Detached Eclipsing Binary,
  V385 Camelopardalis
    Ronald G. Samec, Daniel B. Caton, and
      Danny R. Faulkner   176
Observations and Preliminary Modeling of the Light
  Curves of Eclipsing Binary Systems NSVS 7322420
  and NSVS 5726288
    Mathew F. Knote, Ronald H. Kaitchuck, and
      Robert C. Berrington   194
Observations of the Suspected Variable Star Ross 114
  (NSV 13523)
    Sriram Gollapudy and Wayne Osborn   3
Photometric Analysis of Two Contact Binary Systems:
  USNO-A2.0 1200-16843637 and V1094 Cassiopeiae
    Surjit S. Wadhwa   40
Photometric Analysis of V633 Virginis
    Surjit S. Wadhwa   138
A Photometric Study of the Contact Binary V384 Serpentis
    Edward J. Michaels, Chlöe M. Lanning, and
      Skyler N. Self   43
A Photometric Study of Five Low Mass Contact Binaries
    Edward J. Michaels   209
Radial Velocities for Four δ Sct Variable Stars
    Elizabeth J. Jeffery et al.   111
Recent Maxima of 85 Short Period Pulsating Stars
    Gerard Samolyk   103
Recent Minima of 200 Eclipsing Binary Stars
    Gerard Samolyk   265
Recent Minima of 242 Eclipsing Binary Stars
    Gerard Samolyk   106
Recent Minima of 267 Recently Discovered Eclipsing
  Variable Stars
    Gerard Samolyk and Vance Petriew   270
The Southern Solar-type, Totally Eclipsing Binary
  PY Aquarii
    Ronald G. Samec, Heather A. Chamberlain, and
      Walter Van Hamme   29
Visual Times of Maxima for Short Period Pulsating Stars V
    Gerard Samolyk   98
VR_cI_c Photometric Study of the Totally Eclipsing Pre-
  W UMa Binary, V616 Camelopardalis: Is it Detached?
    Ronald G. Samec et al.   157

**PHOTOELECTRIC PHOTOMETRY
[See PHOTOMETRY, PHOTOELECTRIC]**

PHOTOMETRY, CCD
Apsidal Motion Analysis of the Eccentric Eclipsing Binary
  V1103 Cassiopeiae (Abstract)
    Gary Billings   130
Automated Data Reduction at a Small College Observatory
    Donald A. Smith and Hollis B. Akins   248
Bright Star Monitor Network (Abstract)
    Michael Nicholas   132
BVR_cI_c Photometric Observations and Analyses of the
  Totally Eclipsing, Solar Type Binary, OR Leonis
    Ronald G. Samec, Daniel B. Caton, and
      Danny R. Faulkner   147
CCD Photometry, Light Curve Deconvolution, Period
  Analysis, Kinematics, and Evolutionary Status of the HADS
  Variable V460 Andromedae
    Kevin B. Alton and Kazimierz Stępień   53
CCD Photometry, Light Curve Modeling, and Period Study
  of the Overcontact Binary Systems V647 Virginis and
  V948 Monocerotis
    Kevin B. Alton   7
CCD Photometry, Period Analysis, and Evolutionary Status
  of the Pulsating Variable V544 Andromedae
    Kevin B. Alton   231
Discovery and Period Analysis of Seven Variable Stars
    Tom Polakis   117
Erratum: KAO-EGYPT J064512.06+341749.2 is a
  Low Amplitude and Multi-Periodic δ Scuti Variable Star
    Ahmed Essam, Mohamed Abdel-Sabour, and
      Gamal Bakr Ali   283
Erratum: Recent Minima of 266 Eclipsing Binary Stars
    Gerard Samolyk   134
The Faint Cataclysmic Variable Star V677 Andromedae
  (Abstract)
    Lewis M. Cook et al.   131
Five Years of CCD photometry (Abstract)
    Damien Lemay   278
High Occurrence Optical Spikes and Quasi-Periodic
  Pulses (QPPs) on X-ray Star 47 Cassiopeiae
    Gary Vander Haagen   141
How-to Hour—Planning an Observing Program (Abstract)
    Kenneth J. Menzies, Michael J. Cook, and Franz-Josef
      (Josch) Hambsch   278
Is sCMOS Really sCMAS? (Abstract)
    Gary Walker   131
KAO-EGYPT J064512.06+341749.2 is a Low Amplitude
  and Multi-Periodic δ Scuti Variable Star
    Ahmed Essam, Mohamed Abdel-Sabour, and
      Gamal Bakr Ali   66

The Long-term Period Changes and Evolution of V1,
  a W Virginis Star in the Globular Cluster M12
    Pradip Karmakar, Horace A. Smith, and
      Nathan De Lee   167
Low Resolution Spectroscopy of Miras—X Octantis
    Bill Rea   70
Measuring and Correcting Non-linearities in ABG Cameras
  (Abstract)
    Richard Wagner   277
Medium-term Variation in Times of Minimum of Algol-type
  Binaries: XZ And, RZ Cas, U Cep, TW Dra, U Sge
    Geoff B. Chaplin   222
Multi-color Photometry of the Hot R Coronae Borealis Star
  and Proto-planetary Nebula V348 Sagittarii
    Arlo U. Landolt and James L. Clem   83
Multi-filter Photometric Analysis of W Ursae Majoris
  (W UMa) Type Eclipsing Binary Stars KID 11405559 and
  V342 Boo
    Tatsuya Akiba et al.   186
New Intense Multiband Photometric Observations of the Hot
  Carbon Star V348 Sagittarii (Abstract)
    Franz-Josef Hambsch and Christopher S. Jeffery   132
Observations and Analysis of a Detached Eclipsing Binary,
  V385 Camelopardalis
    Ronald G. Samec, Daniel B. Caton, and
      Danny R. Faulkner   176
Observations and Preliminary Modeling of the Light
  Curves of Eclipsing Binary Systems NSVS 7322420
  and NSVS 5726288
    Mathew F. Knote, Ronald H. Kaitchuck, and
      Robert C. Berrington   194
Observations of the Suspected Variable Star Ross 114
  (NSV 13523)
    Sriram Gollapudy and Wayne Osborn   3
Photometric Accuracy of CMOS and CCD Cameras
  (Preliminary Results) (Abstract)
    Kenneth T. Menzies   276
Photometric Analysis of Two Contact Binary Systems:
  USNO-A2.0 1200-16843637 and V1094 Cassiopeiae
    Surjit S. Wadhwa   40
Photometric Analysis of V633 Virginis
    Surjit S. Wadhwa   138
A Photometric Study of the Contact Binary V384 Serpentis
    Edward J. Michaels, Chlöe M. Lanning, and
      Skyler N. Self   43
A Photometric Study of Five Low Mass Contact Binaries
    Edward J. Michaels   209
Radial Velocities for Four δ Sct Variable Stars
    Elizabeth J. Jeffery et al.   111
Recent Maxima of 85 Short Period Pulsating Stars
    Gerard Samolyk   103
Recent Minima of 200 Eclipsing Binary Stars
    Gerard Samolyk   265
Recent Minima of 242 Eclipsing Binary Stars
    Gerard Samolyk   106
Recent Minima of 267 Recently Discovered Eclipsing
  Variable Stars
    Gerard Samolyk and Vance Petriew   270

The Southern Solar-type, Totally Eclipsing Binary
  PY Aquarii
    Ronald G. Samec, Heather A. Chamberlain, and
      Walter Van Hamme   29
A Study of RR Lyrae Variable Stars in a Professor-Student
  Collaboration using the American Public University System
  (APUS) Observatory (Abstract)
    David Syndergaard and Melanie Crowson   277
VR_cI_c Photometric Study of the Totally Eclipsing Pre-
  W UMa Binary, V616 Camelopardalis: Is it Detached?
    Ronald G. Samec et al.   157

PHOTOMETRY, CMOS

Is sCMOS Really sCMAS? (Abstract)
    Gary Walker   131
Photometric Accuracy of CMOS and CCD Cameras
  (Preliminary Results) (Abstract)
    Kenneth T. Menzies   276

PHOTOMETRY, DSLR

A New Candidate δ Scuti Star in Centaurus: HD 121191
    Roy A. Axelsen   173

PHOTOMETRY, PHOTOELECTRIC

Medium-term Variation in Times of Minimum of Algol-type
  Binaries: XZ And, RZ Cas, U Cep, TW Dra, U Sge
    Geoff B. Chaplin   222
Multi-color Photometry of the Hot R Coronae Borealis Star
  and Proto-planetary Nebula V348 Sagittarii
    Arlo U. Landolt and James L. Clem   83

PHOTOMETRY, PHOTOGRAPHIC

DASCH Project Update (Abstract)
    Edward Los   275
Medium-term Variation in Times of Minimum of Algol-type
  Binaries: XZ And, RZ Cas, U Cep, TW Dra, U Sge
    Geoff B. Chaplin   222
Observations of the Suspected Variable Star Ross 114
  (NSV 13523)
    Sriram Gollapudy and Wayne Osborn   3

PHOTOMETRY, VISUAL

The Contribution of A. W. Roberts' Observations
  to the AAVSO International Database
    Tim Cooper   254
How-to Hour—Planning an Observing Program (Abstract)
    Kenneth J. Menzies, Michael J. Cook, and Franz-Josef
      (Josch) Hambsch   278
Medium-term Variation in Times of Minimum of Algol-type
  Binaries: XZ And, RZ Cas, U Cep, TW Dra, U Sge
    Geoff B. Chaplin   222
Visual Times of Maxima for Short Period Pulsating Stars V
    Gerard Samolyk   98

PLANETARY NEBULAE

Multi-color Photometry of the Hot R Coronae Borealis Star
  and Proto-planetary Nebula V348 Sagittarii
    Arlo U. Landolt and James L. Clem   83

New Intense Multiband Photometric Observations of the Hot
  Carbon Star V348 Sagittarii (Abstract)
    Franz-Josef Hambsch and Christopher S. Jeffery   132

PLANETS
The Rotating Frame (Abstract)
    Christa Van Laerhoven   278

PLANETS, EXTRASOLAR (EXOPLANETS)
Camera Characterization and First Observation after Upgrade
  of Feder Observatory (Abstract)
    Isobel Snellenberger and Adam Kline   132
Contributions of the AAVSO to Year 1 of TESS (Abstract)
    Dennis Conti and Stella Kafka   275
Stepping Stones to TFOP: Experience of the Saint Mary's
  College Geissberger Observatory (Abstract)
    Ariana Hofelmann and Brian Hill   129

POETRY, THEATER, DANCE, SOCIETY
Canada's Role in the James Webb Space Telescope
  (Abstract)
    Nathalie Ouellette   279
Canada's Role in the Maunakea Spectroscopic Explorer
  (Abstract)
    Alan McConnachie   279
Citizen Science
    John R. Percy   1
Dorner Telescope Museum (Abstract)
    Rudolph Dorner and Randall Rosenfeld   280
First Man—The Life of Neil Armstrong (Abstract)
    James R. Hansen   282
The Importance of Religion and Art in Sowing the Seeds
  of Success of the U.S. Space Program in the 1940s and
  1950s (Abstract)
    Catherine L. Newell   279
On the Construction of the Heav'ns: A Multi-media,
  Interdisciplinary Concert and Outreach Event (Abstract)
    James E. Hesser, David Lee, and W. John McDonald   280
The Past, Present and Future of Youth Outreach at the RASC
  (Abstract)
    Jenna Hinds   281
Public Outreach Through the Cronyn Observatory
  (Abstract)
    Jan Cami   279
RASC Robotic Telescope Project (Abstract)
    Paul Mortfield   281
Sky Brightness Measurements and Ways to Mitigate
  Light Pollution in Kirksville, Missouri
    Vayujeet Gokhale et al.   241
A Story of Spectroscopy: How Stellar Composition is Made
  Clear through Deceptively Simple Tools (Abstract)
    Julien da Silva   281
Things You May Not Know about the Apollo 11 Mission
  (Abstract)
    Randy Attwood   281
Toronto's Astronomical Heritage (Abstract)
    John R. Percy   280

Variable Stars and Cultural Astronomy (Abstract)
    Kristine Larsen   130
Who doth not see the measures of the Moon/ Which thirteen
  times she danceth every year?/ And ends her pavan thirteen
  times as soon (Abstract)
    David H. Levy and R. A. Rosenfeld   280
The Zooniverse (Abstract)
    L. Clifton Johnson   278

PROFESSIONAL-AMATEUR COLLABORATION
[See ASTRONOMERS, AMATEUR]

PULSATING VARIABLES
Long-Term Changes in the Variability of Pulsating Red
  Giants (and One RCB Star)
    John R. Percy and Arthur Lei Qiu   76
Period Analysis of All-Sky Automated Survey for
  Supernovae (ASAS-SN) Data on Pulsating Red Giants
    John R. Percy and Lucas Fenaux   202
Radial Velocities for Four δ Sct Variable Stars
    Elizabeth J. Jeffery et al.   111
Studies of Pulsating Red Giants using AAVSO and ASAS-SN
  Data (Abstract)
    John R. Percy and Lucas Fenaux   277
Visual Times of Maxima for Short Period Pulsating Stars V
    Gerard Samolyk   98

R CORONAE BOREALIS VARIABLES
[See also VARIABLE STARS (GENERAL)]
The Contribution of A. W. Roberts' Observations
  to the AAVSO International Database
    Tim Cooper   254
Long-Term Changes in the Variability of Pulsating Red
  Giants (and One RCB Star)
    John R. Percy and Arthur Lei Qiu   76
Multi-color Photometry of the Hot R Coronae Borealis Star
  and Proto-planetary Nebula V348 Sagittarii
    Arlo U. Landolt and James L. Clem   83
New Intense Multiband Photometric Observations of the Hot
  Carbon Star V348 Sagittarii (Abstract)
    Franz-Josef Hambsch and Christopher S. Jeffery   132

RADIO ASTRONOMY; RADIO OBSERVATIONS
Cold War Spy in the Sky now Provides an Eye on the Cosmos
  (Abstract)
    Ken Steiner   130

RED VARIABLES
[See IRREGULAR, MIRA, SEMIREGULAR VARIABLES]

REMOTE OBSERVING
Bright Star Monitor Network (Abstract)
    Michael Nicholas   132

REVIEW ARTICLE
Flare Stars: A Short Review (Abstract)
    Krstinja Dzombeta and John R. Percy   282

RR LYRAE STARS
[See also VARIABLE STARS (GENERAL)]

Automated Data Reduction at a Small College Observatory
  Donald A. Smith and Hollis B. Akins — 248

The Contribution of A. W. Roberts' Observations to the AAVSO International Database
  Tim Cooper — 254

Recent Maxima of 85 Short Period Pulsating Stars
  Gerard Samolyk — 103

A Study of RR Lyrae Variable Stars in a Professor-Student Collaboration using the American Public University System (APUS) Observatory (Abstract)
  David Syndergaard and Melanie Crowson — 277

Visual Times of Maxima for Short Period Pulsating Stars V
  Gerard Samolyk — 98

RS CVN STARS [See ECLIPSING BINARIES; see also VARIABLE STARS (GENERAL)]

RV TAURI STARS
[See also VARIABLE STARS (GENERAL)]

High Occurrence Optical Spikes and Quasi-Periodic Pulses (QPPs) on X-ray Star 47 Cassiopeiae
  Gary Vander Haagen — 141

SATELLITE OBSERVATIONS

Camera Characterization and First Observation after Upgrade of Feder Observatory (Abstract)
  Isobel Snellenberger and Adam Kline — 132

Comparison of North-South Hemisphere Data from AAVSO Visual Observers and the SDO Satellite Computer-Generated Wolf Numbers (Abstract)
  Rodney Howe — 129

Contributions of the AAVSO to Year 1 of TESS (Abstract)
  Dennis Conti and Stella Kafka — 275

Discoveries for δ Scuti Variable Stars in the NASA Kepler 2 Mission (Abstract)
  Joyce A. Guzik, Jorge Garcia, and Jason Jackiewicz — 129

Flare Stars: A Short Review (Abstract)
  Krstinja Dzombeta and John R. Percy — 282

Multi-filter Photometric Analysis of W Ursae Majoris (W UMa) Type Eclipsing Binary Stars KID 11405559 and V342 Boo
  Tatsuya Akiba et al. — 186

SATELLITES; SATELLITE MISSIONS
[See also COORDINATED OBSERVATIONS]

The BRITE Constellation Mission (Abstract)
  Gregg Wade — 279

Canada's Role in the James Webb Space Telescope (Abstract)
  Nathalie Ouellette — 279

Contributions of the AAVSO to Year 1 of TESS (Abstract)
  Dennis Conti and Stella Kafka — 275

Discoveries for δ Scuti Variable Stars in the NASA Kepler 2 Mission (Abstract)
  Joyce A. Guzik, Jorge Garcia, and Jason Jackiewicz — 129

SCIENTIFIC WRITING, PUBLICATION OF DATA

Life, Variable Stars, and Everything
  Nancy D. Morrison — 137

SEMIREGULAR VARIABLES
[See also VARIABLE STARS (GENERAL)]

The Contribution of A. W. Roberts' Observations to the AAVSO International Database
  Tim Cooper — 254

Long-Term Changes in the Variability of Pulsating Red Giants (and One RCB Star)
  John R. Percy and Arthur Lei Qiu — 76

Observations of the Suspected Variable Star Ross 114 (NSV 13523)
  Sriram Gollapudy and Wayne Osborn — 3

Period Analysis of All-Sky Automated Survey for Supernovae (ASAS-SN) Data on Pulsating Red Giants
  John R. Percy and Lucas Fenaux — 202

SSA Analysis and Significance Tests for Periodicity in S, RS, SU, AD, BU, KK, and PR Persei
  Geoff B. Chaplin — 17

Studies of Pulsating Red Giants using AAVSO and ASAS-SN Data (Abstract)
  John R. Percy and Lucas Fenaux — 277

SEQUENCES, COMPARISON STAR [See CHARTS]

SOFTWARE [See COMPUTERS]

SOLAR

Cold War Spy in the Sky now Provides an Eye on the Cosmos (Abstract)
  Ken Steiner — 130

Comparison of North-South Hemisphere Data from AAVSO Visual Observers and the SDO Satellite Computer-Generated Wolf Numbers (Abstract)
  Rodney Howe — 129

Conducting the Einstein Gravitational Deflection Experiment (Abstract)
  Richard L. Berry — 133

Shadow Chasers 2017 (Abstract)
  David Shuman and Paul Simard — 280

SPECTRA, SPECTROSCOPY

Five Years of CCD photometry (Abstract)
  Damien Lemay — 278

The Fun of Processing a Stellar Spectrum—the Hard Way (Abstract)
  Stanley A. Gorodenski — 131

Low Resolution Spectroscopy of Miras—X Octantis
  Bill Rea — 70

Radial Velocities for Four δ Sct Variable Stars
  Elizabeth J. Jeffery et al. — 111

A Simple Objective Prism Spectrograph for Educational Purposes (Abstract)
  John Thompson — 277

A Story of Spectroscopy: How Stellar Composition is Made
  Clear through Deceptively Simple Tools (Abstract)
    Julien da Silva   281

SPECTROSCOPIC ANALYSIS

β Cepheid and Mira Variable Stars: A Spectral Analysis
  (Abstract)
    Jesse D'Shawn Harris and Lucian Undreiu   131
The Fun of Processing a Stellar Spectrum—the Hard Way
  (Abstract)
    Stanley A. Gorodenski   131
Low Resolution Spectroscopy of Miras—X Octantis
    Bill Rea   70
Radial Velocities for Four δ Sct Variable Stars
    Elizabeth J. Jeffery et al.   111

STATISTICAL ANALYSIS

Apsidal Motion Analysis of the Eccentric Eclipsing Binary
  V1103 Cassiopeiae (Abstract)
    Gary Billings   130
BVR_cI_c Photometric Observations and Analyses of the
  Totally Eclipsing, Solar Type Binary, OR Leonis
    Ronald G. Samec, Daniel B. Caton, and
      Danny R. Faulkner   147
CCD Photometry, Light Curve Deconvolution, Period
  Analysis, Kinematics, and Evolutionary Status of the HADS
  Variable V460 Andromedae
    Kevin B. Alton and Kazimierz Stępień   53
CCD Photometry, Light Curve Modeling, and Period Study
  of the Overcontact Binary Systems V647 Virginis and
  V948 Monocerotis
    Kevin B. Alton   7
CCD Photometry, Period Analysis, and Evolutionary Status
  of the Pulsating Variable V544 Andromedae
    Kevin B. Alton   231
Comparison of North-South Hemisphere Data from AAVSO
  Visual Observers and the SDO Satellite Computer-Generated
  Wolf Numbers (Abstract)
    Rodney Howe   129
Conducting the Einstein Gravitational Deflection Experiment
  (Abstract)
    Richard L. Berry   133
Discoveries for δ Scuti Variable Stars in the NASA
  Kepler 2 Mission (Abstract)
    Joyce A. Guzik, Jorge Garcia, and Jason Jackiewicz   129
The Eclipsing Triple Star System b Persei (HD 26961)
  (Abstract)
    Donald F. Collins   276
The Eye is Mightier than the Algorithm: Lessons Learned
  from the Erroneous Classification of ASAS Variables as
  "Miscellaneous" (Abstract)
    Kristine Larsen   275
High Occurrence Optical Spikes and Quasi-Periodic
  Pulses (QPPs) on X-ray Star 47 Cassiopeiae
    Gary Vander Haagen   141
Long-Term Changes in the Variability of Pulsating Red
  Giants (and One RCB Star)
    John R. Percy and Arthur Lei Qiu   76

The Long-term Period Changes and Evolution of V1,
  a W Virginis Star in the Globular Cluster M12
    Pradip Karmakar, Horace A. Smith, and
      Nathan De Lee   167
Medium-term Variation in Times of Minimum of Algol-type
  Binaries: XZ And, RZ Cas, U Cep, TW Dra, U Sge
    Geoff B. Chaplin   222
Multi-color Photometry of the Hot R Coronae Borealis Star
  and Proto-planetary Nebula V348 Sagittarii
    Arlo U. Landolt and James L. Clem   83
Multi-filter Photometric Analysis of W Ursae Majoris
  (W UMa) Type Eclipsing Binary Stars KID 11405559 and
  V342 Boo
    Tatsuya Akiba et al.   186
Observations and Analysis of a Detached Eclipsing Binary,
  V385 Camelopardalis
    Ronald G. Samec, Daniel B. Caton, and
      Danny R. Faulkner   176
Observations and Preliminary Modeling of the Light
  Curves of Eclipsing Binary Systems NSVS 7322420
  and NSVS 5726288
    Mathew F. Knote, Ronald H. Kaitchuck, and
      Robert C. Berrington   194
Period Analysis of All-Sky Automated Survey for
  Supernovae (ASAS-SN) Data on Pulsating Red Giants
    John R. Percy and Lucas Fenaux   202
Photometric Analysis of Two Contact Binary Systems:
  USNO-A2.0 1200-16843637 and V1094 Cassiopeiae
    Surjit S. Wadhwa   40
Photometric Analysis of V633 Virginis
    Surjit S. Wadhwa   138
A Photometric Study of the Contact Binary V384 Serpentis
    Edward J. Michaels, Chlöe M. Lanning, and
      Skyler N. Self   43
A Photometric Study of Five Low Mass Contact Binaries
    Edward J. Michaels   209
The Rotating Frame (Abstract)
    Christa Van Laerhoven   278
SSA Analysis and Significance Tests for Periodicity in S, RS,
  SU, AD, BU, KK, and PR Persei
    Geoff B. Chaplin   17
Sky Brightness at Zenith During the January 2019 Total
  Lunar Eclipse
    Jennifer J. Birriel and J. Kevin Adkins   94
The Southern Solar-type, Totally Eclipsing Binary
  PY Aquarii
    Ronald G. Samec, Heather A. Chamberlain, and
      Walter Van Hamme   29
Studies of Pulsating Red Giants using AAVSO and ASAS-SN
  Data (Abstract)
    John R. Percy and Lucas Fenaux   277
VR_cI_c Photometric Study of the Totally Eclipsing Pre-
  W UMa Binary, V616 Camelopardalis: Is it Detached?
    Ronald G. Samec et al.   157

SU URSAE MAJORIS STARS
[See CATACLYSMIC VARIABLES]

SUN [See SOLAR]

SUNSPOTS, SUNSPOT COUNTS
Comparison of North-South Hemisphere Data from AAVSO Visual Observers and the SDO Satellite Computer-Generated Wolf Numbers (Abstract)
Rodney Howe ... 129

SX PHOENICIS VARIABLES
[See also VARIABLE STARS (GENERAL)]
CCD Photometry, Period Analysis, and Evolutionary Status of the Pulsating Variable V544 Andromedae
Kevin B. Alton ... 231

UNKNOWN; UNSTUDIED VARIABLES
Discovery and Period Analysis of Seven Variable Stars
Tom Polakis ... 117
The Eye is Mightier than the Algorithm: Lessons Learned from the Erroneous Classification of ASAS Variables as "Miscellaneous" (Abstract)
Kristine Larsen ... 275

VARIABLE STAR OBSERVING ORGANIZATIONS
Bringing Astronomy Alive—Student Engagement at John Abbott College (Abstract)
Karim Jaffer ... 281
The Contribution of A. W. Roberts' Observations to the AAVSO International Database
Tim Cooper ... 254
Contributions of the AAVSO to Year 1 of TESS (Abstract)
Dennis Conti and Stella Kafka ... 275
Dorner Telescope Museum (Abstract)
Rudolph Dorner and Randall Rosenfeld ... 280
Life, Variable Stars, and Everything
Nancy D. Morrison ... 137
The Past, Present and Future of Youth Outreach at the RASC (Abstract)
Jenna Hinds ... 281
Public Outreach Through the Cronyn Observatory (Abstract)
Jan Cami ... 279
RASC Robotic Telescope Project (Abstract)
Paul Mortfield ... 281
Science of the Stars: The National Heritage of the Dominion Observatory (Abstract)
Sharon Odell ... 281
So Long, and Thanks for All the Manuscripts
John R. Percy ... 135

VARIABLE STAR OBSERVING
[See also INSTRUMENTATION]
Bright Star Monitor Network (Abstract)
Michael Nicholas ... 132
Camera Characterization and First Observation after Upgrade of Feder Observatory (Abstract)
Isobel Snellenberger and Adam Kline ... 132
Cold War Spy in the Sky now Provides an Eye on the Cosmos (Abstract)
Ken Steiner ... 130
Contributions of the AAVSO to Year 1 of TESS (Abstract)
Dennis Conti and Stella Kafka ... 275
Five Years of CCD photometry (Abstract)
Damien Lemay ... 278
How-to Hour—Planning an Observing Program (Abstract)
Kenneth J. Menzies, Michael J. Cook, and Franz-Josef (Josch) Hambsch ... 278
Photometric Accuracy of CMOS and CCD Cameras (Preliminary Results) (Abstract)
Kenneth T. Menzies ... 276
A Robotic Observatory System that Anyone Can Use for Education, Astro-Imaging, Research, or Fun! (Abstract)
David Lane ... 275
The Rotating Frame (Abstract)
Christa Van Laerhoven ... 278
A Simple Objective Prism Spectrograph for Educational Purposes (Abstract)
John Thompson ... 277
A Study of RR Lyrae Variable Stars in a Professor-Student Collaboration using the American Public University System (APUS) Observatory (Abstract)
David Syndergaard and Melanie Crowson ... 277

VARIABLE STARS (GENERAL)
The BRITE Constellation Mission (Abstract)
Gregg Wade ... 279
The Eye is Mightier than the Algorithm: Lessons Learned from the Erroneous Classification of ASAS Variables as "Miscellaneous" (Abstract)
Kristine Larsen ... 275
Small Observatory Operations: 2018 Highlights from the West Mountain Observatory (Abstract)
Michael Joner ... 129
Variable Stars and Cultural Astronomy (Abstract)
Kristine Larsen ... 130
Variable Stars in the College Classroom (Abstract)
Kristine Larsen ... 277

VARIABLE STARS (INDIVIDUAL); OBSERVING TARGETS
[XZ And] Medium-term Variation in Times of Minimum of Algol-type Binaries: XZ And, RZ Cas, U Cep, TW Dra, U Sge
Geoff B. Chaplin ... 222
[V460 And] CCD Photometry, Light Curve Deconvolution, Period Analysis, Kinematics, and Evolutionary Status of the HADS Variable V460 Andromedae
Kevin B. Alton and Kazimierz Stępień ... 53
[V544 And] CCD Photometry, Period Analysis, and Evolutionary Status of the Pulsating Variable V544 Andromedae
Kevin B. Alton ... 231
[V677 And] The Faint Cataclysmic Variable Star V677 Andromedae (Abstract)
Lewis M. Cook et al. ... 131

[S Aps] The Contribution of A. W. Roberts' Observations to the AAVSO International Database
    Tim Cooper   254

[PY Aqr] The Southern Solar-type, Totally Eclipsing Binary PY Aquarii
    Ronald G. Samec, Heather A. Chamberlain, and Walter Van Hamme   29

[S Ara] The Contribution of A. W. Roberts' Observations to the AAVSO International Database
    Tim Cooper   254

[V342 Boo] Multi-filter Photometric Analysis of W Ursae Majoris (W UMa) Type Eclipsing Binary Stars KID 11405559 and V342 Boo
    Tatsuya Akiba et al.   186

[V385 Cam] Observations and Analysis of a Detached Eclipsing Binary, V385 Camelopardalis
    Ronald G. Samec, Daniel B. Caton, and Danny R. Faulkner   176

[V616 Cam] VR_cI_c Photometric Study of the Totally Eclipsing Pre-W UMa Binary, V616 Camelopardalis: Is it Detached?
    Ronald G. Samec et al.   157

[U Cnc] Long-Term Changes in the Variability of Pulsating Red Giants (and One RCB Star)
    John R. Percy and Arthur Lei Qiu   76

[VZ Cnc] Radial Velocities for Four δ Sct Variable Stars
    Elizabeth J. Jeffery et al.   111

[AD CMi] Radial Velocities for Four δ Sct Variable Stars
    Elizabeth J. Jeffery et al.   111

[X Car] The Contribution of A. W. Roberts' Observations to the AAVSO International Database
    Tim Cooper   254

[R Cas] Long-Term Changes in the Variability of Pulsating Red Giants (and One RCB Star)
    John R. Percy and Arthur Lei Qiu   76

[T Cas] Long-Term Changes in the Variability of Pulsating Red Giants (and One RCB Star)
    John R. Percy and Arthur Lei Qiu   76

[Z Cas] Long-Term Changes in the Variability of Pulsating Red Giants (and One RCB Star)
    John R. Percy and Arthur Lei Qiu   76

[RZ Cas] Medium-term Variation in Times of Minimum of Algol-type Binaries: XZ And, RZ Cas, U Cep, TW Dra, U Sge
    Geoff B. Chaplin   222

[V1094 Cas] Photometric Analysis of Two Contact Binary Systems: USNO-A2.0 1200-16843637 and V1094 Cassiopeiae
    Surjit S. Wadhwa   40

[V1103 Cas] Apsidal Motion Analysis of the Eccentric Eclipsing Binary V1103 Cassiopeiae (Abstract)
    Gary Billings   130

[47 Cas] High Occurrence Optical Spikes and Quasi-Periodic Pulses (QPPs) on X-ray Star 47 Cassiopeiae
    Gary Vander Haagen   141

[R Cen] The Contribution of A. W. Roberts' Observations to the AAVSO International Database
    Tim Cooper   254

[RR Cen] The Contribution of A. W. Roberts' Observations to the AAVSO International Database
    Tim Cooper   254

[DY Cen] Multi-color Photometry of the Hot R Coronae Borealis Star and Proto-planetary Nebula V348 Sagittarii
    Arlo U. Landolt and James L. Clem   83

[U Cep] Medium-term Variation in Times of Minimum of Algol-type Binaries: XZ And, RZ Cas, U Cep, TW Dra, U Sge
    Geoff B. Chaplin   222

[U CrA] The Contribution of A. W. Roberts' Observations to the AAVSO International Database
    Tim Cooper   254

[TW Dra] Medium-term Variation in Times of Minimum of Algol-type Binaries: XZ And, RZ Cas, U Cep, TW Dra, U Sge
    Geoff B. Chaplin   222

[V338 Dra] A Photometric Study of Five Low Mass Contact Binaries
    Edward J. Michaels   209

[S Hor] The Contribution of A. W. Roberts' Observations to the AAVSO International Database
    Tim Cooper   254

[Z Hya] The Contribution of A. W. Roberts' Observations to the AAVSO International Database
    Tim Cooper   254

[KZ Hya] Radial Velocities for Four δ Sct Variable Stars
    Elizabeth J. Jeffery et al.   111

[DE Lac] Radial Velocities for Four δ Sct Variable Stars
    Elizabeth J. Jeffery et al.   111

[V457 Lac] Automated Data Reduction at a Small College Observatory
    Donald A. Smith and Hollis B. Akins   248

[RR Leo] Visual Times of Maxima for Short Period Pulsating Stars V
    Gerard Samolyk   98

[SS Leo] Visual Times of Maxima for Short Period Pulsating Stars V
    Gerard Samolyk   98

[TV Leo] Visual Times of Maxima for Short Period Pulsating Stars V
    Gerard Samolyk   98

[WW Leo] Visual Times of Maxima for Short Period Pulsating Stars V
    Gerard Samolyk   98

[OR Leo] BVR_cI_c Photometric Observations and Analyses of the Totally Eclipsing, Solar Type Binary, OR Leonis
    Ronald G. Samec, Daniel B. Caton, and Danny R. Faulkner   147

[SZ Lyn] Visual Times of Maxima for Short Period Pulsating Stars V
    Gerard Samolyk   98

[RR Lyr] Automated Data Reduction at a Small College Observatory
    Donald A. Smith and Hollis B. Akins   248

[RZ Lyr] Visual Times of Maxima for Short Period Pulsating Stars V
    Gerard Samolyk   98

[V474 Mon] Radial Velocities for Four δ Sct Variable Stars
　Elizabeth J. Jeffery et al.　　　　　　　　　111
[V948 Mon] CCD Photometry, Light Curve Modeling, and Period Study of the Overcontact Binary Systems V647 Virginis and V948 Monocerotis
　Kevin B. Alton　　　　　　　　　　　　　　7
[R Oct] Long-Term Changes in the Variability of Pulsating Red Giants (and One RCB Star)
　John R. Percy and Arthur Lei Qiu　　　　　　76
[X Oct] Low Resolution Spectroscopy of Miras— X Octantis
　Bill Rea　　　　　　　　　　　　　　　　70
[V2802 Ori] A Photometric Study of Five Low Mass Contact Binaries
　Edward J. Michaels　　　　　　　　　　　209
[κ Pav] The Contribution of A. W. Roberts' Observations to the AAVSO International Database
　Tim Cooper　　　　　　　　　　　　　　254
[AV Peg] Visual Times of Maxima for Short Period Pulsating Stars V
　Gerard Samolyk　　　　　　　　　　　　98
[S Per] SSA Analysis and Significance Tests for Periodicity in S, RS, SU, AD, BU, KK, and PR Persei
　Geoff B. Chaplin　　　　　　　　　　　　17
[RS Per] SSA Analysis and Significance Tests for Periodicity in S, RS, SU, AD, BU, KK, and PR Persei
　Geoff B. Chaplin　　　　　　　　　　　　17
[SU Per] SSA Analysis and Significance Tests for Periodicity in S, RS, SU, AD, BU, KK, and PR Persei
　Geoff B. Chaplin　　　　　　　　　　　　17
[AD Per] SSA Analysis and Significance Tests for Periodicity in S, RS, SU, AD, BU, KK, and PR Persei
　Geoff B. Chaplin　　　　　　　　　　　　17
[BU Per] SSA Analysis and Significance Tests for Periodicity in S, RS, SU, AD, BU, KK, and PR Persei
　Geoff B. Chaplin　　　　　　　　　　　　17
[KK Per] SSA Analysis and Significance Tests for Periodicity in S, RS, SU, AD, BU, KK, and PR Persei
　Geoff B. Chaplin　　　　　　　　　　　　17
[PR Per] SSA Analysis and Significance Tests for Periodicity in S, RS, SU, AD, BU, KK, and PR Persei
　Geoff B. Chaplin　　　　　　　　　　　　17
[b Per] The Eclipsing Triple Star System b Persei (HD 26961) (Abstract)
　Donald F. Collins　　　　　　　　　　　276
[V Pup] The Contribution of A. W. Roberts' Observations to the AAVSO International Database
　Tim Cooper　　　　　　　　　　　　　　254
[l_2 Pup] The Contribution of A. W. Roberts' Observations to the AAVSO International Database
　Tim Cooper　　　　　　　　　　　　　　254
[U Sge] Medium-term Variation in Times of Minimum of Algol-type Binaries: XZ And, RZ Cas, U Cep, TW Dra, U Sge
　Geoff B. Chaplin　　　　　　　　　　　222
[MV Sgr] Multi-color Photometry of the Hot R Coronae Borealis Star and Proto-planetary Nebula V348 Sagittarii
　Arlo U. Landolt and James L. Clem　　　　　83

[V348 Sgr] Multi-color Photometry of the Hot R Coronae Borealis Star and Proto-planetary Nebula V348 Sagittarii
　Arlo U. Landolt and James L. Clem　　　　　83
[V348 Sgr] New Intense Multiband Photometric Observations of the Hot Carbon Star V348 Sagittarii (Abstract)
　Franz-Josef Hambsch and Christopher S. Jeffery　132
[RS Sct] Erratum: Recent Minima of 266 Eclipsing Binary Stars
　Gerard Samolyk　　　　　　　　　　　134
[V384 Ser] A Photometric Study of the Contact Binary V384 Serpentis
　Edward J. Michaels, Chlöe M. Lanning, and Skyler N. Self　　　　　　　　　　　　43
[V1377 Tau] A Photometric Study of Five Low Mass Contact Binaries
　Edward J. Michaels　　　　　　　　　　　209
[T TrA] The Contribution of A. W. Roberts' Observations to the AAVSO International Database
　Tim Cooper　　　　　　　　　　　　　　254
[RV UMa] Visual Times of Maxima for Short Period Pulsating Stars V
　Gerard Samolyk　　　　　　　　　　　　98
[R UMi] Long-Term Changes in the Variability of Pulsating Red Giants (and One RCB Star)
　John R. Percy and Arthur Lei Qiu　　　　　　76
[S UMi] Long-Term Changes in the Variability of Pulsating Red Giants (and One RCB Star)
　John R. Percy and Arthur Lei Qiu　　　　　　76
[U UMi] Long-Term Changes in the Variability of Pulsating Red Giants (and One RCB Star)
　John R. Percy and Arthur Lei Qiu　　　　　　76
[Z UMi] Long-Term Changes in the Variability of Pulsating Red Giants (and One RCB Star)
　John R. Percy and Arthur Lei Qiu　　　　　　76
[V633 Vir] Photometric Analysis of V633 Virginis
　Surjit S. Wadhwa　　　　　　　　　　　138
[V647 Vir] CCD Photometry, Light Curve Modeling, and Period Study of the Overcontact Binary Systems V647 Virginis and V948 Monocerotis
　Kevin B. Alton　　　　　　　　　　　　　7
[BW Vul] β Cepheid and Mira Variable Stars: A Spectral Analysis (Abstract)
　Jesse D'Shawn Harris and Lucian Undreiu　　131
[200 eclipsing binary stars] Recent Minima of 200 Eclipsing Binary Stars
　Gerard Samolyk　　　　　　　　　　　265
[242 eclipsing binary stars] Recent Minima of 242 Eclipsing Binary Stars
　Gerard Samolyk　　　　　　　　　　　106
[37 pulsating red giants] Long-Term Changes in the Variability of Pulsating Red Giants (and One RCB Star)
　John R. Percy and Arthur Lei Qiu　　　　　　76
[267 recently discovered eclipsing variable stars] Recent Minima of 267 Recently Discovered Eclipsing Variable Stars
　Gerard Samolyk and Vance Petriew　　　　270

[85 short period pulsators] Recent Maxima of 85 Short Period
Pulsating Stars
  Gerard Samolyk 103
[3UC345-013290] VR$_c$I$_c$ Photometric Study of the Totally
Eclipsing Pre-W UMa Binary, V616 Camelopardalis: Is it
Detached?
  Ronald G. Samec et al. 157
[ASAS-SN-V J045337.64-691811.2] Period Analysis
of All-Sky Automated Survey for Supernovae (ASAS-SN)
Data on Pulsating Red Giants
  John R. Percy and Lucas Fenaux 202
[ASAS-SN-V J053035.52-685923.2] Period Analysis
of All-Sky Automated Survey for Supernovae (ASAS-SN)
Data on Pulsating Red Giants
  John R. Percy and Lucas Fenaux 202
[ASAS-SN-V J054110.62-693804.1] Period Analysis
of All-Sky Automated Survey for Supernovae (ASAS-SN)
Data on Pulsating Red Giants
  John R. Percy and Lucas Fenaux 202
[ASAS-SN-V J054606.99-694202.8] Period Analysis
of All-Sky Automated Survey for Supernovae (ASAS-SN)
Data on Pulsating Red Giants
  John R. Percy and Lucas Fenaux 202
[ASAS-SN-V J185653.55-392537.4] Period Analysis
of All-Sky Automated Survey for Supernovae (ASAS-SN)
Data on Pulsating Red Giants
  John R. Percy and Lucas Fenaux 202
[GSC 00790-00941] Discovery and Period Analysis of Seven
Variable Stars
  Tom Polakis 117
[GSC 01224-00315] Discovery and Period Analysis of Seven
Variable Stars
  Tom Polakis 117
[GSC 01299-01898] Discovery and Period Analysis of Seven
Variable Stars
  Tom Polakis 117
[GSC 01347-00934] Discovery and Period Analysis of Seven
Variable Stars
  Tom Polakis 117
[GSC 01845-00905] Discovery and Period Analysis of Seven
Variable Stars
  Tom Polakis 117
[GSC 04963-01164] Discovery and Period Analysis of Seven
Variable Stars
  Tom Polakis 117
[GSC 4547 0771] VR$_c$I$_c$ Photometric Study of the Totally
Eclipsing Pre-W UMa Binary, V616 Camelopardalis: Is it
Detached?
  Ronald G. Samec et al. 157
[HD 26961] The Eclipsing Triple Star System b Persei
(HD 26961) (Abstract)
  Donald F. Collins 276
[HD 121191] A New Candidate δ Scuti Star in Centaurus:
HD 121191
  Roy A. Axelsen 173
[HV 2671] Multi-color Photometry of the Hot R Coronae
Borealis Star and Proto-planetary Nebula V348 Sagittarii
  Arlo U. Landolt and James L. Clem 83

[KAO-EGYPT J064512.06+341749.2] Erratum: KAO-
EGYPT J064512.06+341749.2 is a Low Amplitude and
Multi-Periodic δ Scuti Variable Star
  Ahmed Essam, Mohamed Abdel-Sabour, and
    Gamal Bakr Ali 283
[KAO-EGYPT J064512.06+341749.2] KAO-EGYPT
J064512.06+341749.2 is a Low Amplitude and Multi-
Periodic δ Scuti Variable Star
  Ahmed Essam, Mohamed Abdel-Sabour, and
    Gamal Bakr Ali 66
[KID 11405559] Multi-filter Photometric Analysis of
W Ursae Majoris (W UMa) Type Eclipsing Binary Stars
KID 11405559 and V342 Boo
  Tatsuya Akiba et al. 186
[Kelt 16b] Camera Characterization and First Observation
after Upgrade of Feder Observatory (Abstract)
  Isobel Snellenberger and Adam Kline 132
[M12] The Long-term Period Changes and Evolution of V1,
a W Virginis Star in the Globular Cluster M12
  Pradip Karmakar, Horace A. Smith, and
    Nathan De Lee 167
[NGC 6218] The Long-term Period Changes and Evolution
of V1, a W Virginis Star in the Globular Cluster M12
  Pradip Karmakar, Horace A. Smith, and
    Nathan De Lee 167
[NSV 5394] A Study of RR Lyrae Variable Stars in a
Professor-Student Collaboration using the American Public
University System (APUS) Observatory (Abstract)
  David Syndergaard and Melanie Crowson 277
[NSV 7465] A Study of RR Lyrae Variable Stars in a
Professor-Student Collaboration using the American Public
University System (APUS) Observatory (Abstract)
  David Syndergaard and Melanie Crowson 277
[NSV 7647] A Study of RR Lyrae Variable Stars in a
Professor-Student Collaboration using the American Public
University System (APUS) Observatory (Abstract)
  David Syndergaard and Melanie Crowson 277
[NSV 13523] Observations of the Suspected Variable Star
Ross 114 (NSV 13523)
  Sriram Gollapudy and Wayne Osborn 3
[NSV 13602] A Study of RR Lyrae Variable Stars in a
Professor-Student Collaboration using the American Public
University System (APUS) Observatory (Abstract)
  David Syndergaard and Melanie Crowson 277
[NSVS 103152] VR$_c$I$_c$ Photometric Study of the Totally
Eclipsing Pre-W UMa Binary, V616 Camelopardalis: Is it
Detached?
  Ronald G. Samec et al. 157
[NSVS 3917713] A Photometric Study of Five Low Mass
Contact Binaries
  Edward J. Michaels 209
[NSVS 5726288] Observations and Preliminary Modeling
of the Light Curves of Eclipsing Binary Systems
NSVS 7322420 and NSVS 5726288
  Mathew F. Knote, Ronald H. Kaitchuck, and
    Robert C. Berrington 194

[NSVS 6133550] A Photometric Study of Five Low Mass Contact Binaries
  Edward J. Michaels   209

[NSVS 7322420] Observations and Preliminary Modeling of the Light Curves of Eclipsing Binary Systems NSVS 7322420 and NSVS 5726288
  Mathew F. Knote, Ronald H. Kaitchuck, and Robert C. Berrington   194

[Ross 114] Observations of the Suspected Variable Star Ross 114 (NSV 13523)
  Sriram Gollapudy and Wayne Osborn   3

[UCAC4 522-042119] Discovery and Period Analysis of Seven Variable Stars
  Tom Polakis   117

[USNO-A2.0 1200-16843637] Photometric Analysis of Two Contact Binary Systems: USNO-A2.0 1200-16843637 and V1094 Cassiopeiae
  Surjit S. Wadhwa   40

[V1-NGC 6218] The Long-term Period Changes and Evolution of V1, a W Virginis Star in the Globular Cluster M12
  Pradip Karmakar, Horace A. Smith, and Nathan De Lee   167

VIDEO

CCD Photometry, Period Analysis, and Evolutionary Status of the Pulsating Variable V544 Andromedae
  Kevin B. Alton   231

NOTES

Made in United States
Troutdale, OR
09/05/2023

12626194R10100